Progress in Mathematical Physics
Volume 66

Editors-in-Chief
Anne Boutet de Monvel, *Université Paris VII Denis Diderot, France*
Gerald Kaiser, *Center for Signals and Waves, Portland, OR, USA*

Editorial Board
C. Berenstein, *University of Maryland, College Park, USA*
Sir M. Berry, *University of Bristol, UK*
P. Blanchard, *University of Bielefeld, Germany*
M. Eastwood, *University of Adelaide, Australia*
A.S. Fokas, *University of Cambridge, UK*
F.W. Hehl, *University of Cologne, Germany*
   University of Missouri, Columbia, USA
D. Sternheimer, *Université de Bourgogne, Dijon, France*
C. Tracy, *University of California, Davis, USA*

For further volumes:
http://www.birkhauser-science.com/series/4813

Bertrand Duplantier • Stéphane Nonnenmacher •
Vincent Rivasseau
Editors

Chaos

Poincaré Seminar 2010

Editors
Bertrand Duplantier
Service de Physique Théorique
CEA Saclay
Gif-sur-Yvette
France

Stéphane Nonnenmacher
Institut de Physique Théorique
CEA Saclay
Gif-sur-Yvette
France

Vincent Rivasseau
Laboratoire de Physique Théorique
Université Paris-Sud
Orsay
France

ISSN 1544-9998   ISSN 2197-1846 (electronic)
ISBN 978-3-0348-0696-1   ISBN 978-3-0348-0697-8 (eBook)
DOI 10.1007/978-3-0348-0697-8
Springer Basel Heidelberg New York Dordrecht London

Library of Congress Control Number: 2013954370

Mathematics Subject Classification (2010): 11M26, 11M50, 37-02, 37D45, 37D50, 37N05, 70K55, 81Q50

© Springer Basel 2013
This work is subject to copyright. All rights are reserved by the Publisher, whether the whole or part of the material is concerned, specifically the rights of translation, reprinting, reuse of illustrations, recitation, broadcasting, reproduction on microfilms or in any other physical way, and transmission or information storage and retrieval, electronic adaptation, computer software, or by similar or dissimilar methodology now known or hereafter developed. Exempted from this legal reservation are brief excerpts in connection with reviews or scholarly analysis or material supplied specifically for the purpose of being entered and executed on a computer system, for exclusive use by the purchaser of the work. Duplication of this publication or parts thereof is permitted only under the provisions of the Copyright Law of the Publisher's location, in its current version, and permission for use must always be obtained from Springer. Permissions for use may be obtained through RightsLink at the Copyright Clearance Center. Violations are liable to prosecution under the respective Copyright Law.
The use of general descriptive names, registered names, trademarks, service marks, etc. in this publication does not imply, even in the absence of a specific statement, that such names are exempt from the relevant protective laws and regulations and therefore free for general use.
While the advice and information in this book are believed to be true and accurate at the date of publication, neither the authors nor the editors nor the publisher can accept any legal responsibility for any errors or omissions that may be made. The publisher makes no warranty, express or implied, with respect to the material contained herein.

Cover design: deblik, Berlin

Printed on acid-free paper

Springer Basel is part of Springer Science+Business Media (www.birkhauser-science.com)

Contents

Foreword ix

Étienne Ghys
The Lorenz Attractor, a Paradigm for Chaos

1 Introduction 1
2 Starting with a few quotations 3
3 The old paradigm of periodic orbits 6
4 Second attempt: hyperbolic dynamics 12
5 The Lorenz attractor 24
6 The topology of the Lorenz attractor 33
7 The attractor as a statistical object 40
8 The butterfly within the global picture 46
  References 50

Stéphan Fauve
Chaotic Dynamos Generated by Fully Turbulent Flows

1 Introduction 55
2 The dynamo effect 58
3 The Karlsruhe and Riga experiments 61
4 The VKS experiment 66
5 Low-dimensional models of field reversals 75
6 A simple model for the dynamics observed in the VKS experiment .. 80
7 Different morphologies for field reversals 84
8 A minimal model for field reversals 87
9 Conclusion 89
  References 90

Uzy Smilansky
Discrete Graphs – A Paradigm Model for Quantum Chaos

1 Introduction	97
2 Spectral statistics and trace formulae	101
3 Eigenfunctions	112
4 Scattering	117
5 Summary and prospects	121
References	122

Paul Bourgade and Jonathan P. Keating
Quantum Chaos, Random Matrix Theory, and the Riemann ζ-function

1 First steps in the analogy	126
2 Quantum chaology	142
3 Macroscopic statistics	152
References	165

Hans-Jürgen Stöckmann
Chaos in Microwave Resonators

1 Introduction	169
2 From classical to quantum mechanics	170
3 Microwave billiards	173
4 Random matrices	175
5 The random plane wave approximation	178
6 Semiclassical quantum mechanics	181
7 Applications	186
References	190

Stéphane Nonnenmacher
Anatomy of Quantum Chaotic Eigenstates

1 Introduction	194
2 What is a quantum chaotic eigenstate?	195
3 Macroscopic description of the eigenstates	206
4 Statistical description	217
5 Nodal structures	225
References	233

Jacques Laskar
Is the Solar System Stable?

1 Historical introduction 239
2 Numerical computations 258
　References 266

Foreword

This book is the twelfth in a series of Proceedings for the *Séminaire Poincaré*, which is directed towards a broad audience of physicists, mathematicians, and philosophers of science.

The goal of this Seminar is to provide up-to-date information about general topics of great interest in physics. Both the theoretical and experimental aspects of the topic are covered, generally with some historical background. Inspired by the *Nicolas Bourbaki Seminar* in mathematics, hence nicknamed *"Bourbaphy"*, the Poincaré Seminar is held twice a year at the Institut Henri Poincaré in Paris, with written contributions prepared in advance. Particular care is devoted to the pedagogical nature of the presentations, so as to be accessible to a large audience of scientists.

This new volume of the Poincaré Seminar Series, *"Chaos"*, corresponds to the fourteenth such seminar, held on June 5, 2010. It presents a complete and interdisciplinary view of the concept of chaos, both in classical mechanics in its deterministic version, and in quantum mechanics. This volume describes recent developments related to the mathematical, physical, theoretical, and experimental aspects of this fascinating concept. It expounds some of the most challenging questions in science, including predictions on the future of our own planetary system.

The first survey, by ÉTIENNE GHYS, entitled *"The Lorenz Attractor, a Paradigm for Chaos"*, offers a broad description of the fundamental mathematical issues at play with **deterministic chaos**, thereby serving in effect as an introductory chapter to the book. After having invoked Laplace, Maxwell, Poincaré and Hadamard's pioneering standpoints on determinism and chaos, the author first presents two past paradigms for chaos which have been superseded by the Lorenz attractor: "(quasi-)periodic dynamics" and "hyperbolic dynamics". He then focusses on describing Lorenz's fundamental 1963 article, which bears the technical title *"Deterministic non periodic flow"*, and stayed largely unnoticed by mathematicians for almost a decade. The 1972 conference by Lorenz, entitled: *"Predictability: does the flap of a butterfly's wings in Brazil set off a tornado in Texas?"*, made the butterfly effect so famous that it finally disseminated to society at large. The reader fascinated by the paradigmatic power contained in this "little ordinary differential equation", originally motivated by meteorological phenomena, will find here the Lorenz butterfly as it is understood today. This includes topological aspects, such as the "Lorenz knots and links", the relation to Hadamard's study of hyperbolic geodesics, as well as the statistical aspects, such as the Sinai–Ruelle–Bowen (SRB) measures which provide a quantitative description of this type of chaotic system. A general picture of dynamical systems finally emerges from this very complete article.

The second article, *"Chaotic Dynamos Generated by Fully Turbulent Flows"*, by STÉPHAN FAUVE, describes a masterpiece experiment, the von Kármán Sodium or VKS experiment, which for the first time established in 2007 the spontaneous generation of a magnetic field in a strongly turbulent sodium flow. The interplay between deterministic chaos in low-dimensional dynamical systems and the many degrees of freedom of hydrodynamic turbulence is magisterally explained, together with the earlier Karlsruhe and Riga experiments. The author then turns to the VKS experimental setup and its spectacular results, reporting the dynamics of the observed magnetic fields, including their reversal, which pertain to the modeling of Earth's magnetic field. These results are compared both to a theoretical model and to numerical simulations.

The research in **quantum chaos** attempts to uncover the fingerprints of classical chaotic dynamics in the corresponding quantum system. In the third contribution, *"Discrete Graphs – A Paradigm Model for Quantum Chaos"*, UZY SMILANSKY describes a simple toy model – the discrete Laplacian on finite d-regular expander graphs – which allows one to grasp the essential ingredients of quantum chaos. Following the pioneering work by Jakobson *et al.*, he reviews the numerical evidence that illustrates the excellent agreement between the spectral statistics of the discrete Laplacian (or its magnetic analogue) on a large d-regular graph, and the predictions of Random Matrix Theory (RMT) for the Gaussian Orthogonal Ensemble (or the Gaussian Unitary Ensemble (GUE)). Theoretical attempts to justify the famed Bohigas–Gianonni–Schmit (BGS) conjecture were mostly based on trace formulae *à la* Gutzwiller, which establish a link between the (quantum) spectral information on one side, and the periodic trajectories of the underlying classical dynamics on the other side. In the present context, the connection between spectral statistics and periodic orbits is directly demonstrated by using explicit trace formulae for d-regular graphs, leading to interesting yet difficult questions on the combinatorics of closed cycles on these graphs. The author takes a "reverse" point of view: assuming that the spectral statistics agrees with RMT, he obtains new (conjectural) expressions for these combinatorics, which are actually well supported by numerical simulations. As the author remarks, this "reverse" viewpoint is similar with that of Keating and Snaith (see the following contribution), who deduced the asymptotic behavior of the high moments of Riemann's ζ-function on the critical line, by assuming that the Riemann zeros satisfy the GUE statistics.

In the thoroughly written article, *"Quantum Chaos, Random Matrix Theory, and the Riemann ζ-function"*, PAUL BOURGADE and JONATHAN P. KEATING illuminate step by step the connections between the distribution of zeros of the Riemann zeta function and the statistics of eigenvalues of random unitary matrices. The classical results on the ζ-function and its relation to the distribution of prime numbers, from Riemann's fundamental idea to apply complex analytic methods, which led to the prime number theorem by Hadamard and de la Vallée-Poussin, to Weil's explicit formula for the zeros of the ζ-function (or "zeta zeros"), are brilliantly summarized. The authors then expound the basics of Dyson's classical ensembles of random matrices. The seminal discovery by Montgomery that, assuming Riemann's Hypothesis, the gaps between the zeta zeros show the same "fermionic" repulsion law as the GUE eigenvalues was the starting point of a flurry of analytic and numerical studies trying to rigorously establish these fascinating connections. The next part provides a digest of spectral statistics in *quantum chaos*: the Berry–Tabor or BGS conjectures, regarding the spectral statistics of quantized integrable or chaotic systems, are analyzed from the viewpoint of Gutwiller's trace formula. The paper ends with a discussion on recent work, in particular by Keating and collaborators, on the macroscopic statistics of zeta zeros and the corresponding conjectures for the moments of the zeta function on the critical axis. This brilliant review convincingly shows how further progress could ultimately provide a *spectral interpretation* for the zeros of the Riemann ζ-function, thus a proof of the celebrated Riemann Hypothesis itself.

HANS-JÜRGEN STÖCKMANN is a pioneer of experimental quantum chaos. In his contribution, *"Chaos in Microwave Resonators"*, he first provides a list of possible physical applications of quantum chaos: nuclear, atomic or mesoscopic physics, but also cavity electromagnetism, acoustics, seismology – in the latter cases one should rather speak of *wave chaos* since the waves are purely classical. This broad field of applications comes from the identity between the time independent Schrödinger equation for a quantum particle in a cavity, and the Helmholtz equation describing stationary (classical) waves in the cavity. The classical motion is the ray dynamics inside the cavity, which may be chaotic for certain very simple shapes (like the famous *stadium billiard*). Such "quantum billiards" are among the simplest and best understood quantum chaotic systems, very much studied at the theoretical and numerical level. After recalling some theoretical predictions of quantum chaos, the author describes in detail the experiments on the propagation of microwaves in 2D or 3D chaotic cavities; these delicate *physical* experiments verified these predictions. This contribution also explains some recent applications of the same type of analysis, e.g., in the framework of constructing efficient microlasers, or towards understanding the "freak (or rogue) wave" phenomenon in oceanic waves.

The contribution of STÉPHANE NONNENMACHER, entitled *"Anatomy of Quantum Chaotic Eigenstates"*, offers a panoramic view of our present knowledge of the *eigenmodes* of quantized chaotic systems, dubbed as "chaotic eigenmodes". Due to the nonseparability of the dynamics, we have no analytic expression at hand for these eigenmodes; as a result, we can only describe them through indirect, unprecise means, or by using statistical methods. The main mathematically rigorous description of chaotic eigenmodes addresses their *macroscopic* localization properties. The central result of this approach, the Quantum Ergodicity theorem (originally formulated by Schnirelman), states that *almost all* the high-energy eigenmodes are *macroscopically equidistributed* over the energy surface, a property viewed as a quantum analogue of the classical ergodicity property. The Quantum Unique Ergodicity conjecture, according to which this asymptotic flatness suffers no exception, has been proven only for very specific systems enjoying extra symmetries, like the case of arithmetic hyperbolic surfaces recently solved by Lindenstrauss. This "macroscopic flatness" does not prevent the eigenfunctions from strongly fluctuating at the *microscopic* scale. Studying these fluctuations on a rigorous footing represents a formidable challenge. A Random Wave Model (RWM) was heuristically proposed to describe these eigenmodes, viewing them locally as random combinations of isoenergetic plane waves. Although a rigorous justification of this statistical model seems totally out of reach, the RWM represents an interesting object by itself, and has recently attracted a lot of attention from probabilists. For instance, one of the most vivid questions consists in understanding the *nodal set* of the random wave, which can be seen as its "microscopic skeleton".

Last, but not least, in *"Is the Solar System Stable?"*, JACQUES LASKAR studies the fundamental question of the stability, hence the fate, of the Solar planetary system we live in, a subject where he made groundbreaking contributions. He offers here a rather complete perspective, both historically and scientifically. From Newton to Laplace to Poincaré, this fascinating problem drove fundamental progress in mathematics and physics throughout the 18th and 19th centuries. Today's scientists, and preeminently the author, assisted by supercomputers, are running extremely long simulations that allow them to establish that the Solar System is indeed chaotic, and has to be dealt with by using a probabilistic approach. Planetary orbits cannot be accurately predicted after a time scale of a few tens of millions of years, and significant probabilities for major planetary collisions or ejections taking place before the end of life of the Sun can be evaluated. These stunning recent results thereby offer a grand ending to this volume, if not to our Solar System!

This book, by the depth of the topics covered in the subject of Chaos, should be of broad interest to mathematicians, physicists and philosophers of science. We further hope that the continued publication of this series of Proceedings will serve the scientific community, at both the professional and graduate levels. We thank the COMMISSARIAT À L'ÉNERGIE ATOMIQUE ET AUX ÉNERGIES ALTERNATIVES (Division des Sciences de la Matière), the DANIEL IAGOLNITZER FOUNDATION, the TRIANGLE DE LA PHYSIQUE FOUNDATION and the ÉCOLE POLYTECHNIQUE for sponsoring this Seminar. Special thanks are due to CHANTAL DELONGEAS for the preparation of the manuscript.

May 2013	BERTRAND DUPLANTIER	VINCENT RIVASSEAU
	STÉPHANE NONNENMACHER	Laboratoire de
	Institut de Physique Théorique	Physique Théorique
	Saclay, CEA, France	Université d'Orsay, France

The Lorenz Attractor, a Paradigm for Chaos

Étienne Ghys

Abstract. It is very unusual for a mathematical or physical idea to disseminate into the society at large. An interesting example is chaos theory, popularized by Lorenz's butterfly effect: *"does the flap of a butterfly's wings in Brazil set off a tornado in Texas?"* A tiny cause can generate big consequences! Mathematicians (and non mathematicians) have known this fact for a long time! Can one adequately summarize chaos theory is such a simple minded way? In this review paper, I would like first of all to sketch some of the main steps in the historical development of the concept of chaos in dynamical systems, from the mathematical point of view. Then, I would like to present the present status of the Lorenz attractor in the panorama of the theory, as we see it Today.

Translation by Stéphane Nonnenmacher *from the original French text*[1]

1. Introduction

The "Lorenz attractor" is the paradigm for chaos, like the French verb "aimer" is the paradigm for the verbs of the 1st type. Learning how to conjugate "aimer" is not sufficient to speak French, but it is doubtlessly a necessary step. Similarly, the close observation of the Lorenz attractor does not suffice to understand all the mechanisms of *deterministic chaos*, but it is an unavoidable task for this aim. This task is also quite pleasant, since this object is beautiful, both from the mathematical and aesthetic points of view. It is not surprising that the "butterfly effect" is one of the few mathematical concepts widely known among non-scientists.

In epistemology, a paradigm is "a dominant theoretical concept, at a certain time, in a given scientific community, on which a certain scientific domain bases the questions to be asked, and the explanations to be given."[2] The Lorenz attractor has indeed played this role in the modern theory of dynamical systems, as I will try

---
[1] http://www.bourbaphy.fr/ghys.pdf
[2] Trésor de la Langue Française.

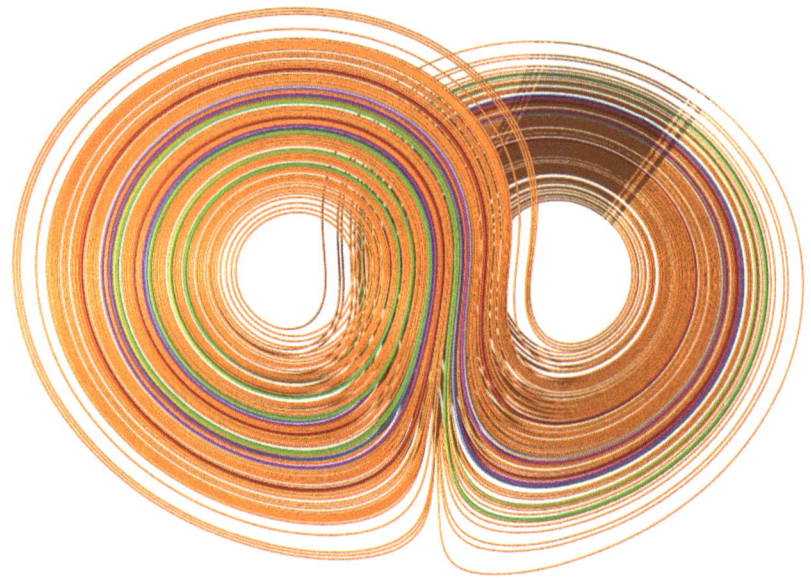

FIGURE 1. The Lorenz attractor

to explain. The Lorenz dynamics features an ensemble of qualitative phenomena which are thought, *today*, to be present in "generic" dynamics.

According to the spirit of this seminar, this text is not written exclusively for mathematicians. The article [81] is another accessible reference for a description of the Lorenz attractor. The nice book "Dynamics beyond uniform hyperbolicity. A global geometric and probabilistic perspective" by Bonatti, Díaz and Viana gives an account of the state of the art on the subject, but is aimed at experts [16].

In a first step, I wish to rapidly present two past paradigms which have been superseded by the Lorenz attractor: the "(quasi-)periodic dynamics" and the "hyperbolic dynamics". Lorenz's article dates back to 1963, but it was really noticed by mathematicians only a decade later, and it took another decade to realize the importance of this example. One could regret this lack of communication between mathematicians and physicists, but this time was also needed for the hyperbolic paradigm to consolidate, before yielding to its nonhyperbolic successors[3]. In a second step I will present the Lorenz butterfly as it is understood today, focussing on the topological and statistical aspects. Then, I will try to sketch the general picture of dynamical systems, in the light of an ensemble of (optimistic) conjectures due to Palis.

Chaos theory is often described from a negative viewpoint: the high sensitivity to initial conditions makes it impossible to practically determine the future evolution of a system, because these initial conditions are never known with total

---

[3]For a historical presentation of chaos theory, see for instance [8].

precision. Yet, the theory would be rather poor if it was limited to this absence of determinism and did not encompass any deductive aspect. On the contrary, I want to insist on the fact that, by asking the good questions, the theory is able to provide rich and nontrivial information, and leads to a real understanding of the dynamics.

There remains a lot of work to do, halfway between mathematics and physics, in order to understand whether this "little ordinary differential equation" can account for meteorological phenomena, which initially motivated Lorenz. Long is the way between these differential equations and the "true" Navier–Stokes partial differential equations at the heart of the physical problem. Due to my incompetence, I will not dwell on this important question.

I prefer to consider the butterfly as a nice gift from physicists to mathematicians!

I thank Aurélien Alvarez, Maxime Bourrigan, Pierre Dehornoy, Jos Leys and Michele Triestino for their help when preparing those notes. I also thank Stéphane Nonnenmacher for his excellent translation of the French version of this paper.

2. Starting with a few quotations

I would like to start by a few quotations, which illustrate the evolution of the opinions on dynamics across the last two centuries.

Let us start with Laplace's famous definition of determinism, in his 1814 "Essai philosophique sur les probabilités" [41]:

> We ought then to consider the present state of the universe as the effect of its previous state and as the cause of that which is to follow. An intelligence that, at a given instant, could comprehend all the forces by which nature is animated and the respective situation of the beings that make it up, if moreover it were vast enough to submit these data to analysis, would encompass in the same formula the movements of the greatest bodies of the universe and those of the lightest atoms. For such an intelligence nothing would be uncertain, and the future, like the past, would be open to its eyes.

The fact that this quotation comes from a (fundamental) book on probability theory shows that Laplace's view on determinism was far from naive [38]. We lack the "vast intelligence" he mentions, so we are forced to use probabilities to understand dynamical systems. Isn't that a modern idea, the first reference to ergodic theory?

In his little book "Matter and Motion" published in 1876, Maxwell insists on the *sensitivity to initial conditions* in physical phenomena: the intelligence mentioned by Laplace must indeed be *infinitely* vast [48]! One should notice that, according to Maxwell, this sensitivity is not the common rule, but rather and exception. This debate is still not really closed today.

> There is a maxim which is often quoted, that "The same causes will always produce the same effects".
>
> To make this maxim intelligible we must define what we mean by the same causes and the same effects, since it is manifest that no event ever happens more that once, so that the causes and effects cannot be the same in *all* respects.
>
> [...]
>
> There is another maxim which must not be confounded with that quoted at the beginning of this article, which asserts "That like causes produce like effects".
>
> This is only true when small variations in the initial circumstances produce only small variations in the final state of the system. In a great many physical phenomena this condition is satisfied; but there are other cases in which a small initial variation may produce a great change in the final state of the system, as when the displacement of the "points" causes a railway train to run into another instead of keeping its proper course.

With his sense of eloquence, Poincaré expresses in 1908 the dependence to initial conditions in a way almost as fashionable as Lorenz's butterfly that we will describe below, including the devastating cyclone [62]:

> Why have meteorologists such difficulty in predicting the weather with any certainty? Why is it that showers and even storms seem to come by chance, so that many people think it quite natural to pray for rain or fine weather, though they would consider it ridiculous to ask for an eclipse by prayer? We see that great disturbances are generally produced in regions where the atmosphere is in unstable equilibrium. The meteorologists see very well that the equilibrium is unstable, that a cyclone will be formed somewhere, but exactly where they are not in a position to say; a tenth of a degree more or less at any given point, and the cyclone will burst here and not there, and extend its ravages over districts it would otherwise have spared. If they had been aware of this tenth of a degree they could have known it beforehand, but the observations were neither sufficiently comprehensive nor sufficiently precise, and that is the reason why it all seems due to the intervention of chance.

Poincaré's second quotation shows that he does not consider chaos as an obstacle to a global understanding of the dynamics [61]. However, the context shows that he is discussing gas kinetics, which depends on a huge number of degrees of freedom (positions and speeds of all atoms). Even though Poincaré has realized the possibility of chaos in celestial mechanics (which depends on much fewer degrees of freedom), he has apparently not proposed to use probabilistic methods to study it[4].

---

[4]Except for the recurrence theorem?

> You are asking me to predict future phenomena. If, quite unluckily, I happened to know the laws of these phenomena, I could achieve this goal only at the price of inextricable computations, and should renounce to answer you; but since I am lucky enough to ignore these laws, I will answer you straight away. And the most astonishing is that my answer will be correct.

Could this type of ideas apply to celestial mechanics as well? In his 1898 article on the geodesics of surfaces of negative curvature, after noticing that "a tiny change of direction of a geodesic [...] is sufficient to cause any variation of the final shape of the curve", Hadamard concludes in a cautious way [34].

> Will the circumstances we have just described occur in other problems of mechanics? In particular, will they appear in the motion of celestial bodies? We are unable to make such an assertion. However, it is likely that the results obtained for these difficult cases will be analogous to the preceding ones, at least in their degree of complexity.
> [...]
> Certainly, if a system moves under the action of given forces and its initial conditions have given values *in the mathematical sense*, its future motion and behavior are exactly known. But, in astronomical problems, the situation is quite different: the constants defining the motion are only *physically* known, that is with some errors; their sizes get reduced along the progresses of our observing devices, but these errors can never completely vanish.

Some people have interpreted these difficulties as a sign of disconnection between mathematics and physics. On the opposite, as Duhem already noticed in 1907, they can be seen as a new challenge for mathematics, namely the challenge to develop what he called the "mathematics of approximation" [23].

> One cannot go through the numerous and difficult deductions of celestial mechanics and mathematical physics without suspecting that many of these deductions are condemned to eternal sterility.
>
>   Indeed, a mathematical deduction is of no use to the physicist so long as it is limited to asserting that a given *rigorously* true proposition has for its consequence the *rigorous* accuracy of some such other proposition. To be useful to the physicist, it must still be proved that the second proposition remains *approximately* exact when the first is only *approximately* true. And even that does not suffice. The range of these two approximations must be delimited; it is necessary to fix the limits of error which can be made in the result when the degree of precision of the methods of measuring the data is known; it is necessary to define the probable error that can be granted the data when we wish to know the result within a definite degree of approximation.
>
>   Such are the rigorous conditions that we are bound to impose on mathematical deduction if we wish this absolutely precise language to be

able to translate without betraying the physicist's idiom, for the terms of this latter idiom are and always will be vague and inexact like the perceptions which they are to express. On these conditions, but only on these conditions, shall we have a mathematical representation of the *approximate*.

But let us not be deceived about it; this "mathematics of approximation" is not a simpler and cruder form of mathematics. On the contrary, it is a more thorough and more refined form of mathematics, requiring the solution of problems at times enormously difficult, sometimes even transcending the methods at the disposal of algebra today.

Later I will describe Lorenz's fundamental article, dating back to 1963, which bears the technical title "Deterministic non periodic flow", and was largely unnoticed by mathematicians during about 10 years [44]. In 1972 Lorenz gave a conference entitled "Predictability: does the flap of a butterfly's wings in Brazil set off a tornado in Texas?", which made famous the *butterfly effect* [45]. The three following sentences, extracted from this conference, seem to me quite remarkable.

If a single flap of a butterfly's wing can be instrumental in generating a tornado, so all the previous and subsequent flaps of its wings, as can the flaps of the wings of the millions of other butterflies, not to mention the activities of innumerable more powerful creatures, including our own species.

If a flap of a butterfly's wing can be instrumental in generating a tornado, it can equally well be instrumental in preventing a tornado.

More generally, I am proposing that over the years minuscule disturbances neither increase nor decrease the frequency of occurrence of various weather events such as tornados; the most they may do is to modify the sequence in which these events occur.

The third sentence in particular is scientifically quite deep, since it proposes that the *statistical* description of a dynamical system could be *unsensitive* to the initial conditions; this idea could be seen as a precursor of the *Sinai–Ruelle–Bowen measures* which, as I will later describe, provide a quantitative description of this type of chaotic system.

3. The old paradigm of periodic orbits

3.1. Some jargon

One should first set up the jargon of the *theory of dynamical systems*. The spaces on which the motion takes place will almost always be the numerical spaces $\mathbb{R}^n$ (and most often $\mathbb{R}^3$). Sometimes more general spaces, like *differentiable manifolds* V (e.g., a sphere or a torus) will represent the phase space of the system. The topology of V and the dynamics can be strongly correlated, this interaction being the main motivation for Poincaré to study topology. However, we will ignore this aspect here ...

The dynamics is generated by an evolution differential equation, or equivalently a *vector field* X (assumed differentiable) on V. Each point x on V is the starting point of a *trajectory* (or an *orbit*) on X. Without any further assumption, this trajectory may not be defined for all time $t \in \mathbb{R}$: it could escape to infinity in finite time. If all trajectories are defined for *for all* $t \in \mathbb{R}$, the field is said to be *complete*: this is always the case if V is compact. In this case, for $x \in V$ and $t \in \mathbb{R}$, we may denote by $\phi^t(x)$ the position at time t of the trajectory starting at x for $t = 0$. For each $t \in \mathbb{R}$ the transformation $\phi^t : x \mapsto \phi^t(x)$ is a differentiable bijection on V (with differentiable inverse ϕ^{-t}): it is a *diffeomorphism* of V. Obviously $\phi^{t_1+t_2}$ is the composition of ϕ^{t_1} and ϕ^{t_2}, and $(\phi^t)_{t \in \mathbb{R}}$ is called the *flow generated by the vector field* X. In other terms, ϕ^t represents the "vast intelligence" Laplace was dreaming of, which would potentially solve all differential equations! When a mathematician writes "Let ϕ^t be the flow generated by X", he claims he has made this dream real... Of course, in most cases we make as if we knew ϕ^t, but in reality knowing only asymptotic behavior when t goes to infinity would be sufficient for us.

One could object – with some ground – that the evolution of a physical system has no reason to be *autonomous*, namely the vector field X could itself depend on time. This objection is, in some way, at the heart of Maxwell's argument, when he notices that the same cause can never occur at two different times, because indeed the times are different. I will nevertheless restrict myself to autonomous differential equations, because they cover a sufficiently vast range of applications, and also because it would be impossible to develop such a rich theory without any assumption on the time dependence of the vector field.

The vector fields on $\mathbb{R}^n$ are not always complete, but they may be transverse to a sphere, such that the orbits of points on the sphere enter the ball and can never exit. The trajectories of points in the ball are then well defined for all $t \geqslant 0$. In this case, ϕ^t is in fact only a *semiflow* of the ball, in the sense that it is defined only for $t \geqslant 0$; this is not a problem if one is only interested in the future of the system. The reader not familiar with differential topology can, as a first step, restrict himself to this particular case.

It is customary to study as well *discrete time* dynamics. One then chooses a diffeomorphism ϕ on a manifold and studies its successive *iterations* $\phi^k = \phi \circ \phi \circ \cdots \circ \phi$ (k times) where the time k is now an integer.

One can often switch between these two points of view. From a diffeomorphism ϕ on V, one can glue together the two boundaries of $V \times [0,1]$ using ϕ, such as to construct a manifold $\tilde{V}$ with one dimension more than V, which carries a natural vector field X, namely $\frac{\partial}{\partial s}$, where s is the coordinate on $[0,1]$ (see Figure 2). Notice that the vector field X is nonsingular. The original manifold V can be considered as a hypersurface in $\tilde{V}$, transverse to X, for instance given by fixing $s = 0$. V is often called a *global section* of X in $\tilde{V}$. The orbits $\phi^t(x)$ of X travel inside $\tilde{V}$ and meet the global section V when t is integral, on the orbits of the diffeomorphism ϕ. Clearly, the continuous time flow ϕ^t on $\tilde{V}$ and the discrete

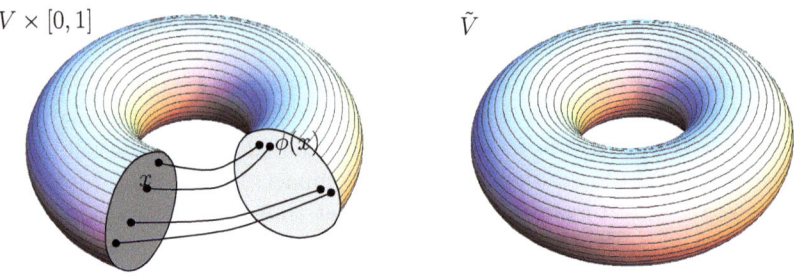

FIGURE 2. A suspension

time dynamics ϕ^k on V are so closely related to each other, that understanding the latter is equivalent with understanding the former. The flow ϕ^t, or the field X, is called the *suspension* of the diffeomorphism ϕ.

Conversely, starting from a vector field X on a manifold $\tilde{V}$, it is often possible to find a hypersurface $V \subset \tilde{V}$ which meets each orbit infinitely often. To each point $x \in V$ one can then associate the point $\phi(x)$, which is the first return point on V along the future orbit of x. This diffeomorphism ϕ on V is the *first return map*. The field X is then the suspension of the diffeomorphism ϕ.[5]

Yet all fields are not suspensions. In particular, vector fields with singularities cannot be suspensions, and we will see that this is the case of the Lorenz equation... Nevertheless, this idea (due to Poincaré) to transform a continuous dynamics into a discrete one is extremely useful, and can be adapted, as we will see later.

3.2. A belief

For a long time, one has focussed on two types of orbits:

- the *singularities* of X are *fixed points* of the flow ϕ^t: these are *equilibrium positions* of the system;
- the *periodic orbits* are the orbits of points x such that, for some $T > 0$ one has $\phi^T(x) = x$.

The old paradigm which entitles this section is the belief that, *in general, after a transient period, the motion evolves into a permanent régime which is either an equilibrium, or a periodic orbit.*

Let us be more precise. For x a point in V, one defines the ω-*limit set* (resp. the α-*limit set*) of x, as the set $\omega_X(x)$ (resp. $\alpha_X(x)$) made of the accumulation points of the orbit $\phi^t(x)$ when $t \to +\infty$ (resp. $t \to -\infty$). The above belief consists in the statement that, for a "generic" vector field X on V, and for each $x \in V$, the set $\omega_X(x)$ reduces either to a single point, or to a periodic orbit. It is not necessary to insist on the ubiquity of periodic phenomena in science (for instance in

---

[5]In general the return time is not necessarily constant, and the flow needs to be reparametrized to obtain a true suspension.

astronomy). The realization that other types of régimes could be expected occurred amazingly late in the history of science.

In fact, the first fundamental results in the theory of dynamical systems, around the end of the nineteenth century, had apparently confirmed this belief. Let us cite for instance the *Poincaré–Bendixson theorem*: for any vector field on the two-dimensional disk, entering the disk on the boundary and possessing (to simplify) finitely many fixed points, the ω-limit sets can only be of three types: a *fixed point*, a *periodic orbit* or a *singular cycle* (that is, a finite number of singular points connected by finitely many regular orbits). See Figure 3.

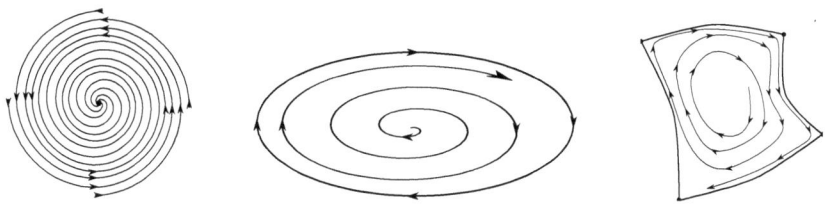

FIGURE 3. A fixed point, a limiting periodic orbit, and a "cycle".

It is easy to show that the case of a singular cycle is "unstable", that is, it does not occur for a "generic" vector field; this remark confirms the above belief, in the case of flows on the disk. Interestingly, Poincaré did not explicitly notice the (rather simple) fact that, among the three asymptotic limits, one is "exceptional", while the two others are "generic". Robadey's thesis [63] discusses the concepts of genericity in Poincaré's work. The explicit investigation of the behavior of generic vector fields started much later, probably with Smale and Thom at the end of the 1950s.

The case of *vector fields on general surfaces* needed more work, and was only completed in 1962 by Peixoto [55]. There is no Poincaré–Bendixson type theorem, and the ω-limit sets can be much more complicated than for the disk. Yet, these complex examples happen to be "rare", and generally the limit sets are indeed equilibrium points or periodic orbits.

We shall now introduce an important concept, which we will later widely generalize. A singularity x_0 of a vector field X (on a manifold V) is said to be *hyperbolic* if the linearized vector field at this point (namely, a matrix) has no eigenvalue on the imaginary axis. The set of points x the orbits of which converge to x_0 is the *stable manifold* of x_0: it is a submanifold $W^s(x_0)$ immersed in V, which contains x_0 and has dimension equal to the number of eigenvalues with negative real parts. Switching to $-X$, one similarly defines the *unstable manifold* $W^u(x_0)$ of x_0, of dimension complementary to that of $W^s(x_0)$ (since the singularity is hyperbolic).

If γ is a periodic orbit containing a point x_0, one can choose a ball B of dimension $n-1$ containing x_0 and transverse to the flow. The flow then allows to define *Poincaré's first return map*, which is a local diffeomorphism in B defined near x_0 and fixing that point. If the linearization of this diffeomorphism at x_0 does not have any eigenvalue of modulus 1, then *the orbit γ is said to be hyperbolic*. The set of points x such that $\omega_X(x)$ coincides with the periodic orbit γ is the *stable manifold of γ*, denoted by $W^s(\gamma)$. As above, it is an immersed submanifold. The unstable manifold $W^u(\gamma)$ is defined similarly.

In 1959 Smale defined the vector fields which are now called of *Morse–Smale type* [72]. These fields are characterized by the following properties:

- X has at most finitely many singularities and periodic orbits, which are all hyperbolic;
- *The α and ω-limit sets of all points are singularities or periodic orbits;*
- *the stable and unstable manifolds of the singularities or the periodic orbits intersect each other transversally.*

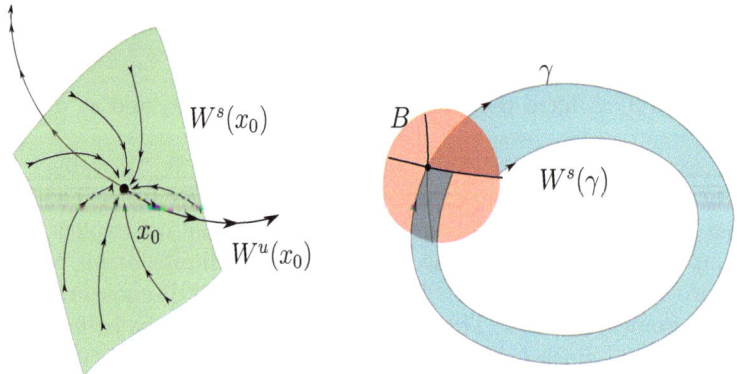

FIGURE 4. Hyperbolic fixed point and periodic orbit

In this article Smale formulated a triple conjecture. Before stating it, I first need to explain the fundamental concept of *structural stability*, introduced in 1937 by Andronov and Pontrjagin [5]. A vector field X is structurally stable if there exists a neighborhood of X (in the C^1 topology on vector fields) such that all fields X' in this neighborhood are *topologically conjugate* to X. This conjugacy means that there exists a (generally non-differentiable) homeomorphism in V which maps the orbits of X to the orbits of X', keeping the time orientations. For such fields X, the topological dynamics is qualitatively insensitive to small perturbations. This concept of structural stability thus belongs to the "mathematics of approximation" called for by Duhem. Andronov and Pontrjagin had shown that certain very simple fields on the two-dimensional disk are structurally stable. The most naive example is given by the radial field $X = -x\partial/\partial x - y\partial/\partial y$, for which all points converge towards the (singular) origin. If one perturbs X, the new field will still have an attracting singularity near the origin: X is thus structurally stable.

I can now state Smale's conjectures:

1. *Given a compact manifold V, the Morse–Smale fields form an* open dense set *in the space of all vector fields on V.*
2. *All Morse–Smale fields are structurally stable.*
3. *All structurally stable fields are of Morse–Smale type.*

The first and third conjecture are *false*, as Smale will soon himself discover. But the second one is true... Smale's motivation for these conjectures was clear: he wanted to prove the (then dominant) belief in a permanent régime – equilibrium point or periodic orbit – for a generic system.

In 1962 Peixoto proved the 3 conjectures if V is a compact orientable *surface* [55].

The first of Smale's conjectures is surprising, since Poincaré or Birkhoff already knew it was false: a flow can have infinitely many periodic orbits in a stable way. About this period, Smale wrote in 1998 [77]:

> It is astounding how important scientific ideas can get lost, even when they are aired by leading scientific mathematicians of the preceding decades.

It was explicitly realized around 1960 that the dynamics of a generic vector field is likely to be much more complicated than that of a Morse–Smale field. The paradigm of the periodic orbits yielded to the next one – that of hyperbolic systems – which I will describe in the next section. But periodic orbits will continue to play a fundamental rôle in dynamics, as Poincaré had explained in 1892 [59, Chap. 3, Sec. 36]:

> In addition, these periodic solutions are so valuable for us because they are, so to say, the only breach by which we may attempt to enter an area heretofore deemed inaccessible.

One more remark before going on towards the Lorenz attractor ... The dynamics we are considering here are not assumed to be *conservative*. One could discuss for instance the case of *Hamiltonian* dynamical systems, which enjoy very different qualitative behaviors. For example, the preservation of phase space volume implies that almost all points are recurrent, as follows from *Poincaré's recurrence theorem*. A conservative field is thus *never* of Morse–Smale type.

In the domain of Hamiltonian vector fields, the old paradigm is *quasi-periodic motion*. One believes that in the ambient manifold, "many orbits" are situated on invariant tori supporting a linear dynamics, generated by a certain number of uncoupled harmonic oscillators with different periods. The typical example is the Keplerian motion of the planets, assuming they do not interact with each other: each planet follows a periodic trajectory, and the whole system moves on a torus of dimension given by the number of planets. The KAM theory allows to show that many of these invariant tori persist if one perturbs a completely integrable Hamiltonian system. On the other hand, there is no reason to find such tori for a "generic" Hamiltonian system.

One should thus keep in mind that we are discussing here *a priori dissipative dynamics*. It is amazing that eminent physicists like Landau and Lifschitz have for a while presented turbulence as an almost periodic phenomenon, with invariant tori of dimensions depending on Reynold's number. Only in the 1971 second edition of their famous treatise on fluid mechanics have they realized that almost periodic functions are "too nice" to describe turbulence.

4. Second attempt: hyperbolic dynamics

4.1. Hadamard and the geodesics on a bull's forehead

In 1898 Hadamard publishes a remarkable article on the dynamical behavior of the geodesics on surfaces of negative curvature [34]. This article can be considered as the starting point of the theory of hyperbolic dynamical systems and of symbolic dynamics. It probably appeared too early since, more than 60 years later, Smale had to follow again the same path followed by Hadamard before continuing much further, as we will see.

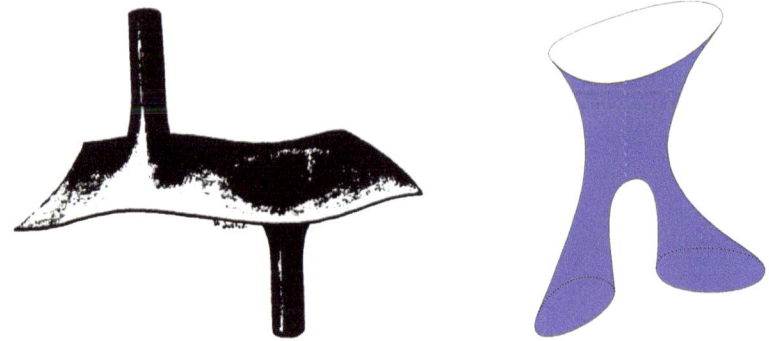

FIGURE 5. Hadamard's pants. Reprinted from [34]. We tried to trace the rights holder. If someone has legitimate claims, please contact the publisher.

Hadamard starts by giving some concrete examples of *negatively curved surfaces* in the ambient space. The left part of Figure 5 is extracted from his article. This surface is diffeomorphic to a plane minus two disks, which allows me to call it "(a pair of) pants", and to draw it like on the right part of Figure 5.

The problem is to understand the dynamical behavior of the geodesics on this surface. That is, a point is constrained to move on P, only constrained by the reaction force. At each moment the acceleration is orthogonal to the surface: the trajectories are *geodesics* of P, at constant speed.

An initial condition consists in a point on P and a tangent vector to P on this point, say of length 1. The set of these initial conditions forms the manifold $V = T^1 P$ of dimension 3, called the *unitary tangent bundle* of P. The geodesic flow

ϕ^t acts on V: one considers the geodesic starting from a point in a certain direction, and follow it during the time t to get another point and another direction.

The main property of the negative curvature used by Hadamard is the following: every continuous path drawn on the surface between two points can be deformed (keeping the boundary points fixed) into a unique geodesic arc joining the two boundary points.

One can find three closed geodesics g_1, g_2, g_3 cutting P into 4 parts. Three of them correspond to the three "ends" of P, and the fourth one is the *convex core*, which is a *compact* surface with boundary given by the three geodesics (see Figure 6).

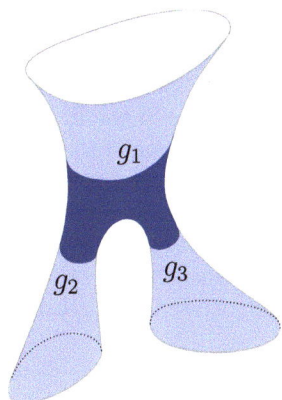

FIGURE 6. The core of the pants

Let us now consider a geodesic $t \in \mathbb{R} \mapsto g(t) \in P$. If the curve g intersects g_1, g_2 or g_3 at time t_0, to exit the core and enter one end, the property I have just mentioned shows that for any $t > t_0$ the curve $g(t)$ remains in this end and cannot come back in the core. In fact, one can check that $g(t)$ goes to infinity in that end. Conversely, if a geodesic enters the core at time t_0, it remains in one of the ends when $t \to -\infty$. One can also show that no geodesic can stay away from the core for ever. There are thus several types of geodesics:

- $g(t)$ *is in the core for all* $t \in \mathbb{R}$;
- $g(t)$ *comes from one end, enters the core and exits into an end;*
- $g(t)$ *comes from one end, enters the core and stays there for all large t;*
- $g(t)$ *is in the core for t sufficiently negative, and exits into an end.*

The most interesting ones are of the first type: they are called nowadays *nonwandering orbits*. Hadamard analyzed them as follows. Let us join g_1 and g_2 by an arc c_3 of minimal length: it is a geodesic arc inside the core, orthogonal to g_1 and g_2. Similarly, let c_1 connect g_2 with g_3, and c_2 connect g_1 with g_3. The arcs c_i are the *seams of the pants*. If one cuts the core along the seams, one obtains two hexagons H_1 and H_2 (see Figure 7).

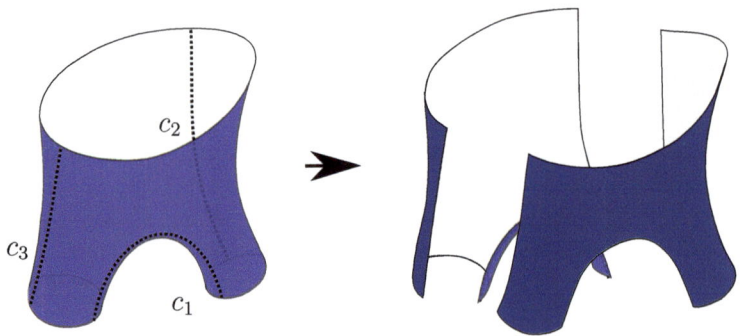

FIGURE 7. Three cuts in the pants leading to two hexagons

Let us now consider a geodesic g of the first type, that is entirely contained in the core. The point $g(0)$ belongs to one of the hexagons, maybe to both if it lies on a seam, but let us ignore this particular case. When one follows the geodesic g from $g(0)$ by increasing the time t, one successively intersects the seams c_1, c_2, c_3 infinitely many times[6]. One can then read an infinite word written in the 3-letter alphabet $\{1, 2, 3\}$. If g intersects the seam c_1, the next seam cannot be c_1: it will be either c_2 or c_3. The word associated to a geodesic will thus never contain two successive identical letters.

Similarly, one can follow g in the negative time direction and obtain a word. These two words make up a single bi-infinite word $m(g)$ associated with the geodesic g.

Actually, Hadamard only treats the closed geodesics, which are associated with the periodic words. In 1923 Morse will complete the theory by coding the nonperiodic geodesics by bi-infinite words [50]. In the sequel I will mix the two articles, and call their union Hadamard–Morse.

The main result of Hadamard–Morse is the following:

For any bi-infinite word m in the alphabet $\{1, 2, 3\}$ without repetition, there exists a geodesic g realizing that word. The geodesic g is unique if one specifies the hexagon containing $g(0)$.

Of course, the uniqueness of the geodesic should be understood as follows: one can move the origin of g into $g(\tau)$ without changing the word, as long as g does not meet any seam between $t = 0$ and $t = \tau$. Uniqueness means that if two geodesics of type 1 are associated with the same word and start from the same hexagon, then they can only differ by such a (short) time shift.

The proof of this result is quite easy (with modern techniques). Starting from a bi-infinite word, one takes the word m_p of length $2p$ obtained by keeping only the p first letters on the right and on the left. One then considers a path $\gamma_p : [-l_p, l_p] \mapsto P$ which "follows" the word m_p and starts from H_1 or H_2. More

---

[6]This fact is due to the negative curvature.

precisely, one chooses a point x_1 in H_1 and x_2 in H_2, and the path γ_p is formed by $2p$ geodesic arcs alternatively joining x_1 and x_2 and crossing the seams as indicated in the word m_p. This arc γ_p can be deformed, keeping the boundaries fixed, into a geodesic arc $\tilde{\gamma}_p : [-\bar{l}_p, \bar{l}_p] \to P$. One then needs to show that this path converges to a geodesic $g : \mathbb{R} \to P$ when $p \to \infty$, which is easy for a modern mathematician (using Ascoli's theorem etc.). Uniqueness is not very difficult to check either.

Here are a few qualitative consequences:

- If two bi-infinite words m and m' coincide from a certain index on, the corresponding geodesics g, g' will approach each other when $t \to +\infty$: they are *asymptotes*. There exists τ such that the distance between $g(t)$ and $g'(t + \tau)$ tends to 0 when $t \to +\infty$. Nowadays we would say that g and g' belong to the same stable manifold. Of course, a similar remark applies when m and m' coincide for all sufficiently negative indices: the geodesics then belong to the same unstable manifold.

- Starting from two infinite words m_+ and m_- without repetitions, one can of course construct a bi-infinite word m coinciding with m_+ for sufficiently positive indices, and with m_- for sufficiently negative indices. Hence, given two nonwandering geodesics, one can always find a third one which is an asymptote of the first one in the future, and an asymptote of the second one in the past. One can even arbitrarily fix any finite number of indices of m. Obviously, this result implies a sensitive dependence to the initial conditions: *an arbitrarily small perturbation of a given geodesic can make it asymptote to two arbitrary geodesics, respectively in the past and in the future*.

- Another interesting property: *any nonwandering geodesic can be approached arbitrarily close by periodic geodesics*. Starting from a bi-infinite word, it suffices to consider a long segment of this word and to repeat it infinitely often to construct a nearby periodic geodesic. Therefore there exist *countably many periodic geodesics, and their union is dense in the set of nonwandering points*. The dynamical behavior of these geodesics is thus much more complicated than for a Morse–Smale flow.

To end this brief account of these two articles, I shall mention some consequences Hadamard–Smale could have "easily" obtained, but did not notice.

- The description of the dynamics of the geodesics does not depend on the choice of metric with negative curvature. One could "almost" deduce that for two metrics of negative curvature on P, the geodesic flows are topologically conjugate. We are close to the structural stability, proved by Anosov in 1962 [3]: *the geodesic flow of a Riemannian metric with negative curvature is structurally stable*. Moreover, Gromov showed that, on a given compact manifold, the geodesic flows of two arbitrary metrics of negative curvature are topologically conjugate [30].[7]

- Another aspect, implicit in these articles, is the introduction of a dynamics on a *space of symbols*. The space of all sequences $(x_i)_{i \in \mathbb{Z}} \in \{1, 2, 3\}^{\mathbb{Z}}$ is compact, and the *shift* of the indices $(x_i) \mapsto (x_{i+1})$ is a homeomorphism which "codes"

---

[7] Actually, the two metrics could be defined on two different manifolds with isomorphic fundamental groups.

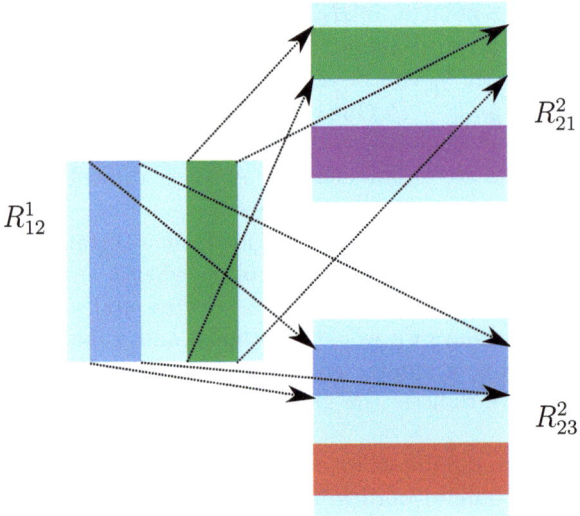

FIGURE 8. Dynamics on the rectangles R_{ij}

the dynamics of the geodesic flow. More precisely, one should take the "subshift of finite type" formed by the sequences without repetitions, and (even more precisely) one should take two copies of this space, corresponding to the two hexagons.

• Finally, by extrapolating a little, one could see in these articles a subliminal construction of Smale's horseshoe, an object I will soon describe. Indeed, consider all unit vectors tangent to P on a point of a seam, heading towards another seam. The geodesic generated by such an initial condition starts from c_i, crosses H_1 or H_2, and lands on c_j. These unit vectors form twelve rectangles: one has to choose c_i, c_j, then H_1 or H_2, which gives twelve possibilities, and for each one we should indicate the starting and arrival points on the seams. These twelve rectangles – denoted by R_{ij}^1, R_{ij}^2 – are embedded in $T^1 P$, the unitary tangent bundle of P, transversely to the geodesic flow.

Strictly speaking, the union R of these rectangles is *not* a global section for the geodesic flow, since they do not meet *all* the geodesics: the seams themselves define geodesics not meeting R. Nevertheless, one can define a "first return" map ϕ, which is not defined in the whole of R, and which is not surjective. Starting from a unit vector x in R, the corresponding geodesic crosses a hexagon and lands on another cut. The tangent vector at the exit point is not necessarily in R, since the geodesic could then exit the second hexagon along g_1, g_2 or g_3 instead of another cut. If this tangent vector is still in R, we denote it by $\phi(x)$. The domain of definition of ϕ is the union of 24 "vertical rectangular" zones, two in each rectangle, and its image is formed by 24 "horizontal rectangular" zones. Each of the 24 vertical rectangular zones is contracted by ϕ along the vertical direction, and expanded along the horizontal direction, and its image is one of the 24 horizontal rectangular zones. Figure 8 displays the two vertical zones in R^1_{12} and their images in R^2_{21} and R^2_{23}.

This simultaneous appearance of *expansion and contraction* is clearly at the heart of the phenomenon, but it was not explicitly noticed by Hadamard–Morse. An easy observation: if one slightly perturbs the geodesic flow (not necessarily into another geodesic flow), R will still be transverse to the flow, and the "return" on R will have the same shape, with slightly deformed rectangular zones; the nonwandering orbits will still be coded by sequences of symbols, accounting for the sequences of rectangles crossed along the evolution. The structural stability (at least of the nonwandering set) can be easily deduced from this observation.

It is surprising that this article of Hadamard, containing so many original ideas, could stay unnoticed so long. Yet, Duhem described this article in such a "colorful" way as to attract the attention even of non-mathematicians [23].

> Imagine the forehead of a bull, with the protuberances from which the horns and ears start, and with the collars hollowed out between these protuberances; but elongate these horns and ears without limit so that they extend to infinity; then you will have one of the surfaces we wish to study. On such a surface geodesics may show many different aspects. There are, first of all, geodesics which close on themselves. There are some also which are never infinitely distant from their starting point even though they never exactly pass through it again; some turn continually around the right horn, others around the left horn, or right ear, or left ear; others, more complicated, alternate, in accordance with certain rules, the turns they describe around one horn with the turns they describe around the other horn, or around one of the ears. Finally, on the forehead of our bull with his unlimited horns and ears there will be geodesics going to infinity, some mounting the right horn, others mounting the left horn, and still others following the right or left ear. [...] If, therefore, a material point is thrown on the surface studied starting from a geometrically given position with a geometrically given velocity, mathematical deduction can determine the trajectory of this point and tell whether this path goes to infinity or not. But, for the physicist, this deduction is forever useless. When, indeed, the data are no longer known geometrically, but are determined by physical procedures as precise as we may suppose, the question put remains and will always remain unanswered.

4.2. Smale and his horseshoe

Smale has given several accounts of his discovery of hyperbolic systems around 1960[8] (see for instance [77]). This discovery is independent of the previous contributions of Poincaré and Birkhoff, which however played a rôle in the subsequent developments of the theory. It seems that Hadamard's article played absolutely

---
[8]On the beach in Copacabana, more precisely in Leme!

no rôle at this time. Nevertheless, Smale constructs a counterexample to his own conjecture, according to which Morse–Smale flows form an open dense set in the space of dynamical systems: the famous *horseshoe*. It is a diffeomorphism ϕ on the two-dimensional sphere $\mathbb{S}^2$, thought of as the plane $\mathbb{R}^2$ with an extra point at infinity. ϕ is assumed to map a rectangle R in the plane as shown on Figure 9, by expanding the vertical directions and contracting the horizontal ones. The intersection $R \cap \phi(R)$ is the union of two rectangles R_1, R_2.

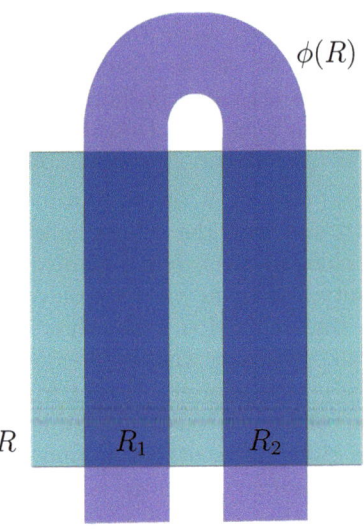

FIGURE 9. Smale's horseshoe

One can assume that the point at infinity is a repulsive fixed point. The points of R which always stay in R, that is $\cap_{k\in\mathbb{Z}}\phi^k(R)$, form a Cantor set, homeomorphic to $\{1,2\}^\mathbb{Z}$. Each of its points is coded by the sequence of rectangles R_1, R_2 successively visited by its orbit. In particular, the periodic points form an infinite countable set, dense in this Cantor set. Smale then establishes that *the horseshoe is structurally stable*, a rather easy fact (nowadays). If one perturbs the diffeomorphism ϕ into a diffeomorphism ϕ', the intersection $\phi'(R) \cap R$ is still made of two "rectangular" zones crossing R all along, and one can still associate a single orbit to each sequence in $\{1,2\}^\mathbb{Z}$, allowing to construct a topological conjugacy between ϕ and ϕ', at least on these invariant Cantor sets. There just remains to extend the conjugacy in the exterior, using the fact that all exterior orbits come from infinity.

Smale publishes this result in the proceedings of a workshop organized in the Soviet Union in 1961 [73]. Anosov tells us about this "hyperbolic revolution" in [4].

> The world turned upside down for me, and a new life began, having read Smale's announcement of "a structurally stable homeomorphism with an infinite number of periodic points", while standing in line to register for

a conference in Kiev in 1961. The article is written in a lively, witty, and often jocular style and is full of captivating observations. [...] [Smale] felt like a god who is to create a universe in which certain phenomena would occur.

Afterwards the theory progresses at a fast pace. The horseshoe is quickly generalized by Smale (see for instance [74]). As I already mentioned, Anosov proves in 1962 that the geodesic flow on a manifold of negative curvature is structurally stable[9]. For this aim, he conceives the concept of what is known today as an *Anosov flow*.

A *nonsingular* flow ϕ^t on a compact manifold V, generated by a vector field X, is an Anosov flow if at each point x one can decompose the tangent space $T_x V$ into three subspaces $\mathbb{R}.X(x) \oplus E_x^u \oplus E_x^s$. The first one $\mathbb{R}.X(x)$ is the line generated by the vector field, and the others are called respectively unstable and stable subspaces. This decomposition should be invariant through the differential $d\phi^t$ of the flow, and there should exist constants $C > 0, \lambda > 0$ such that for all $t \in \mathbb{R}$ and $v_u \in E_x^u$, $v_s \in E_x^s$:

$$\|d\phi^t(v_u)\| \geqslant C^{-1}\exp(\lambda t)\|v_u\| \quad ; \quad \|d\phi^t(v_s)\| \leqslant C\exp(-\lambda t)\|v_s\|$$

(here $\|\cdot\|$ is an auxiliary Riemannian metric).

Starting from the known examples of structurally stable systems (Morse–Smale, horseshoe, geodesic flow on a negatively curved manifold and a few others), Smale cooked up in 1965 the fundamental concept of dynamical systems satisfying the *Axiom A* (by contraction, *Axiom A systems*).

Consider a compact set $\Lambda \subset V$ invariant through a diffeomorphism ϕ. Λ is said to be a *hyperbolic set* if the tangent space to V restricted to points of Λ admits a continuous decomposition into a direct sum: $T_\Lambda V = E^u \oplus E^s$, invariant through the differential $d\phi$, and such that the vectors in E^u are expanded, while those in E^s are contracted. Precisely, there exists $C > 0, \lambda > 0$ such that, for all $v_u \in E^u, v_s \in E^s, k \in \mathbb{Z}$, one has:

$$\|d\phi^k(v_u)\| \geqslant C^{-1}\exp(k\lambda)\|v_u\| \quad ; \quad \|d\phi^k(v_s)\| \leqslant C\exp(-k\lambda)\|v_s\|.$$

A point in V is called *wandering* if it has an open neighborhood U disjoint from all its iterates: $\phi^k(U) \cap U = \emptyset$ for all $k \neq 0$. The set of *nonwandering points* is an invariant closed set, traditionally denoted by $\Omega(\phi)$[10].

By definition, ϕ is *Axiom A* if $\Omega(\phi)$ is hyperbolic and if the set of periodic points is dense in $\Omega(\phi)$.

Under this assumption, if x is a nonwandering point, the set $W^s(x)$ (*resp.* $W^u(x)$) of the points y such that the distance between $\phi^k(y)$ and $\phi^k(x)$ goes to 0 when k tends to $+\infty$ (resp. $-\infty$) is a submanifold immersed in V: it is the stable (resp. unstable) manifold of the point x. An Axiom A diffeomorphism satisfies the *strong transversality assumption* if the stable manifolds are transverse to the unstable ones.

---

[9]Surprisingly, he does not seem to know Hadamard's work.
[10]Although the notation looks like one of an open set.

An analogous definition can be given for vector fields X: one then needs to continuously decompose $T_\Lambda V$ into a sum $\mathbb{R}.X \oplus E^u \oplus E^s$. In particular, this definition implies that the singular points of X are isolated in the nonwandering set, otherwise the dimensions of the decomposition would be discontinuous.

Smale then states three conjectures, parallel to the ones he had formulated nine years earlier:

1. *Given a compact manifold V, the Axiom A diffeomorphisms satisfying the strong transversality condition form an open dense set in the set of diffeomorphisms of V.*
2. *The Axiom A diffeomoprhisms satisfying the strong transversality condition are structurally stable*
3. *The structurally stable diffeomorphisms are Axiom A and satisfy the strong transversality condition.*

The second conjecture is correct, as shown by Robbin in 1971 and by Robinson in 1972 (assuming diffeomorphisms of class C^1). One had to wait until 1988 and Mañé's proof to check the third conjecture.

However, the first conjecture is wrong, as Smale himself will show in 1966 [75]. As we will see, a counterexample was actually "available" in Lorenz's article four years earlier, but it took time to the mathematics community to notice this result. Smale's counterexample is of different nature: he shows that a defect of transversality between the stable and unstable manifolds can sometimes not be "repaired" through a small perturbation. However, Smale introduces a weaker notion than structural stability, called Ω-*stability*: it requires that, after a perturbation of the diffeomorphism ϕ into ϕ', the *restrictions* of ϕ to $\Omega(\phi)$ and of ϕ' to $\Omega(\phi')$ be conjugate through a homeomorphism. He conjectures that property to be generic.

Smale's 1967 article *Differential dynamical systems* represents an important step for the theory of dynamical systems [76], a "masterpiece of mathematical literature" according to Ruelle [69].

But, already in 1968, Abraham and Smale found a counterexample to this new conjecture, also showing that Axiom A is not generic [1]. In 1972, Shub and Smale experiment another concept of stability [70], which will lead Meyer to the following comment:

> In the never-ending quest for a solution of the yin-yang problem more and more general concepts of stability are proffered.

Bowen's 1978 review article is interesting on several points [18]. The theory of Axiom A systems has become solid and, although difficult open questions remain, one has a rather good understanding of their dynamics, both from the topological and ergodic points of view. Even if certain "dark swans" have appeared in the landscape, destroying the belief in the genericity of Axiom A, these are still studied, at that time, "as if they were hyperbolic". Here are the first sentences of Bowen's article, illustrating this point of view.

> These notes attempt to survey the results about Axiom A diffeomorphisms since Smale's well-known paper of 1967. In that paper, Smale

defined these diffeomorphisms and set up a program for dynamical systems centered around them. These examples are charming in that they display complicated behavior but are still intelligible. This means that there are many theorems and yet some open problems. [...] The last sections deal with certain non-Axiom A systems that have received a good deal of attention. These systems display a certain amount of Axiom A behavior. One hopes that further study of these examples will lead to the definition of a new and larger class of diffeomorphisms, with the Axiom A class of prototype.

It is not possible here to seriously present the theory of hyperbolic systems, the reader may try [36]. I will still describe a bit later some of the most important theorems, which give a more precise account of the ergodic behavior of these systems.

Still, I have to cite one of the major results – due to Bowen – which allows to understand the dynamics à la Hadamard. Assume one covers the nonwandering set $\Omega(\phi)$ by finitely many "boxes" B_i ($i = 1, 2, \ldots, l$), assumed compact with disjoint interiors. Let us construct a finite graph, with vertices indexed by i, and such that each oriented link connects i to j if the interior of $\phi(B_i) \cap B_j$ is nonempty (as a subset of $\Omega(\phi)$). Let us denote by Σ the closure of the set of sequences $\{1, 2, \ldots, l\}^{\mathbb{Z}}$ corresponding to the infinite paths on the graph, namely the sequences of indices $\sigma(p)_{p \in \mathbb{Z}}$ such that two consecutive indices are connected in the graph. One says that Σ is a *subshift of finite type*. The collection of boxes is a *Markov partition* if this subshift faithfully codes the dynamics of ϕ. One requires that for each sequence $\sigma \in \Sigma$, there exists a unique point $x = \pi(\sigma)$ in $\Omega(\phi)$, the orbit of which precisely follows the itinerary σ: for each k, one has $\phi^k(x) \in B_{\sigma(k)}$. But one also requires that the coding $\pi : \Sigma \to \Omega(\phi)$ be as injective as possible: each fiber $\pi^{-1}(x)$ is finite and, for x a generic point (in Baire's sense in $\Omega(\phi)$), it contains a single itinerary. Bowen establishes that the nonwandering sets of hyperbolic sets can be covered by this type of Markov partition. The usefulness of this result is clear, since it transfers a dynamical question into a combinatorial one.

Nowadays, Axiom A systems seem to occupy a much smaller place as was believed at the end of the 1970s. The hyperbolic paradigm has abandoned its dominant position ... Anosov's following quotation could probably be expanded beyond the mathematical world [4].

> Thus the grandiose hopes of the 1960s were not confirmed, just as the earlier naive conjectures were not confirmed.

For a more detailed description of the "hyperbolic history" one can also read the introduction of [36] or of [54]. See also "What is ... a horseshoe" by one of the actors of the subject [69].

4.3. How about Poincaré's infinitely thin tangle?

I barely dare to set a doubt on Poincaré's legacy in this seminar bearing his name. Many authors claim that Poincaré is at the origin of chaos theory. His rôle is doubtlessly very important, but maybe not as much as is often claimed. Firstly, the idea that physical phenomena can be very sensitive to initial conditions was not totally new, as we have seen with Maxwell. Second, Poincaré's contributions on these questions seem to have been largely forgotten in the subsequent years, and have only played a minor rôle in the further development of the theory. Most of Poincaré's results were rediscovered. More essentially, the central idea that I wish to present here is that chaos theory cannot be restricted to the "powerless" statement that the dynamics is complicated: the theory must encompass methods allowing to explain the internal mechanisms of this dynamics. The following famous quotation of Poincaré illustrates this powerlessness facing the complexity of the dynamics [59, Chap. 33, Sec. 397]:

> When we try to represent the figure formed by these two curves and their infinitely many intersections, each corresponding to a doubly asymptotic solution, these intersections form a type of trellis, tissue, or grid with infinitely fine mesh. Neither of the two curves must ever cut across itself again, but it must bend back upon itself in a very complex manner in order to cut across all of the meshes in the grid an infinite number of times.
>
> The complexity of this figure is striking, and I shall not even try to draw it. Nothing is more suitable for providing us with an idea of the complex nature of the three-body problem, and of all the problems of dynamics in general, where there is no uniform integral and where the Bohlin series are divergent.

Poincaré's major contribution[11] was to realize the crucial importance of *homoclinic orbits*. Let's consider a diffeomorphism ϕ on the plane, with a hyperbolic fixed point at the origin, with stable and unstable subspaces of dimension 1. A point x (different from the origin) is *homoclinic* if it belongs to the intersection $W^s \cap W^u$ of the stable and unstable manifolds of the origin. The orbit of x converges to the origin both in the future and in the past.

It is not necessary to recall here in detail how this concept appeared in 1890. On this topic, I recommend the reading of [12], which relates the beautiful story of Poincaré's mistake in the first version of his manuscript when applying for King Oscar's prize [12]. The book [11] also discusses this prize, insisting on the historical and sociological aspects rather than the mathematical ones.

Poincaré's setup was inspired by celestial mechanics, so ϕ preserves the area, and he had first thought it was impossible for W^s and W^u to cross each other without being identical. He deduced from there a stability theorem in the three body problem. When he realized his error, he understood that these homoclinic points where W^s meets W^u transversely are not only possible, but actually the rule. He also understood that the existence of such a point of intersection implies that the geometry of the curves W^s and W^u must indeed constitute an "infinitely thin tangle". Poincaré did not try to draw the tangle, but to illustrate the complexity of the situation he shows that the existence of a transverse homoclinic

---
[11]in this domain!

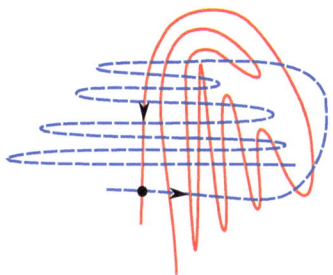

FIGURE 10. A homoclinic orbit

intersection induces that of infinitely many other homoclinic orbits: the arc of W^s situated between x and $\phi(x)$ crosses the unstable manifold W^u infinitely often.

In [68] Shil'nikov wonders why Poincaré does not attach more importance to Hadamard's article, where one may obviously detect the presence of homoclinic orbits: an infinite sequence of $1, 2, 3$ without repetitions and taking the value 1 only finitely many times defines a geodesic on the pants, which converges towards the periodic orbit g_1 both in the future and in the past. But Poincaré thinks that [60]:

> The three-body problem should not be compared to the geodesics on surfaces of opposite curvatures; on the opposite, it should be compared to the geodesics on convex surfaces.

The opposition hyperbolic/elliptic, negative/positive curvature is not new.

The next step in the analysis of homoclinic orbits is due to George Birkhoff in 1935: through a cute geometrical argument, he establishes that one can find periodic points arbitrarily close to a transverse homoclinic point [13].

Then comes Smale's rediscovery 25 years later. If one considers a rectangle R near the stable manifold, containing both the origin and the homoclinic point, like in Figure 11, then a sufficient iterate ϕ^l on R acts like a *horseshoe*. One thus

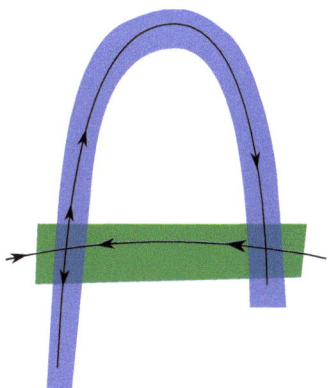

FIGURE 11. A homoclinic point and its horseshoe

obtains a description of the "internal mechanism" of this thin tangle, for instance using sequences of symbols 0, 1 as we have seen above. The presence of infinitely many periodic orbits becomes obvious. One can read in [68] an account on the evolution of the ideas around these homoclinic orbits.

5. The Lorenz attractor

How could one explain the lack of communication in the 1960s and 1970s between a theoretical physicist like Lorenz and a mathematician like Smale? The first one was working on the East Coast and the second one on the West Coast of the United States ... According to Williams [82], one reason would be the journal where Lorenz published his article[12]:

> Though many scientists, especially experimentalists, knew this article, it is not too surprising that most mathematicians did not, considering for example where it was published. Thus, when Ruelle–Takens proposed (1971) specifically that turbulence was likely an instance of a "strange attractor", they did so without specific solutions of the Navier–Stokes equations, or truncated ones, in mind. This proposal, controversial at first, has gained much favor.

It seems that Smale had very few physical motivations when cooking up his theory of hyperbolic systems, while physics itself does not seem to encompass many hyperbolic systems. This is at least Anosov's point of view [4]:

> *One gets the impression that the Lord God would prefer to weaken hyperbolicity a bit rather than deal with restrictions on the topology of an attractor that arise when it really is "1960s-model" hyperbolic.*

Even nowadays, it is not easy to find physical phenomena with strictly hyperbolic dynamics (see however [35, 39]). In my view, one of the main challenges of this part of mathematics is to "restore contact" with physics.

5.1. Lorenz and his butterfly

Lorenz's 1963 article [44] is magnificent. Lorenz had been studying for a few years simplified models describing the motion of the atmosphere, in terms of ordinary differential equations depending on few variables. For instance, in 1960 he describes a system he can explicitly solve using elliptic functions: the solutions are "still" *quasiperiodic in time* [42]. His 1962 article analyzes a differential equation in a space of dimension 12, in which he numerically detects a sensitive dependence to initial conditions [43]. But it is the 1963 paper which – for good reasons – lead him to fame. The aim of the paper is clear:

> In this study we shall work with systems of deterministic equations which are idealizations of hydrodynamical systems.

---

[12] One may also wonder whether the prestigious journal where Williams published his paper [82] is accessible to physicists.

After all, the atmosphere is made of finitely many particles, so one indeed needs to solve an ordinary differential equation in a space of "huge" dimension. But of course, such equations are "highly intractable", and one must treat them through partial differential equations; in turn, the latter must be discretized on a finite grid, leading to new ordinary differential equations depending on fewer variables, and probably more useful than the original ones. Lorenz discusses the type of differential equations he wants to study.

> In seeking the ultimate behavior of a system, the use of conservative equations is unsatisfactory [...]. This difficulty may be obviated by including the dissipative processes, thereby making the equations non conservative, and also including external mechanical or thermal forcing, thus preventing the system from ultimately reaching a state at rest.

A typical differential equation presenting both *viscosity* and *forcing* has the following form:

$$\frac{dx_i}{dt} = \sum_{j,k} a_{ijk} x_j x_k - \sum_j b_{ij} x_j + c_i$$

where $\sum a_{ijk} x_i x_j x_k$ vanishes identically[13] and $\sum b_{ij} x_i x_j$ is positive definite. The quadratic terms $a_{ijk} x_j x_k$ represent the *advection*, the linear terms $\sum b_{ij} x_j$ correspond to the *friction* and the constant terms to the *forcing*. Lorenz observes that under these conditions, the vector field is transverse to spheres of large radii (that is, of high energy), so that the trajectories entering a large ball will stay there forever. He can then discuss diverse notions of stability, familiar to contemporary mathematicians; periodicity, quasiperiodicity, stability in the sense of Poisson, etc. The bibliographic references in Lorenz's article include one article of Poincaré, but it is the famous one from 1881 [57]. In this article, founding the theory of dynamical systems, Poincaré introduces the limit cycles, and shows particular cases of the Bendixson–Poincaré theorem, introduces the first return maps etc., but there is no mention of chaotic behavior yet; chaos will be studied starting from the 1890 memoir we have discussed above, which Lorenz seems to have overlooked. Another bibliographic reference is a dynamical systems book by Birkhoff published in 1927. Again, this reference precedes Birkhoff's works in which he "almost" obtains a horseshoe...

Then, Lorenz considers as example the phenomenon of *convection*. A thin layer of a viscous fluid is placed between two horizontal planes, set at two different temperatures, and one wants to describe the resulting motion. The high parts of the fluid are colder, therefore denser; they have thus a tendency to go down due to gravity, and are then heated when they reach the lower regions. The resulting circulation of the fluid is complex. Physicists know well the Bénard and Rayleigh experiments. Assuming the solutions are periodic in space, expanding in Fourier

---

[13] This condition expresses the fact that the "energy" $\sum x_i^2$ is invariant through the quadratic part of the field.

series and truncating these series to keep only few terms, Salzman had just obtained an ordinary differential equation describing the evolution. Simplifying again this equation, Lorenz obtained "his" equation:

$$dx/dt = -\sigma x + \sigma y$$
$$dy/dt = -xz + rx - y$$
$$dz/dt = xy - bz.$$

Here x represents the intensity of the convection, y represents the temperature difference between the ascending and descending currents, and z is proportional to the "distortion of the vertical temperature profile from linearity, a positive value indicating that the strongest gradients occur near the boundaries". Obviously, one should not seek in this equation a faithful representation of the physical phenomenon ... The constant σ is the *Prandtl number*. Guided by physical considerations, Lorenz is lead to choose the numerical values $r = 28, \sigma = 10, b = 8/3$; it was a good choice, and these values remain traditional. He could then numerically solve these equations, and observe a few orbits. The electronic computer Royal McBee LGP-30 was rather primitive: according to Lorenz, it computed (only!) 1000 times faster than by hand ...

But Lorenz's observations are nevertheless remarkably fine. He first observes the famous *sensitivity to initial conditions*[14]. More importantly, he notices that these sensitive orbits still seem to accumulate on a complicated compact set, which is itself *insensitive* to initial conditions. He observes that this invariant compact set approximately resembles a surface presenting a "double" line along which two leaves meet each other.

> Thus within the limits of accuracy of the printed values, the trajectory is confined to a pair of surfaces which appear to merge in the lower portion. [...] It would seem, then, that the two surfaces merely appear to merge, and remain distinct surfaces. [...] Continuing this process for another circuit, we see that there are really eight surfaces, etc., and we finally conclude that there is an infinite complex of surfaces, each extremely close to one or the other of the two merging surfaces.

Figure 12 is reprinted from Lorenz's article. Starting from an initial condition, the orbit rapidly approaches this "two-dimensional object" and then travels "on" this surface. The orbit then turns around the two holes, left or right, in a seemingly random way. Notice the analogy with Hadamard's geodesics turning around the bull's horns.

---

[14]The anecdote is quite well known "I started the computer again and went out for a cup of coffee"... It was told in the conference Lorenz gave on the occasion of the 1991 Kyoto prize, "A scientist by choice", which contains many other interesting things. In particular, he discusses there his relations with mathematics. In 1938 Lorenz is a graduate student in Harvard and works under the guidance of G. Birkhoff "on a problem in mathematical physics". He does not mention any influence of Birkhoff on his conception of chaos. A missed encounter? On the other hand, Lorenz mentions that Birkhoff "was noted for having formulated a theory of aesthetics". Almost all Lorenz's works, including a few unpublished ones, can be downloaded on http://eapsweb.mit.edu/research/Lorenz/publications.htm.

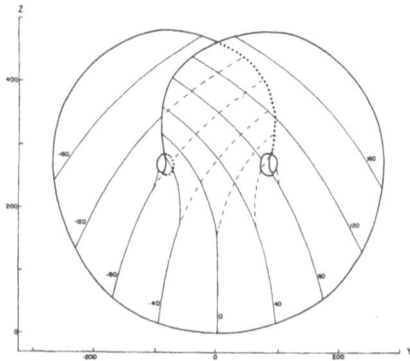

FIGURE 12. Lorenz's diagram. From [44, p. 38] with permission from © 1963 American Meteorological Society.

Besides, Lorenz studies the way the orbits come back to the "branching line" between the two leaves, which can be parametrized by an interval $[0, 1]$. Obviously, this interval is not very well defined, since the two leaves do not really come in contact, although they coincide "within the limits of accuracy of the printed values". Starting from a point on this interval, one can follow the future trajectory and observe its return onto the interval. For this first return map $[0, 1] \to [0, 1]$, each point has one image but two preimages. This corresponds to the fact that, to go back in time and describe the past trajectory of a point in $[0, 1]$, one should be able to see two copies of the interval; these copies are undistinguishable on the figure, so that two different past orbits emanate from the "same point" of the interval. But of course, if there are two past orbits starting from "one" point, there are four, then eight, etc., which is what Lorenz expresses in the above quotation. Numerically, the first return map is featured on the left part of Figure 13. Working by analogy, Lorenz compares this application to the (much simpler) following one: $F(x) = 2x$ if $0 \leqslant x \leqslant 1/2$ and $F(x) = 2 - 2x$ if $1/2 \leqslant x \leqslant 1$ (right part of Figure 13). Nowadays the chaotic behavior of this "tent map" is well known, but this was much less classical in 1963 ... In particular, the *periodic points* of F are exactly the rational numbers with odd denominators, which are dense in $[0, 1]$. Lorenz does not hesitate to claim that the same property applies to the iterations of the "true" return map. The periodic orbits of the Lorenz attractor are "thus" dense. What an intuition!

> There remains the question as to whether our results really apply to the atmosphere. One does not usually regard the atmosphere as either deterministic or finite, and the lack of periodicity is not a mathematical certainty, since the atmosphere has not been observed forever.

To summarize, this article contains *the first example of a dissipative and physically relevant dynamical system presenting all the characteristics of chaos. The orbits are unstable but their asymptotic behavior seems relatively insensitive to initial*

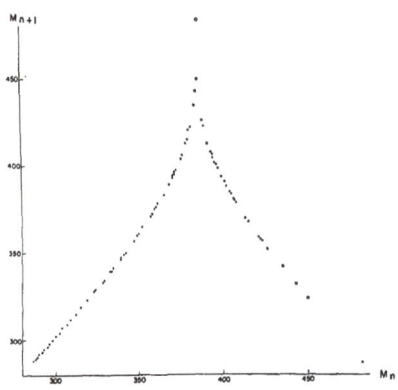
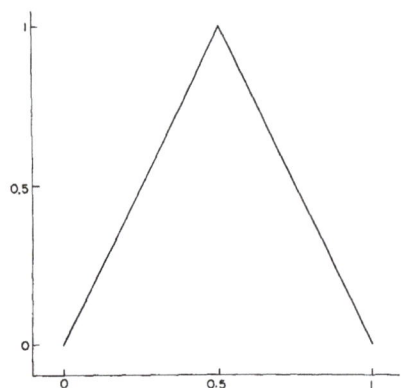

FIG. 4. Corresponding values of relative maximum of Z (abscissa) and subsequent relative maximum of Z (ordinate) occurring during the first 6000 iterations.

FIG. 5. The function $M_{n+1}=2M_n$ if $M_n<\frac{1}{2}$, $M_{n+1}=2-2M_n$ if $M_n>\frac{1}{2}$, serving as an idealization of the locus of points in Fig. 4.

FIGURE 13. Lorenz's graphs of first return maps. From [44, p. 38] with permission from © 1963 American Meteorological Society.

conditions. None of the above assertions is justified, at least in the mathematical sense. How frustrating!

Very surprisingly, an important question is not addressed in Lorenz's article. The observed behavior happens to be *robust*: if one slightly perturbs the differential equation, for instance by modifying the values of the parameters, or by adding small terms, then the new differential equation will feature the same type of attractor with the general aspect of a surface. This property will be rigorously established later, as we will see.

Everybody has heard of the "butterfly effect". The terminology seems to have appeared in the best-selling book of Gleick [29], and be inspired by the title of a conference by Lorenz in 1972 [45].

5.2. Guckenheimer, Williams and their template

The Lorenz equation pops up in mathematics in the middle of the 1970s. According to Guckenheimer [32], Yorke mentioned to Smale and his students the existence of this equation, which was not encompassed by their studies. The well-known 1971 article by Ruelle and Takens on turbulence [67] still proposes hyperbolic attractors as models, but in 1975 Ruelle observes that "Lorenz's work was unfortunately overlooked" [65]. Guckenheimer and Lanford are among the first people to show some interest in this equation (from a mathematical point of view) [31, 40]. Then the object will be fast appropriated by mathematicians, and it is impossible to give an exhaustive account of all their works. As soon as 1982 a whole book is devoted to the Lorenz equation, although it mostly consists in a list of open problems for mathematicians [79].

I will only present here the fundamental works of Guckhenheimer and Williams, who constructed the *geometric Lorenz models* [33, 82] (independently from Afraimovich, Bykov and Shil'nikov [2]). The initial problem consists in justifying the phenomena *observed* by Lorenz on his equation. We have seen that this equation is itself a rough approximation of the physical phenomenon. Proving that the precise Lorenz equation satisfies the observed properties is thus not the most interesting issue. Guckenheimer and Williams have another aim: they consider the behaviors observed by Lorenz as an *inspiration*, in order to construct vector fields, called geometric Lorenz models, satisfying the following properties:

- *for each or these fields, the set of nonwandering points is not hyperbolic, since it contains both nonsingular points and a singular one*
- *the fields are not structurally stable;*
- *the fields form an open set in the space of vector fields.*

Let us first consider a linear vector field:

$$\frac{dx}{dt} = ax, \quad \frac{dy}{dt} = -by, \quad \frac{dz}{dt} = -cz$$

with $0 < c < a < b$. Let C be the square $[-1/2, 1/2] \times [-1/2, 1/2] \times \{1\} \subset \mathbb{R}^3$. The orbit starting from a point $(x_0, y_0, 1)$ in this square is $(x_0 \exp(at), y_0 \exp(-bt), \exp(-ct))$. Call $T_\pm$ the triangular zones where these orbits intersect the planes given by $\{x = \pm 1\}$. They are defined by the equations $x = \pm 1$, $|y| \leq \frac{1}{2}|z|^{b/c}$ and $z > 0$, so they are "triangles" with their lower corner being a cusp. One then considers the zone ("box") B swept by the orbits starting from C until they reach $T_\pm$, to which one adds the future orbits of the points in $\{x = 0; -1/2 \leq y \leq 1/2; z = 1\}$ (which never intersect $T_\pm$), as well as the wedge $\{-1 \leq x \leq 1; y = 0; z = 0\}$ (see Figure 14).

The "Lorenz vector field" we will construct coincides with this linear vector field inside the box B. Outside one proceeds such that the orbits exiting from $T_\pm$

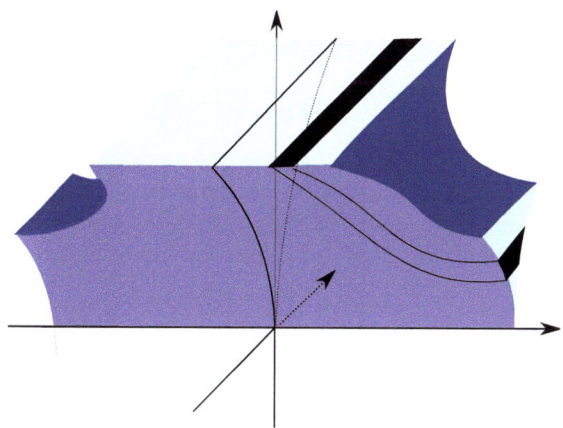

FIGURE 14. The box B

FIGURE 15. A Lorenz geometric model

come back inside the square C. One then obtains a vector field defined only inside a certain domain D sketched in Figure 15.

The main objective is to understand the dynamics inside D, but one can also extend the vector field outside this domain to get a globally defined field. Such a vector field X is a *geometric Lorenz model*.

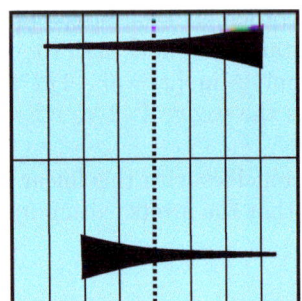 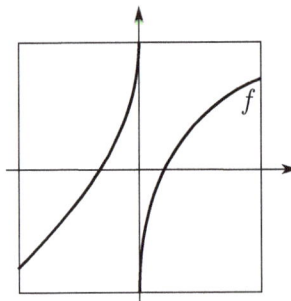

FIGURE 16. Return maps

An important remark is that not all the points of the triangles originate from the square: the tips do not, since they come from the singular point at the origin. There are several ways to organize the return from $T_\pm$ onto C; one can make sure that the Poincaré return map $F : C \to C$ has the following form:

$$F(x,y) = (f(x), H(x,y)).$$

Technically, one requires that $H(x,y) > 1/4$ for $x > 0$ and $H(x,y) < 1/4$ for $x < 0$. Furthermore, that the map $f : [-1/2, 1/2] \to [-1/2, 1/2]$ satisfies the following conditions:

1. $f(0^-) = 1/2$, $f(0^+) = -1/2$;
2. $f'(x) > \sqrt{2}$ for all x in $[-1/2, 1/2]$.

The second conditions implies that for all interval J contained in $[-1/2, 1/2]$, there exists an integer $l > 0$ such that $f^l(J) = [-1/2, 1/2]$[15]

To describe the structure of the orbits inside the box, Williams introduces the concept of *template*. Figure 17 is reprinted from [15][16]: We are dealing with a

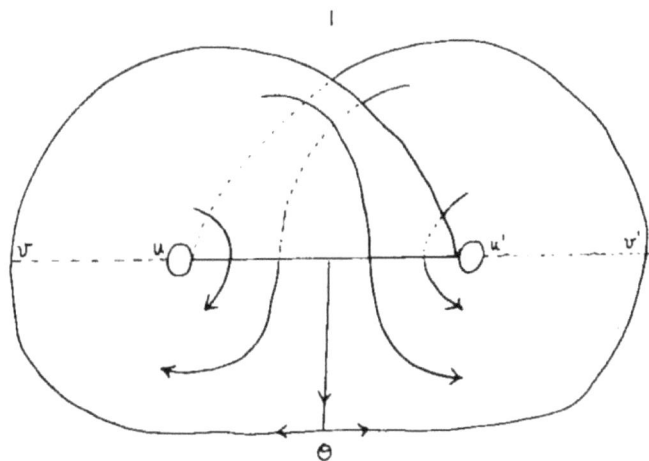

FIGURE 17. The template. From [15, p. 52] with permission from © 1983 Elsevier.

branched surface $\Sigma \subset \mathbb{R}^3$ embedded in space, on which one can define a *semiflow* ψ^t ($t \geqslant 0$). A semiflow means that $\psi^t : \Sigma \to \Sigma$ is defined only for $t \geqslant 0$ and that $\psi^{t_1+t_2} = \psi^{t_1} \circ \psi^{t_2}$ for all t_1, t_2. The trajectories of the semiflow are sketched on the figure: a point in Σ has a future but has no past, precisely because of the two leaves which meet along an interval. The first return map on this interval is chosen to be the map f defined above. The dynamics of the semiflow is easy to understand: the orbits turn on the surface, either on the left or on the right wing, according to the signs of the iterates $f^k(x)$

We shall now construct a flow starting from the semiflow using a well-known method, the *projective limit*. One considers the abstract space $\hat{\Sigma}$ of the curves $c : \mathbb{R} \to \Sigma$ which are trajectories of ψ^t in the following sense: for all $s \in \mathbb{R}$ and $t \in \mathbb{R}^+$, one has $\phi^t(c(s)) = c(s+t)$. Given a point x on Σ, to choose such a curve c with $c(0) = x$ amounts to "selecting a past" for x: on goes backwards in time along the semiflow, and at each crossing of the interval, one chooses one of the two possible preimages. The map $c \in \hat{\Sigma} \mapsto c(0) \in \Sigma$ thus has totally discontinuous fibers, which are Cantor sets. The space $\hat{\Sigma}$ is an abstract compact set equipped with a *flow* $\hat{\psi}^t$ defined by $\hat{\psi}^t(c)(s) = c(s+t)$, which now makes sense for all $t \in \mathbb{R}$.

---

[15]The fact that the graph of f does not resemble Figure 13 is due to a different choice of notations.
[16]Incidentally, this figure shows that the quality of an article does not depend on that of its illustrations . . .

Let us fix a Lorenz model X generating a flow ϕ^t and associated with a first return map f. Williams shows in [82] that: *There exists a compact Λ contained in the box D, such that:*

- *the ω-limit set of each point in D is contained in Λ;*
- *Λ is invariant through ϕ^t, and the restriction of ϕ^t on Λ is topologically conjugate with $\hat{\psi}^t$ acting on $\hat{\Sigma}$;*
- *Λ is topologically transitive: it contains an orbit which is dense in Λ;*
- *the union of all periodic orbits is dense in Λ.*

We have now justified Lorenz's intuition, according to which the attractor Λ behaves as a surface "within the limits of accuracy of the printed values".

We also see that the topological dynamics of the vector field is completely determined by that of the map f. To understand it, one uses the notion of *kneading sequence* introduced around the same time in a more general context. Take the two sequences $\alpha_k, \beta_k = \pm$ given by the signs of $f^k(0^-)$ et $f^k(0^+)$ for $k \geqslant 0$ (the sign of 0 is defined to be +). Obviously, two applications f of the above type which are conjugated by a homeomorphism (preserving the orientation) define the same sequences α_k, β_k, and the converse is true. Back to the vector field, these two sequences can be obtained by considering the two unstable separatrices starting from the origin, which will intersect the square C infinitely many times, either in the part $x > 0$ or in the part $x < 0$. The two sequences precisely describe the forward evolution of these two separatrices. It should be clear that these fields are structurally unstable, since a slight perturbation can change the sequences α_k, β_k.

Guckenheimer and Williams prove that the two sequences contain all the topological information on the vector field. Precisely, the establish the following theorem:

There exists an open $\mathcal{U}$ in the space of vector fields in $\mathbb{R}^3$ and a continuous map π from $\mathcal{U}$ into a two-dimensional disk, such that two fields X, Y on $\mathcal{U}$ are conjugate through a homeomorphism close to the identity if and only if $\pi(X) = \pi(Y)$.

We won't dwell on the proofs, but will insist on an important point: the Lorenz flow, although nonhyperbolic, is still "singular hyperbolic". Namely, at each point x of the attractor Λ, one can decompose the tangent space into a direct sum of a line E_x^s and a plane F_x, such that the following properties hold:

- E_x^s and F_x depend continuously of the point $x \in \Lambda$, and are invariants through the differential of the flow ϕ^t;
- the vectors in E_x^s are contracted by the flow: there exists $C > 0, \lambda > 0$ such that for all $t > 0$ and all $v \in E_x^s$, one has $\|d\phi^t(v)\| \leqslant C\exp(-\lambda t)\|v\|$;
- the vectors in F_x are not necessarily expanded by the flow, but they cannot be contracted as much as the vectors in E^s. Precisely, if $u \in F_x$ and $v \in E_x^s$ are unitary, one has $\|d\phi^t(u)\| \geqslant C^{-1} \exp(\lambda t)\|d\phi^t(v)\|$ for all $t > 0$;
- the flow uniformly expands the two-dimensional volume along

$$F_x : \det(d\phi^t|F_x) \geqslant \exp(\lambda' t)$$

for some $\lambda' > 0$ and all $t > 0$.

For each $x \in \Lambda$ one may consider the set $W^s(x)$ of the points $y \in \mathbb{R}^3$ such that the distance $\|\phi^t(x) - \phi^t(y)\|$ goes to zero faster than $\exp(-\lambda t)$. It is a smooth curve with tangent in x given by E_x^s. The collection of these curves defines a foliation of an open neighborhood of Λ, justifying the terminology "*attractor*": all the points in this neighborhood are attracted by Λ and their trajectories are asymptotes of trajectories in Λ. The branched surface Σ is constructed from the local leaves of this foliation. Of course, one needs to show that all these structures exist, and that they persist upon a perturbation of the vector field.

Hence, the open set in the space of vector fields detected by Guckenheimer and Williams trespasses the hyperbolic systems, but the dynamics of this type of fields can still be understood (at list qualitatively). The specificity of these fields is that they are, in some sense, suspensions of maps f on the interval, but also contain a singular point in their nonwandering set. In some sense, their dynamics can be translated into a discrete time dynamics, but the presence of the singular point shows that the first return time on C is unbounded. It would be too naive to believe that this type of phenomenon, together with hyperbolic systems, suffice to understand "generic dynamics". We will see later some other types of phenomena (different from hyperbolicity) happening in a stable manner. But the "Lorenz phenomenon" I have just described is doubtlessly one of the few phenomena representing a generic situation. We will come back later to this question.

As we have mentioned, the geometric models for the Lorenz attractor have been *inspired* by the original Lorenz equation, but it wasn't clear whether the Lorenz equation indeed behaves like a geometric model. This question was not really crucial, since Lorenz could clearly have made other choices to cook up his equation, which resulted from somewhat arbitrary truncations of Fourier series. Lorenz himself never claimed that his equation had any physical sense. Nevertheless, the question of the connection between the Lorenz equation and the Guckenheimer-Williams dynamics was natural, and Smale chose it as one of the "mathematical problems for the next century" in 1998 [78]. The problem was positively solved by Tucker [80] (before the "next century"!). The goal was to construct a square C adapted to the original Lorenz equation, the first return map on C, and to check that they have the properties required by the geometric model. The proof uses a computer, and one needs to bound from above the errors. The major difficulty – which makes the problem quite delicate – is due to the presence of the singular point, and the fact that the return time may become very large ... For a brief description of the method used by Tucker, see for instance [81].

6. The topology of the Lorenz attractor

6.1. Birman, Williams and their "can of worms"

We have seen that the periodic orbits of a geometric Lorenz model are dense in the attractor. To better understand the topology of the attractor, Birman and Williams had the idea to consider these periodic orbits as knots. A *knot* is a closed oriented curve embedded in space without double points. A topologist will consider that two knots are identical (the technical term is *isotopic*) if it is possible to continuously deform the former into the latter without any double point. Now the questions are: which knots are represented by at least one periodic orbit of the Lorenz model? Can a single knot be represented by infinitely many periodic orbits? Beyond knots, one can also consider the *links*, which are unions of finitely many disconnected knots. Each collection of finitely many periodic orbits defines a link. Since the periodic orbits are dense in the attractor, one can hope to approximate the latter by a link containing a huge number of periodic orbits. The article [15] nicely mixes topology and dynamics.

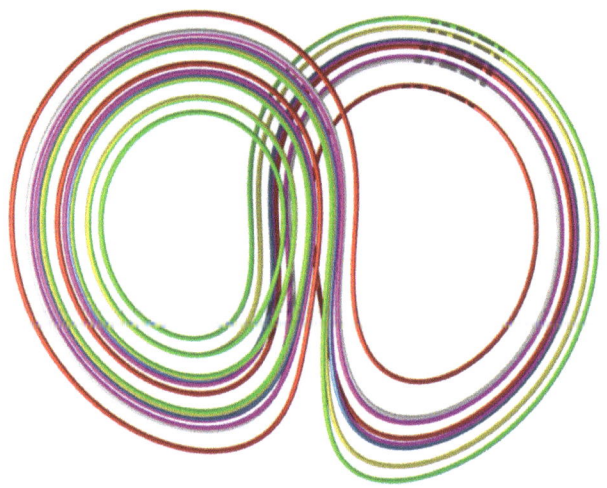

FIGURE 18. A few periodic orbits

A priori, this study of knots and links should be performed for each geometric Lorenz model, that is for each return map f, or more precisely for each choice of a pair of kneading sequences α, β. However, we may restrict ourselves to the particular case of the "multiplication by two" case, by setting $f_0(x) = 2x + 1$ for $x \in [-1/2, 0[$ and $f_0(x) = 2x - 1$ for $x \in [0, 1/2]$. A point in $[-1/2, 1/2]$ is fully determined by the sequence of signs of its iterates by f_0, and each sequence of signs corresponds to a point in $[-1/2, 1/2]$. This is nothing but the dyadic decomposition of numbers in $[-1/2, 1/2]$. If f is the return map on $[-1/2, 1/2]$ for a given geometric Lorenz model, one can associate to each point $x \in [-1/2, 1/2]$ the unique point $h(x) \in [-1/2, 1/2]$ such that for each $k \geqslant 0$ the numbers $f^k(x)$ and $f_0^k(h(x))$ have the same sign. This defines an injection $h : [-1/2, 1/2] \to [-1/2, 1/2]$ such that $h \circ f = f_0 \circ h$, and one can thus think that f_0 "contains" all the one-dimensional dynamics we are interested in. Of course, the map h is not always a bijection, depending on the specific geometric Lorenz model. We will study the maximal case of f_0, since it contains all the others. Strictly speaking, f_0 cannot be a first return map of a geometric Lorenz model, since its derivative does not blow up near $x = 0$. However, starting from f_0 one may construct, like in the previous section, a *topological* semiflow on the branched surface, and then a flow through a projective limit. This topological flow is perfectly adapted to our problem, which is the nature of knots and links formed by the periodic orbits of the geometric Lorenz models.

Each periodic orbit of the Lorenz flow can be projected onto the template, so it is associated with a periodic orbit of f. One could fear that this projection could modify the topology of the knot through the appearance of double points. But this does not happen because the projection on the template occurs along the (one-dimensional) stable manifolds, and clearly a stable manifold can meet a periodic orbit on at most one point (two different points cannot simultaneously be periodic and asymptotes of each other).

The periodic points of f_0 are easy to determine: they are the rational numbers with denominators of the form $2(2l+1)$, with $l \in \mathbb{Z}$. If x is such a periodic point of period k, one constructs a braid as follows: inside the square $[-1/2, 1/2] \times [0, 1]$, for $i = 0, \ldots, k-1$ one connects the points $(f^i(x), 0)$ and $(f^{i+1}(x), 1)$ by a segment, such that the segments "climbing to the right" are below those "climbing to the left", like on Figure 19. Then, one closes the braid as usual in topology, to obtain

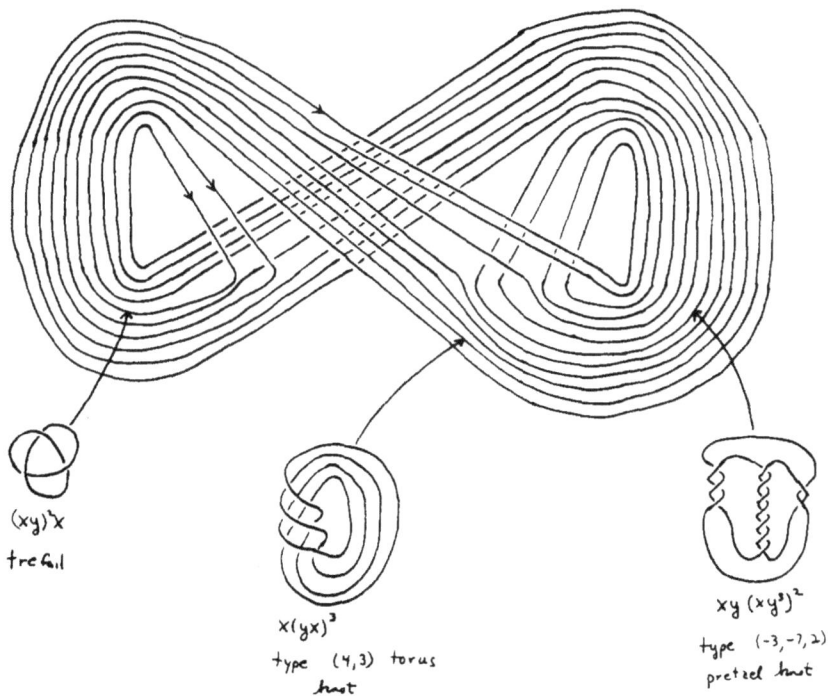

FIGURE 19. A Lorenz link

a knot $k(x)$. Starting from a finite number of periodic points $x_1, x_2, \ldots, x_l$, one obtains a link. Those are called the "Lorenz knots and links".

Before describing some of the results obtained in [15, 83], I should remind some definitions related with knots. An oriented knot is always the boundary of a oriented surface embedded in the three-dimensional space (called its Seifert surface). The minimal genus of such a surface is the *genus* of the knot. If a knot cannot be obtained as the connected sum of two nontrivial knots, it is said to be *prime*. A knot is said to be *chiral* if it cannot be continuously deformed into its mirror image. Given a knot k embedded in the three-dimensional sphere $\mathbb{S}^3$ (union of $\mathbb{R}^3$ and a point at infinity), it is said to be *fibered* if the complement $\mathbb{S}^3 \setminus k$ fibres on the circle; this means that there exists a family of surfaces with boundaries, parametrized by an angle $\theta \in \mathbb{S}^1$, which all share the same boundary k but do not intersect each other outside their boundary, and which altogether cover the whole sphere. Near the knot, the surfaces looks like the pages of a book near the binding. Finally, given two disjoint knots k_1 and k_2, one can define as follows their *linking number*: choose an oriented surface with boundary k_1, and count the algebraic intersection number of k_2 with this surface. This integer $\mathrm{enl}(k_1, k_2)$ is independent of the chosen surface, and $\mathrm{enl}(k_1, k_2) = \mathrm{enl}(k_2, k_1)$.

Here are a few properties of the Lorenz links:

- *The genus of a Lorenz knot can be arbitrary large.*
- *The linking number of two Lorenz knots is always positive.*
- *The Lorenz knots are prime.*
- *The Lorenz links are fibered.*
- *The nontrivial Lorenz knots are chiral.*

The Lorenz knots are very particular. For instance, using a computer and the tables containing the 1 701 936 prime knots representable by plane diagrams with less than 16 crossings, one can show that only 21 of them are Lorenz knots [28]. More information on the Lorenz knots and the recent developments on the subject can be found in [22, 14]. Surprisingly, Ghrist has shown that if one embeds the branched surface in 3-space by twisting one of the wings by a half-turn, the resulting flow is *universal*: *all* the links are represented by (finite collections of) periodic orbits of that flow [25].

6.2. The right-handed attractor...

Try to see the attractor as a "topological limit" of its periodic orbits, and take into account the fact that each finite union of periodic orbits defines a fibered link; how do these circle fibrations behave when the number of components of the link tends to infinity? In a way, one would like to see the complement of the attractor itself as a fibered object ... In [27] I propose a global description of these fibrations, based on the concept of *right-handed* vector field.

Consider a *nonsingular* vector field X on the sphere $\mathbb{S}^3$, generating a flow ϕ^t. Let us call $\mathcal{P}$ the convex compact set formed by the probability measures invariant by the flow. This space contains, for instance, the probability measures equidistributed on the periodic orbits (if any), as well as their convex combinations. A general invariant probability measure can be seen as a generalized periodic orbit (a *foliated cycle* in Sullivan's terminology). We have seen that two knots in the sphere always define a linking number; I show that one can also define a linking quadratic form enl : $\mathcal{P} \times \mathcal{P} \to \mathbb{R}$. The flow is then called right-handed if enl only takes positive values. For the Lorenz flow, enl is only nonnegative, but it is positive when restricted to invariant measures which do not charge the singular point. This fact can be directly "read" on Williams's template, since two arcs of trajectory only have positive intersections when projected onto the plane. The Lorenz flow is thus not strictly right-handed.

One of the main results in [27] is that a right-handed flow is always *fibered*, in the following sense. One can find a positive *Gauss linking form* on the flow. Precisely, there exists a $(1,1)$-differential form Ω on $\mathbb{S}^3 \times \mathbb{S}^3$ minus the diagonal, such that:

- if $\gamma_1, \gamma_2 : \mathbb{S}^1 \to \mathbb{S}^3$ *are two disjoint closed curves, their linking number is given by the Gauss integral* $\int\int \Omega_{\gamma_1(t_1),\gamma_2(t_2)}(\frac{d\gamma_1}{dt_1}, \frac{d\gamma_2}{dt_2}) dt_1 dt_2$;
- *if x_1, x_2 are two distinct points, then* $\Omega_{x_1,x_2}(X(x_1), X(x_2)) > 0$.

Interestingly, such a Gauss form directly provides a fibration on the complement of each periodic orbit. Indeed, for μ any invariant probability measure, the integral

$$\omega_x(v) = \int \Omega_{x,y}(v, X(y))\, d\mu(y)$$

defines a *closed and nonsingular* 1-form on the complement of the support of μ. If μ is supported on a periodic orbit, the form ω admits a multivalued primitive, which defines a fibration on the circle of the complement of the orbit.

Even though the Lorenz flow does not strictly belong to this frame, I show in [27] how this construction allows to better understand the results of Birman and Williams I have described above.

6.3. From Lorenz back to Hadamard...

Hadamard studies the geodesics on negatively curved pants, and shows that they can be described by the sequences of symbols enumerating the crossed seams. Williams analyzes the trajectories of the Lorenz attractor by a semiflow on the "Lorenz template", itself described by two sequences of symbols, following the wings (left or right) successively crossed by two limiting trajectories. It is thus not surprising to find a connection between these two dynamics. Such a connection was indeed exhibited in [26].

Following a nonwandering geodesic on the pants P, after each crossing with a seam one may consider to turn right or turn left to reach the next seam. It is thus possible to associate to each nonwandering geodesic a bi-infinite sequence of "left/right" symbols. Yet, this new coding is not perfect, because it is not bijective. Assume the pants is embedded symmetrically in space, meaning that it is invariant

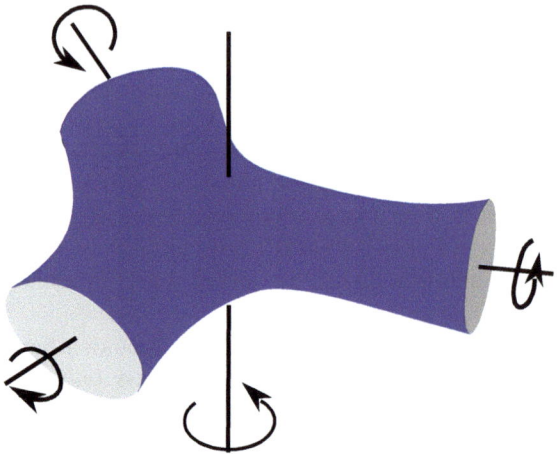

FIGURE 20. Symmetries of the pants

through six rotations (the identity, two rotations of order three, and three rotations of order two), as in Figure 20.

Obviously, each of these rotations maps a geodesic of P into another geodesic, which shares the same "left-right" sequence. One is thus brought to consider the quotient $\hat{P}$ of P by this group of order six. This quotient is not exactly a surface, due to the intersection points of the rotation axes with P. Ignoring this "detail", the geodesics of $\hat{P}$ are bijectively coded by the sequences "left-right", like the orbits of the Lorenz flow.

We now recognize in $\hat{P}$ a variant of the *modular surface*, quotient of the Poincaré half-plane $\mathbb{H} = \{z \in \mathbb{C} | \Im z > 0\}$ through the action of $\Gamma = \mathrm{PSL}(2, \mathbb{Z})$ by Moebius transformations. The subgroup $\Gamma[2]$ formed by the matrices congruent to the identity modulo 2, acts without fixed points on $\mathbb{H}$, and the quotient is a sphere minus three points. The group $\Gamma/\Gamma[2]$ of order six acts on this punctured sphere similarly as it acts on P. *A geodesic of the modular surface is thus coded by a sequence of "left-right" symbols, like an orbit of the Lorenz flow.*

To go further, one should compare the spaces supporting these two dynamics. For the Lorenz equation, it is the usual space $\mathbb{R}^3$. For the modular geodesic flow, it is the quotient $\mathrm{PSL}(2, \mathbb{R})/\mathrm{PSL}(2, \mathbb{Z})$: this is due to the fact that $\mathrm{PSL}(2, \mathbb{R})$ can be identified with the unitary tangent bundle to $\mathbb{H}$. Figure 21 displays a fundamental domain for the action of $\mathrm{PSL}(2, \mathbb{Z})$ on the Poincaré half-plane. Three copies of this domain form a circular triangle, with corners sitting on the real axis. Two copies of this triangle form a fundamental domain for $\Gamma[2]$.

In [26] I use the topological identification (through the classical modular forms) of $\mathrm{PSL}(2, \mathbb{R})/\mathrm{PSL}(2, \mathbb{Z})$ with the complement in the sphere $\mathbb{S}^3$ of the trefoil knot, in order to represent the modular geodesics on the sphere. I can then "identify" the topological dynamics of the Lorenz flow and the modular geodesic

FIGURE 21. A fundamental domain for the action of $\mathrm{PSL}(2,\mathbb{Z})$ on the half-plane

flow. For instance, I show that the periodic geodesics on the modular surface, seen as knots on the sphere, can be deformed into the Lorenz knots. The left part of Figure 22 shows one of the periodic orbits of the modular geodesic flow, embedded in the complement of (a version of) the Lorenz template. The right part shows the Lorenz orbit with the same symbolic coding. The central part illustrates a step in the deformation. For more explanations, see [26, 28].

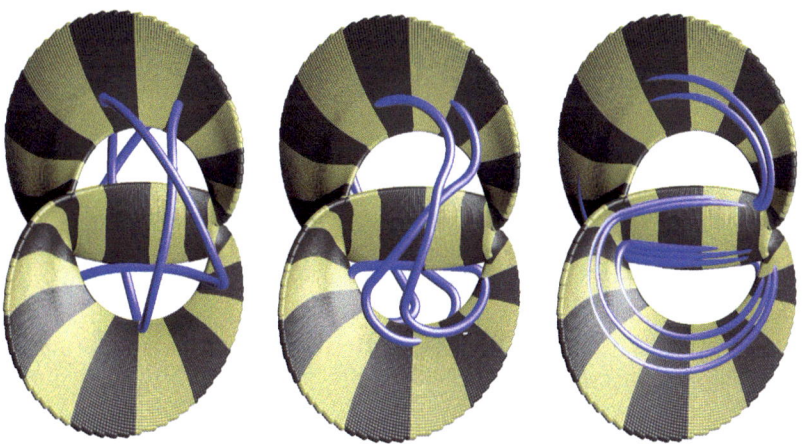

FIGURE 22. Deformation of a modular knot into a Lorenz knot

The analogy between a fluid motion and a geodesic flow is not new. In a remarkable 1966 article, Arnold showed that the Euler equation for perfect fluids is nothing but the equation for the geodesics on the infinite-dimensional group of volume preserving diffeomorphisms [7]. Besides, he shows that the sectional curvatures are "often" negative, inducing him to propose a behavior à la Hadamard for the solutions of the Euler equation. Of course, Lorenz does not consider perfect fluids, he starts from the Navier–Stokes equation instead of the Euler equation; still, the motivations are similar.

7. The attractor as a statistical object

7.1. Physical measures

As we have seen, one of the major ideas to "escape" the sensitivity to initial conditions is to introduce probabilities.

Lorenz is one of the people who have most clearly expressed this idea: "over the years minuscule disturbances neither increase nor decrease the frequency of occurrence of various weather events such as tornados; the most they may do is to modify the sequence in which these events occur."

A flow ϕ^t on a compact manifold V preserves at least one probability measure. This results, for instance, from the fact that this flow acts on the convex compact set of probability measures on V (in an affine way); such an action of an abelian group always admits a fixed point. This mathematical existence theorem will certainly not impress any physicist! But, once we have "selected" an invariant probability measure μ, we may invoke *Birkhoff's ergodic theorem*: for each function $u : V \to \mathbb{R}$ integrable w.r.t. the measure μ, the asymptotic time average

$$\tilde{u}(x) = \lim_{T \to +\infty} \frac{1}{T} \int_0^T u(\phi^t(x)) \, dt$$

exists for μ-almost all point x. The function $\tilde{u}$, defined almost everywhere, is obviously constant along the orbits of the flow. One can choose[17] the measure μ to be ergodic, so that any invariant function is constant almost everywhere. This ergodic theorem could be considered as fulfilling Lorenz's wish: the *frequency* of an event does not depend on the initial conditions. However, the difficulty lies in the fact that a given flow ϕ^t usually preserves infinitely many measures, and the sets of full measure for two invariant measures μ_1 and μ_2 may be disjoint. The question is to determine whether there exists a "physically meaningful" invariant measure.

In the early 1970s, Sinai, Ruelle and Bowen have discovered a fundamental concept to answer this question [71, 66, 17].

---

[17] An invariant probability measure is ergodic if any invariant subset has measure zero or one. The ergodic measures are the extremal points of the convex compact set of invariant probability measures, showing that there exist such measures (this existence proof is rather abstract).

Let us recall that on a manifold there is no intrinsic notion of Lebesgue measure, yet the Lebesgue negligible sets are well defined, that is, they do not depend on the choice of coordinate system. For this reason, we may say that a measure is absolutely continuous with respect to "the" Lebesgue measure. Lebesgue negligible sets could be considered to be negligible for a physicist as well...

It is time to give a precise definition of the word *attractor* I have used many times. *A compact set Λ in V is an* attractor *if:*

- *Λ is invariant by the flow ϕ^t;*
- *there exists a point in Λ the orbit of which is dense in Λ;*
- *the basin of attraction $B(\Lambda) = \{x \in V | \omega(x) \subset \Lambda\}$ has nonzero Lebesgue measure.*

A probability measure μ invariant by ϕ^t is an SRB *measure (for Sinai–Ruelle–Bowen) if, for each* continuous *function $u : V \to \mathbb{R}$, the set of points x such that*

$$\lim_{T \to +\infty} \frac{1}{T} \int_0^T u(\phi^t(x))\, dt = \int u\, d\mu$$

has nonzero Lebesgue measure.

This set of points is called the *basin* of μ, and denoted by $B(\mu)$.

Of course there are similar definitions for discrete time dynamical systems $\phi : V \to V$ (not necessarily bijective).

A simple example, yet not generic in our context, is that of a flow preserving a probability measure absolutely continuous w.r.t. the Lebesgue measure, and which is ergodic. Indeed, in that case the ergodic theorem precisely means that the time averages converge almost surely towards the spatial average. But I have already mentioned that our discussion mostly concerns the dissipative dynamics... An SRB measure is an invariant measure which "best remembers the Lebesgue measure".

Before giving more examples, let us start by a counterexample, due to Bowen. Let us consider a vector field admitting two saddle connections, as on Figure 23. For an initial condition x inside the region bounded by the separatrices, the future orbit will accumulate on the two singular points. Since the dynamics is slow near

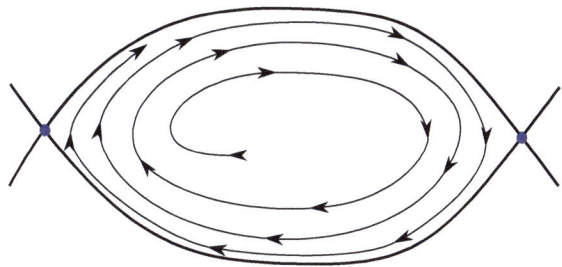

FIGURE 23. Bowen's counterexample

the singular points, the trajectories are made of a sequence of intervals during which the point is almost at rest, near one of the singular points, separated by short phases of fast transitions. The slow intervals become longer and longer. As a result, the time averages "hesitate" between the two Dirac masses on the singular points. The statistics do not converge to any invariant measure; there is no SRB measure.

This counterexample may seem a bit artificial, and in fact no significantly different counterexample is known. As we will see, *it is conjectured that the existence of an SRB measure holds generically among all dynamical systems* (see also [84]).

Before providing some significative examples, an important word of caution. Until now we have only considered the topological equivalence between flows: a homeomorphism mapping the orbits of one flow to those of another one. Of course, such a homeomorphism is often *singular* w.r.t. the Lebesgue measure, so it does not map an SRB measure of the first flow to an SRB measure of the second flow...

The simplest example is given by an *expanding map on the circle*, which stands as a model for all further developments. It is a map ψ, say infinitely differentiable, from the circle to itself, with derivative always greater than 1. Of course, such a map cannot be bijective. One can show that ψ is *topologically* conjugate to $x \in \mathbb{R}/\mathbb{Z} \mapsto d.x \in \mathbb{R}/\mathbb{Z}$ for a certain integer d ($d \neq \pm 1$) but, as we have just noticed, this does not imply any special property regarding the Lebesgue measure.

The *Distortion Lemma* is a simple, yet fundamental idea. Consider a branch F_k of the inverse of ψ^k, defined on an interval I of the circle. This branch can be written as $f_{k-1} \circ f_{k-2} \circ \cdots \circ f_0$, where each f_i is one of the inverse branches of ψ. To evaluate the difference $(\log F_k')(x_0) - (\log F_k')(y_0)$, let us set $x_i = f_i \circ \cdots \circ f_0(x_0)$ and $y_i = f_i \circ \cdots \circ f_0(y_0)$. We observe that $|x_i - y_i| \leqslant \delta^i |x_0 - y_0|$ for some $\delta < 1$, which allows to write:

$$|(\log F_k')(x) - (\log F_k')(y)| \leqslant \sum_{i=0}^{k-1} |(\log f_i')(x_i) - (\log f_i')(y_i)|$$

$$\leqslant C \sum_{i=0}^{k-1} \delta^i |x_0 - y_0|$$

$$\leqslant \frac{C}{1-\delta} |x_0 - y_0|,$$

where C is an upper bound for the derivative of $(\log f_i')$. This shows that the quotients $F_k'(x)/F_k'(y)$ are bounded uniformly w.r.t. k and the chosen branch. This is the statement of the lemma.

Now let Leb be the Lebesgue measure, and Leb_k its push-forward through ψ^k. The density of Leb_k w.r.t. Leb is the sum of the derivatives of all the inverse branches of ψ^k. This density is therefore bounded from below and from above by two positive constants, independently of the number k of iterations.

It is quite easy to check that any weak limit of the Birkhoff averages $\left(\sum_1^k \text{Leb}_i\right)/k$ is an measure invariant through ψ which is absolutely continuous w.r.t. Leb, with density bounded by the same constants. We have thus shown *the existence of an invariant measure absolutely continuous w.r.t. the Lebesgue measure.*

One can then show that ψ is *ergodic* w.r.t. the Lebesgue measure: if $A \subset \mathbb{R}/\mathbb{Z}$ is invariant through ψ (meaning that $\psi^{-1}(A) = A$), then the Lebesgue measure of A or of its complement vanishes. This property is also a consequence of the Distortion Lemma. Assume the Lebesgue measure of A is nonzero; one can then find intervals I_k such that $\operatorname{Leb}(I_k \cap A)/\operatorname{Leb}(I_k)$ goes to 1 when $k \to \infty$, provided we allow the lengths of the I_k to be arbitrary small. If we now set $J_k = \psi^{n_k}(I_k)$ for some appropriate n_k, the Distortion Lemma shows that $\operatorname{Leb}(J_k \cap A)/\operatorname{Leb}(J_k)$ also goes to 1, but this time we may assume that the lengths of the J_k are bounded from below. Hence, we can find intervals of lengths bounded from below, which contain an arbitrary large proportion of A. In the limit, we obtain a nonempty interval J_∞ such that $\operatorname{Leb}(J_\infty \cap A) = \operatorname{Leb}(J_\infty)$, implying that A has a nonempty interior, up to a negligible set. If we expand once more, using the fact that ψ has a dense orbit, we finally show that A has full measure.

We have just shown that *there exists a unique invariant measure absolutely continuous w.r.t. the Lebesgue measure*, and the Birkhoff ergodic theorem then implies that this measure is SRB.

This proof is quite easy, and it has been generalized in many ways, under less and less hyperbolicity assumptions.

The easiest generalization concerns an *expanding map* $\psi : V \to V$, where V may have dimension greater than 1. One shows analogously that there exists an invariant measure absolutely continuous w.r.t. the Lebesgue measure, and ergodic. For this it suffices to replace the estimates on the derivatives by estimates on Jacobian determinants.

The first really different case is that of a hyperbolic attractor. The situation is as follows: a diffeomorphism ϕ preserves a compact set Λ and the tangent bundle $T_\Lambda V$ above Λ splits into the sum $E^s \oplus E^u$ of a fibre made of uniformly contracted vectors, and another fibre made of uniformly expanded ones. We assume that Λ is an attractor, meaning that for each $x \in \Lambda$, the unstable manifold $W^u(x)$ – made of the points whose *past* orbit is asymptotic to that of x – is fully contained in Λ. Each point y in an open neighborhood U of Λ then lies on the stable manifold of a certain point $x \in \Lambda$: the future orbit of y is asymptotic to that of x, and thus accumulates in Λ.

The set Λ is foliated by unstable manifolds, which are uniformly expanded through ϕ. The idea is then to copy the Distortion Lemma, but only along the expanding directions. One starts from a probability measure supported in a small ball contained in some unstable manifold, and which is absolutely continuous w.r.t. the Lebesgue measure *on the unstable manifold*. One then successively pushes forward this measure through the diffeomorphism, and tries to show a weak convergence by time averaging the obtained sequence of measures. We don't need to estimate the full Jacobian determinants, but rather the unstable Jacobians, computed along E^u.

The result is as follows: one obtains an invariant measure which is not necessarily absolutely continuous w.r.t. the Lebesgue measure, but which is absolutely continuous *when restricted on the unstable manifolds*. The last statement needs some explanations. Locally, near a point $x \in \Lambda$, the situation is that of a product $B^u \times B^s$ of an open ball in the unstable manifold $W^u(x)$ and an open ball in the stable manifold $W^s(x)$. Two points with the same first coordinate (resp. the same second coordinate) lie in the same stable manifold (resp. unstable manifold). A measure μ can then be "disintegrated" in these local coordinates: if $A \subset B^u \times B^s$,

$$\mu(A) = \int_T \mu_t(A \cap (B^u \times \{t\})) \, d\nu(t),$$

where ν is a measure on B^s and μ_t is a measure on B^u defined for ν-almost all $t \in B^s$. The measures constructed above by time averaging are such that μ_t is absolutely continuous w.r.t. the Lebesgue measure on the unstable leaves, for ν-almost all t. Technically, μ is called an (unstable) *Gibbs state*.

There remains to show that these Gibbs states are SRB measures. If we consider the basin $B(\mu)$, namely the set of points for which, for any continuous test function, the time average is equal to the μ-average, this basin is clearly a union of stable manifolds: if two points have asymptotic future orbits, their time averages coincide. Besides, the defining property of Gibbs states implies that a set of full μ-measure intersects almost all the unstable manifolds along parts of full Lebesgue measure. Does this indeed imply that the basin has positive Lebesgue measure in V? To show it, one needs to establish that a subset of $B^u \times B^s$ which is a union of vertical balls, and meets each horizontal ball on a set of full Lebesgue measure in B^u, has positive full Lebesgue measure in $B^u \times B^s$. This would be the case if the parametrization of a neighborhood of x through $B^u \times B^u$ were differentiable, unfortunately it is generally not the case. On the other hand, one fundamental technical result states that the "holonomies" of these stable and unstable foliations are absolutely continuous w.r.t. the Lebesgue measure: this means that, in the coordinates $B^u \times B^s$, the projection on the first coordinate allows to identify different unstable balls, and these identifications all preserve the sets of zero Lebesgue measure. One can then apply Fubini's theorem to conclude that a Gibbs state is indeed an SRB measure: its basin $B(\mu)$ has full Lebesgue measure in the open neighborhood U of Λ. These results make up most of [71, 66, 17].

To go further, one may relax the hyperbolicity assumptions. For instance, one may assume that the diffeomorphism ϕ is *partially hyperbolic*: it admits an invariant decomposition $E^u \oplus E$. The vectors in E^u are still uniformly expanded, but we don't assume any more that the vectors in E are contracted. On the other hand, if a vector in E is expanded, it should be so less strongly than the vectors in E^u: one speaks of a *dominated decomposition*. Under this hypothesis, it is not difficult to copy the previous reasoning and show the existence of unstable Gibbs states. On the other hand, this hypothesis is not sufficient to show that the Gibbs states are SRB measures: the condition ensuring the existence of sufficiently many stable manifolds, such that almost every point of the open U is asymptotic to a point in the attractor, is lacking. To show that a Gibbs state μ is an SRB measure, it is enough for instance to require that for μ-almost every point $x \in \Lambda$ the vectors in E_x are exponentially contracted.

The quest for the weakest conditions allowing to show the existence of SRB measures is summarized in Chapter 11 of the book [16]. This question is fundamental since, as we will see, one hopes that "almost all" diffeomorphisms admit SRB measures.

7.2. Ergodic theory of the Lorenz attractor

The geometric Lorenz models are not hyperbolic, hence are not encompassed by the works of Sinai, Ruelle and Bowen we have just summarized. On the other hand, the Lorenz flow is *singular hyperbolic*. At each point x of the attractor Λ, the tangent space can be split into a direct sum of a line E_x^s and a plane F_x, such that the vectors of E^s are uniformly contracted, while a vector in F_x is "less contracted" than the vectors in E^s. Of course the plane F_x contains the Lorenz vector field. Notice that a vector field is invariant through the flow it generates, so at each nonsingular point it provides a nonexpanding direction. One expects to find expansion in F_x, in the direction transverse to the field, but this expansion may not be uniform, due to the defect of hyperbolicity. For each point $x \in \Lambda$ we have a stable manifold $W^s(x)$, which is a curve "transverse" to the attractor.

We have also studied the structure of the first return map on the square C: it has all the characteristics of a hyperbolic map. This allows, e.g., to show the existence of an invariant measure absolutely continuous w.r.t. the Lebesgue measure for the map f on the interval. The return map of an orbit on the square diverges (logarithmically) when going to the discontinuity of f. This poses technical difficulties (but, quite strangely, sometimes also simplifies the work). Most of these technical difficulties have been overcome, and we now understand quite well the ergodic behavior of the Lorenz flows.

One of the first results, obtained in the 1980s, is that *the Lorenz attractor supports a unique SRB measure* [21, 56].

Moreover, *the unstable Lyapunov exponent for this measure μ is positive*. This means that for μ-almost every point $x \in \Lambda$ one may find a line E_x^u contained in the plane F_x, such that for $v \in E_x^u$,

$$\lim_{T \to \infty} \frac{1}{T} \log \|d\phi^T(v)\| = \lambda > 0,$$

with λ independent of x. In other words, the dynamics exponentially expands almost all the vectors. Of course, this expansion is not uniform.

One can then apply Pesin's theory for the Lyapunov exponents, which allows to construct unstable manifolds. Precisely, for μ-almost all point $x \in \Lambda$, the set of points y with past orbits asymptotic to that of x is a smooth curve W_x^u contained in Λ. In a way, μ is hyperbolic almost everywhere, with a splitting of the tangent space into three directions: E_x^s (which continuously depends on x), E_x^u (defined almost everywhere) and the direction of the vector field (defined everywhere except on the singular point).

The fact that this SRB measure is *mixing* has been shown only recently [46]: if A, B are two measurable subsets of U, one has

$$\lim_{t \to +\infty} \mu(\phi^t(A) \cap B) = \mu(A)\mu(B).$$

The most complete result is that obtained recently by Holland and Melbourne [37]. Consider a smooth function u defined on the open set U, and to simplify the statement, assume that its μ-average vanishes. Then, there exists a Brownian motion W_T of variance $\sigma^2 \geq 0$ and $\epsilon > 0$ such that, μ-almost surely,

$$\int_0^T u \circ \phi^t dt = W_T + O(T^{\frac{1}{2}-\varepsilon}) \quad \text{when} \quad t \to +\infty.$$

That is, the Birkhoff sums oscillate like a Brownian motion around their average value. This implies a *Central Limit Theorem* and a *Law of the iterated logarithm*.

Let us also mention the preprint [24] which analyzes in detail the statistics of the recurrence times in a small ball.

8. The butterfly within the global picture

8.1. Singular hyperbolicity

The Lorenz attractor is mainly an *example* featuring a certain number of properties:

- it is complicated and chaotic, yet not hyperbolic;
- its dynamics can nevertheless be described reasonably well, both from the topological and ergodic points of view;
- it admits "some" hyperbolicity;
- it is robust: any perturbation of a Lorenz model is another Lorenz model.

The central question is to determine whether it constitutes a "significative" example. The aim is to find sufficiently general qualitative properties on a dynamics, allowing to show that it resembles a Lorenz attractor.

Before stating some recent results in this direction, I shall first explain a precursory theorem of Mañé, dating back to 1982 [47]. Let ϕ be a diffeomorphism on a compact manifold V, and $\Lambda \subset V$ a compact invariant set. Λ is said to be *transitive* if it contains a dense orbit (w.r.t. the induced topology in Λ). It is said to be *maximal* if there exists an open neighborhood U such that for each point x outside Λ, the orbit of x will leave U in the past or in the future. Finally, Λ is said to be *robustly transitive* if these properties persist under perturbation: there exists a neighborhood $\mathcal{V}$ of ϕ in the space of C^1 diffeomorphisms, such that if $\psi \in \mathcal{V}$, the compact set $\Lambda_\psi = \cap_{k \in \mathbb{Z}} \psi^k(U)$ is nonempty and transitive. Mañé's theorem is the first one showing that hyperbolicity can result from dynamical conditions of this kind. It states that *in dimension 2, the robustly transitive sets are exactly the hyperbolic sets*. Such a result cannot be generalized to higher dimension, where one should try to obtain some partial hyperbolicity. Chapter 7 in [16] is entirely devoted to describing the recent results in this direction.

In the case of vector fields in dimension 3, the main result has been obtained by Morales, Pacifico and Pujals in 2004 [49]. Let ϕ^t be a flow on a manifold V, preserving a compact set Λ. Let us recall a definition given in Section 5.2: Λ is said to be *singular hyperbolic* if – like in the case of the Lorenz model – the tangent space $T_\Lambda V$ decomposes into $E^s \oplus F$, such that:

- E^s_x and F_x depend continuously on the point $x \in \Lambda$ and are invariant through the differential of ϕ^t;
- the vectors in E^s_x are uniformly contracted by the flow;
- the decomposition $E^s_x \oplus F_x$ is dominated: if $u \in F_x$ and $v \in E^s_x$ are unitary, one has $\|d\phi^t(u)\| \geqslant C \exp(\lambda t) \|d\phi^t(v)\|$ for all $t > 0$;
- the flow expands the bidimensional volume on F_x: $\det(d\phi^t|F_x) \geqslant C \exp(\lambda t)$ for all $t > 0$.

The main theorem in [49] states that *if a compact invariant set Λ of a three-dimensional flow is robustly transitive, then:*

- Λ *is singular hyperbolic;*
- *if Λ does not contain any singular point, it is a hyperbolic set;*
- *all the field singularities contained in Λ are of Lorenz type; the three eigenvalues $a_1 < a_2 < a_3$ of the linearization satisfy $a_1 < 0 < a_3$ and either $-a_3 < a_2 < 0$ (index 2) or $0 < a_2 < -a_1$ (index 1);*
- *all the singularities have the same index and Λ is an attractor or a repeller, (i.e., an attractor for the reversed field) depending on the value 2, resp. 1, of the index.*

Hence, this theorem provides the Lorenz attractors (more precisely, the dynamics of the same family) with a new status: they are *singular hyperbolic attractors*. They represent, in dimension 3, a qualitative phenomenon: robust transitivity. One can read a sketch of proof, as well as some complements, in Chapter 9 of [16].

One still needs to generalize the results available for the Lorenz attractor to all singular hyperbolic attractors. The topological understanding of these objects is still incomplete, although we know, for instance, that periodic orbits are dense in the attractor [9]. The ergodic understanding is now much more advanced, and rather close to the case of the Lorenz attractors; for instance, *there exists an SRB measure, the basin of which contains almost all the points of an open neighborhood of the attractor* [6].

8.2. Palis's big picture

The history of dynamical systems seems to be punctuated by a long sequence of hopes, soon to be abandoned ... A Morse–Smale world, replaced by a world of hyperbolic attractors, in turn destroyed by an abundance of examples like Lorenz's model. Yet, dynamicists do not lack optimism, and they do not hesitate to brush a world view according to their present belief, (naively) hoping that their view will not soon be obsolete. Palis has formulated such a vision in three articles in 1995,

2005 and 2008 [51, 52, 53]. He states there an ensemble of conjectures describing the dynamics of "almost all" diffeomorphisms or flows. These conjectures are necessarily technical, and it is not useful to describe them in detail here. I will only give a sketch of the general spirit.

The first difficulty – which is not specific to this domain – is to give a meaning to "almost all" dynamics. The initial idea from the 1960s was to describe an *open dense* set in the space of dynamical systems, or at least, a countable intersection of open dense sets, in order to use *Baire genericity*. Yet, this notion has proved too strict. It is well known that no notion of "Lebesgue measure" can be defined on a space of infinite dimension, but a substitute – *prevalence* – seems to impose itself little by little. Let $\mathcal{P}$ be a subset of the set $\mathrm{Diff}(V)$ of diffeomorphisms (of a certain regularity) on a manifold V. We then consider a family of diffeomorphisms ϕ_ν on V depending on a parameter ν, which for instance takes values in a ball in $\mathbb{R}^p$. One can then evaluate the Lebesgue measure of the set of ν such that ϕ_ν belongs to $\mathcal{P}$. Of course, if the family ϕ_ν is degenerate, for instance if it does not depend on ν, it can completely avoid $\mathcal{P}$. The set $\mathcal{P}$ is said to be *prevalent* if, for a generic family ϕ_ν (in the Baire sense within the space of families), the set of ν such that ϕ_ν belongs to $\mathcal{P}$ has full Lebesgue measure (within the ball covered by the parameter ν).

Palis's *finiteness conjecture* reads as follows:

The following properties are prevalent among the diffeomorphisms or the flows on a given compact manifold:

- *there exist finitely many attractors $\Lambda_1, \ldots, \Lambda_k$;*
- *each attractor admits an SRB measure;*
- *the union of the basins of attraction of the SRB measures covers almost all the manifold, in the sense of the Lebesgue measure.*

As for the lost hope of generic structural stability, it has been replaced by *stochastic stability*. Consider a family of diffeomorphisms ϕ_ν depending on a parameter ν in a neighborhood of the origin in $\mathbb{R}^p$, with $\phi_0 = \phi$. Choose a point x_0 in an attractor Λ, a small $\varepsilon > 0$, and consider the random sequence of points

$$x_0 \quad ; \quad x_1 = \phi_{\nu_1}(x_0) \quad ; \quad x_2 = \phi_{\nu_2}(x_1) \quad ; \quad x_k = \phi_{\nu_k}(x_{k-1}) \quad ; \quad \ldots,$$

where the parameters ν_i are randomly chosen in the ball of radius ε. The SRB measure μ on the attractor Λ is said to be stochastically stable if, for any neighborhood $\mathcal{U}$ of μ in the weak topology on measures, and for any family ϕ_ν, one can find ε such that, for almost all choices of the above random sequence, the average over the Dirac masses

$$\frac{1}{k+1}(\delta_{x_0} + \delta_{x_1} + \cdots + \delta_{x_k})$$

has a limit when $k \to \infty$, and the limit belongs to $\mathcal{U}$. In other words, through a stochastic perturbation, the statistical behavior "does not change too much"...

A second part of Palis's conjectures states that *stochastic stability is prevalent*.

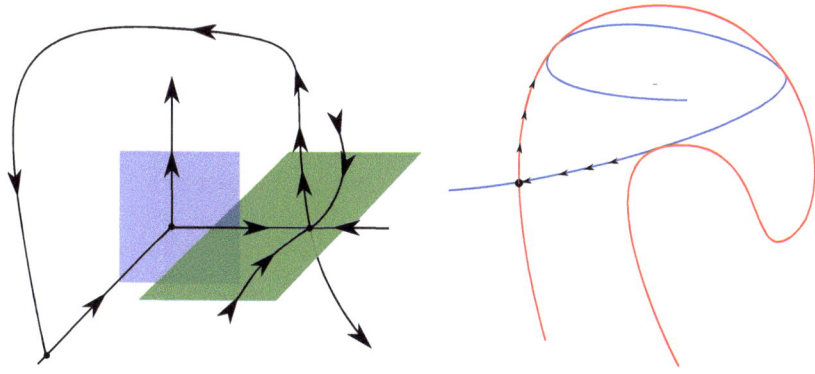

FIGURE 24. A heteroclinic orbit and a homoclinic tangency

These conjectures come with a general strategy of proof. The idea is to explain the non-density of hyperbolicity through the appearance of a few paradigmatic phenomena, one of which is the Lorenz type attractor.

So far only a few nonhyperbolic phenomena have been identified:

• *heterodimensional cycles*

Such a cycle consists in two periodic hyperbolic points x, y in the same transitive set, the unstable manifolds of which have different dimensions, and such that the stable manifold of each point transversally intersects the unstable manifold of the other. It is, of course, an obstruction to hyperbolicity. For diffeomorphisms, this type of cycle only appears for dimensions ≥ 3.

• *homoclinic tangencies*

A homoclinic tangency is a nontransverse intersection between the stable and unstable manifolds of a same periodic point. Once more, it is an obstruction to hyperbolicity.

• *singular cycles*

In the case of flows, the possible presence of singularities offers new possibilities. A *cycle* is given by a finite sequence $\gamma_1, \ldots, \gamma_l$ of hyperbolic singularities or periodic orbits, which are cyclically connected by regular orbits: there are points $x_1, \ldots, x_l$ such that the α-limit set of x_i is γ_i and the ω-limit set of x_i is γ_{i+1} (where $\gamma_{l+1} = \gamma_1$). The cycle is singular provided at least one of the γ_i is a singularity. These singular cycles appear densely in the family of Lorenz models: the two unstable curves starting from the singular point spiral in the attractor, and can be localized in the stable manifold of this singular point. This happens if the point 0 is periodic for the map f on $[-1/2, 1/2]$ we have studied.

Palis proposes that *on a given compact manifold, the set of vectors fields which are either hyperbolic or display one of these phenomena is dense (in the C^r topology)*.

In dimension 3, he conjectures that *the union of the vector fields which are either hyperbolic, or admit a homoclinic tangency, or have attractors (or repellers) of Lorenz type, is dense (in the C^r topology)*.

Arroyo and Rodriguez–Hertz have made progresses in this direction, by showing that a three-dimensional field can be approached, in the C^1 topology, by fields which are either hyperbolic, or present a homoclinic tangency or a singular cycle [10].

In any case, the Lorenz attractor displays phenomena which could be characteristic of "typical chaos". At least in the framework of mathematical chaos, since the relevance of the Lorenz model to describe meteorological phenomena remains widely open [64].

8.3. A long and beautiful sentence

To finish, here is a quotation by Buffon dating back to 1783, and showing that his view of an ergodic and mixing world was quite close from Lorenz's one [20].

> [...] tout s'opère, parce qu'à force de temps tout se rencontre, et que dans la libre étendue des espaces et dans la succession continue du mouvement, toute matière est remuée, toute forme donnée, toute figure imprimée; ainsi tout se rapproche ou s'éloigne, tout s'unit ou se fuit, tout se combine ou s'oppose, tout se produit ou se détruit par des forces relatives ou contraires, qui seules sont constantes, et se balançant sans se nuire, animent l'Univers et en font un théâtre de scènes toujours nouvelles, et d'objets sans cesse renaissants.

References

[1] R. Abraham and S. Smale, *Nongenericity of Ω-stability*. Global Analysis (Proc. Sympos. Pure Math., Vol. XIV, Berkeley, Calif., 1968) pp. 5–8 Amer. Math. Soc., Providence, R.I. 1970.

[2] V.S. Afraimovich, V.V. Bykov and L.P. Shil'nikov, The origin and structure of the Lorenz attractor. (En russe), *Dokl. Akad. Nauk SSSR* **234** no. 2, 336–339 (1977).

[3] D.V. Anosov, Roughness of geodesic flows on compact Riemannian manifolds of negative curvature. (En russe), Dokl. Akad. Nauk SSSR **145**, 707–709 (1962).

[4] D.V. Anosov, *Dynamical systems in the 1960s: the hyperbolic revolution*. Mathematical events of the twentieth century, 1–17, Springer, Berlin, 2006.

[5] A. Andronov. and L. Pontrjagin, Systèmes grossiers. Dokl. Akad. Nauk SSSR **14** (1937).

[6] V. Araújo, E.R. Pujals, M.J. Pacifico and M. Viana, Singular-hyperbolic attractors are chaotic. Transactions of the A.M.S. **361**, 2431–2485 (2009).

[7] V. Arnold, Sur la géométrie différentielle des groupes de Lie de dimension infinie et ses applications à l'hydrodynamique des fluides parfaits. Annales de l'institut Fourier **16** no. 1, 319–361 (1966).

[8] D. Aubin and A. Dahan Dalmedico, Writing the history of dynamical systems and chaos: longue durée and revolution, disciplines and cultures. Historia Math. **29** no. 3, 273–339 (2002).

[9] A. Arroyo and E. Pujals, Dynamical properties of singular-hyperbolic attractors. Discrete Contin. Dyn. Syst. **19** no. 1, 67–87 (2007).

[10] A. Arroyo and F. Rodriguez Hertz, Homoclinic bifurcations and uniform hyperbolicity for three-dimensional flows. Ann. Inst. H. Poincaré Anal. Non Linéaire **20** no. 5, 805–841 (2003).

[11] J. Barrow-Green, *Poincaré and the three body problem*. History of Mathematics, **11**. American Mathematical Society, Providence, RI; London Mathematical Society, London, 1997. xvi+272 pp.

[12] F. Béguin, *Le mémoire de Poincaré pour le prix du roi Oscar*. In L'héritage scientifique de Poincaré, éd. Charpentier, Ghys, Lesne, éditions Belin 2006.

[13] G.D. Birkhoff, Nouvelles recherches sur les systèmes dynamiques. Mem. Pont. Acad. Sci. Nov. Lyncaei **53**, 85–216 (1935).

[14] J. Birman and I. Kofman, A new twist on Lorenz links. J. Topol. **2** no. 2, 227–248 (2009).

[15] J. Birman and R.F. Williams, Knotted periodic orbits in dynamical systems, I. Lorenz's equations. Topology **22** no. 1, 47–82 (1983).

[16] C. Bonatti, L. Díaz and M. Viana, *Dynamics beyond uniform hyperbolicity. A global geometric and probabilistic perspective*. Encyclopaedia of Mathematical Sciences, **102**. Mathematical Physics, III. Springer-Verlag, Berlin, 2005.

[17] R. Bowen, *Equilibrium states and the ergodic theory of Anosov diffeomorphisms*. Lecture Notes in Mathematics **470**, Springer-Verlag, Berlin, 1975.

[18] R. Bowen, *On Axiom A diffeomorphisms*. Regional Conference Series in Mathematics, No. 35. American Mathematical Society, Providence, R.I., 1978. vii+45 pp.

[19] R. Bowen and D. Ruelle, The ergodic theory of Axiom A flows. Invent. Math. **29** no. 3, 181–202 (1975).

[20] G.L. Buffon, *Du Soufre*. Histoire Naturelle des minéraux, tome 2, 1783.

[21] L.A. Bunimovich, *Statistical properties of Lorenz attractors*. Nonlinear dynamics and turbulence, Pitman, p. 7192, 1983.

[22] P. Dehornoy, *Les nœuds de Lorenz*. to be published in *L'Enseignement mathématique*, arXiv:0904.2437

[23] P. Duhem, *La théorie physique; son objet, sa structure*. 1906. English transl. by P.P. Wiener, *The aim and structure of physical theory*, Princeton University Press, 1954.

[24] S. Galatolo and M.J. Pacifico, Lorenz-like flows: exponential decay of correlations for the Poincaré map, logarithm law, quantitative recurrence. Ergod. Th. Dynam. Sys. **30**, 1703–1737 (2010).

[25] R. Ghrist, Branched two-manifolds supporting all links. Topology **36** no. 2, 423–448 (1997).

[26] E. Ghys, *Knots and dynamics*. International Congress of Mathematicians. Vol. I, pp. 247–277, Eur. Math. Soc., Zürich, 2007.

[27] E. Ghys, Right-handed vector fields & the Lorenz attractor. Jpn. J. Math **4** no. 1, 47–61 (2009).

[28] E. Ghys and J. Leys, *Lorenz and modular flows: a visual introduction.* American Math. Soc. Feature Column, November 2006, www.ams.org/featurecolumn/archive/lorenz.html

[29] J. Gleick, *Chaos: Making a New Science.* Viking Penguin, 1987.

[30] M. Gromov, Three remarks on geodesic dynamics and fundamental group. Enseign. Math. **46** no. 3–4, 391–402 (2002).

[31] J. Guckenheimer, A strange, strange attractor. In *The Hopf Bifurcation*, Marsden and McCracken, eds. Appl. Math. Sci., Springer-Verlag, (1976).

[32] J. Guckenheimer, The Lorenz Equations: Bifurcations, Chaos, and Strange Attractors. Amer. Math. Monthly **91** no. 5, 325–326 (1984).

[33] J. Guckenheimer and R.F. Williams, Structural stability of Lorenz attractors. Inst. Hautes études Sci. Publ. Math. **50**, 59–72 (1979).

[34] J. Hadamard, Les surfaces à courbures opposées et leurs lignes géodésiques. Journal de mathématiques pures et appliquées 5e série **4**, 27–74 (1898).

[35] J.T. Halbert and J.A. Yorke, Modeling a chaotic machine's dynamics as a linear map on a "square sphere" . http://www.math.umd.edu/ halbert/taffy-paper-1.pdf

[36] B. Hasselblatt, *Hyperbolic dynamical systems.* Handbook of dynamical systems, Vol. 1A, pp. 239–319, North-Holland, Amsterdam, 2002.

[37] M. Holland and I. Melbourne, Central limit theorems and invariance principles for Lorenz attractors. J. Lond. Math. Soc. (2) **76** no. 2, 345–364 (2007).

[38] J. P. Kahane, Hasard et déterminisme chez Laplace. Les Cahiers Rationalistes **593** – mars-avril 2008.

[39] S. Kuznetsov, Plykin-type attractor in nonautonomous coupled oscillators. Chaos **19**, 013114 (2009).

[40] O. Lanford, *An introduction to the Lorenz system.* Papers from the Duke Turbulence Conference (Duke Univ., Durham, N.C., 1976), Paper No. 4, i+21 pp. Duke Univ. Math. Ser., Vol. III, Duke Univ., Durham, N.C., 1977.

[41] P.S. Laplace, *Essai philosophique sur les probabilités.* 1814. English transl. by A.I. Dale, *Philosophical essay on probabilities*, Springer, 1995

[42] E.N. Lorenz, Maximum simplification of the dynamic equations. Tellus **12**, 243–254 (1960).

[43] E.N. Lorenz, *The statistical prediction of solutions of dynamic equations.* Proc. Internat. Sympos. Numerical Weather Prediction, Tokyo, pp. 629–635.

[44] E.N. Lorenz, Deterministic non periodic flow. J. Atmosph. Sci. **20**, 130–141 (1963).

[45] E.N. Lorenz, *Predictability: does the flap of a butterfly's wings in Brazil set off a tornado in Texas?* 139th Annual Meeting of the American Association for the Advancement of Science (29 Dec. 1972), in *Essence of Chaos* (1995), Appendix 1, 181.

[46] S. Luzzatto, I. Melbourne and F. Paccaut, The Lorenz attractor is mixing. Comm. Math. Phys. **260** no. 2, 393–401 (2005).

[47] R. Mañé, An ergodic closing lemma. Ann. of Math. **116** no. 3, 503–540 (1982).

[48] J.C. Maxwell, *Matter and Motion.* 1876, rééd. Dover (1952).

[49] C.A. Morales, M.J. Pacifico and E.R. Pujals, Robust transitive singular sets for 3-flows are partially hyperbolic attractors or repellers. Ann. of Math. **160** no. 2, 375–432 (2004).

[50] H.M. Morse, A one-to-one representation of geodesics on a surface of negative curvature. Amer. J. Math. **43** no. 1, 33–51 (1921).

[51] J. Palis, *A global view of dynamics and a conjecture on the denseness of finitude of attractors.* Géométrie complexe et systèmes dynamiques (Orsay, 1995), *Astérisque* No. 261 (2000), xiii–xiv, pp. 335–347.

[52] J. Palis, A global perspective for non-conservative dynamics. Ann. Inst. H. Poincaré Anal. Non Linéaire **22** no. 4, 485–507 (2005).

[53] J. Palis, Open questions leading to a global perspective in dynamics. Nonlinearity **21** no. 4, T37–T43 (2008).

[54] J. Palis and F. Takens, *Hyperbolicity and sensitive chaotic dynamics at homoclinic bifurcations.* Cambridge Studies in Advanced Mathematics, 35. Cambridge University Press, Cambridge, 1993. x+234 pp.

[55] M. Peixoto, Structural stability on two-dimensional manifolds. Topology **1**, 101–120 (1962).

[56] Ya.B. Pesin, Dynamical systems with generalized hyperbolic attractors: hyperbolic, ergodic and topological properties. Ergodic Theory Dynam. Systems **12** no. 1, 123–151 (1992).

[57] H. Poincaré, Mémoire sur les courbes définies par une équation différentielle. Journal de mathématiques pures et appliquées **7**, 375–422 (1881).

[58] H. Poincaré, Sur le problème des trois corps et les équations de la dynamique. Acta Mathematica **13**, 1–270 (1890).

[59] H. Poincaré, *Les méthodes nouvelles de la mécanique céleste.* Gauthier-Villars, Paris, vol. 3, (1892–1899). English transl. *New methods of celestial mechanics,* D.L. Goroff (ed.), American Institute of Physics, 1993

[60] H. Poincaré, Sur les lignes géodésiques des surfaces convexes. Trans. Amer. Math. Soc. **6**, 237–274 (1905).

[61] H. Poincaré, Le hasard. Revue du Mois **3**, 257–276 (1907).

[62] H. Poincaré, *Science et méthode.* Flammarion, (1908). English transl. by F. Maitland, *Science and Method,* T. Nelson and Sons, London, 1914.

[63] A. Robadey, *Différentes modalités de travail sur le général dans les recherches de Poincaré sur les systèmes dynamiques.* Thèse, Paris 7, 2005.

[64] R. Robert, L'effet papillon n'existe plus! Gaz. Math. **90**, 11–25 (2001).

[65] D. Ruelle, *The Lorenz attractor and the problem of turbulence.* Turbulence and Navier–Stokes equations (Proc. Conf., Univ. Paris-Sud, Orsay, 1975), pp. 146–158, Lecture Notes in Math., 565, Springer, Berlin, (1976).

[66] D. Ruelle, A measure associated with axiom-A attractors, Amer. J. Math. **98** no. 3, 619–654 (1976).

[67] D. Ruelle and F. Takens, On the nature of turbulence. Commun. Math. Phys. **20**, 167–192 (1971).

[68] L.P. Shil'nikov, *Homoclinic trajectories: from Poincaré to the present.* Mathematical events of the twentieth century, 347–370, Springer, Berlin, 2006.

[69] M. Shub, What is ... a horseshoe?, Notices Amer. Math. Soc. **52** no. 5, 516–517 (2005).

[70] M. Shub, et S. Smale, Beyond hyperbolicity. Ann. of Math. (2) **96**, 587–591(1972).

[71] Ja.G. Sinai, Gibbs measures in ergodic theory. (En russe), Uspehi Mat. Nauk **27**, no. 4(166), 21–64 (1972).

[72] S. Smale, On dynamical systems. Bol. Soc. Mat. Mexicana **5**, 195–198 (1960).

[73] S. Smale, *A structurally stable differentiable homeomorphism with an infinite number of periodic points*. Qualitative methods in the theory of non-linear vibrations (Proc. Internat. Sympos. Non-linear Vibrations, Vol. II, 1961) pp. 365–366 Izdat. Akad. Nauk Ukrain. SSR, Kiev 1963.

[74] S. Smale, *Diffeomorphisms with many periodic points*. Differential and Combinatorial Topology (A Symposium in Honor of Marston Morse) pp. 63–80, Princeton Univ. Press, Princeton, N.J. 1965.

[75] S. Smale, Structurally stable systems are not dense. Amer. J. Math. **88**, 491–496 (1966).

[76] S. Smale, Differentiable dynamical systems. Bull. Amer. Math. Soc. **73**, 747–817 (1967).

[77] S. Smale, Finding a horseshoe on the beaches of Rio. Math. Intelligencer **20** no. 1, 39–44 (1998).

[78] S. Smale, Mathematical problems for the next century. Math. Intelligencer **20** no. 2, 7–15 (1998).

[79] C. Sparrow, *The Lorenz equations: bifurcations, chaos, and strange attractors*. Applied Mathematical Sciences, 41. Springer-Verlag, New York-Berlin, (1982).

[80] W. Tucker, A rigorous ODE solver and Smale's 14th problem. Found. Comput. Math. **2**, no. 1, 53–117 (2002).

[81] M. Viana, What's new on Lorenz strange attractors? Math. Intelligencer **22** no. 3, 6–19 (2000).

[82] R.F. Williams, The structure of Lorenz attractors. Inst. Hautes Études Sci. Publ. Math. **50**, 73–99 (1979).

[83] R.F. Williams, Lorenz knots are prime. Ergodic Theory Dynam. Systems **4** no. 1, 147–163 (1984).

[84] L.S. Young, What are SRB measures, and which dynamical systems have them? . Dedicated to David Ruelle and Yasha Sinai on the occasion of their 65th birthdays. J. Statist. Phys. **108** no. 5–6, 733–754 (2002).

Étienne Ghys
CNRS – ÉNS Lyon
UMPA, 46 Allée d'Italie
F-69364 Lyon, France

Chaotic Dynamos Generated by Fully Turbulent Flows

Stéphan Fauve

Abstract. We report experimental results on the generation of a magnetic field by the motion of an electrically conducting liquid metal. We recall that this dynamo effect is a canonical example of an instability process that occurs on a strongly turbulent flow. Although the magnetic field is driven by a turbulent velocity field that involves a wide range of interacting scales, we observe that its dynamics results from a small number of interacting modes. We present a model that describes both periodic and random reversals of the magnetic field and compare it with the experimental results and direct numerical simulations.

1. Introduction

Until the seventies, it was believed that an erratic temporal behavior of some quantity $x(t)$ in a statistically stationary state had to result from the interaction of many degrees of freedom. For instance, a qualitative description of the transition to turbulence of fluid flows given by Landau & Lifshitz (1959) was ascribing the temporal complexity to the arbitrary phases resulting from a large number of successive Hopf bifurcations. Most quantitative models of systems with an erratic behavior were based on the theory of stochastic processes, assuming more or less explicitly that their randomness was resulting from a large number of uncontrolled parameters or initial conditions. It was of course known since Poincaré that low-dimensional Hamiltonian systems can display complicated (non periodic) solutions, iterations of simple maps were used to generate random numbers, and erratic behaviors were observed in simple electric devices, but the concept of low-dimensional deterministic chaos was far from being considered among physicists who were used to problems either governed by linear equations, or constrained by extremum principles. Modeling an erratic temporal behavior with a low-dimensional dynamical system, i.e., a small number of coupled nonlinear differential equations, has been made first in the sixties, one of the most famous example being the Lorenz system

(Lorenz 1963), which certainly triggered a large part of the studies on temporal chaos that were performed in the seventies and the eighties. Ruelle and Takens (1971) proposed a qualitative mechanism to get a chaotic regime after a small number of bifurcations from a stationary state, in contrast to Landau's picture. A chaotic velocity field in a Couette–Taylor flow, generated from a stationary state after a small number of transitions, was reported by Swinney & Gollub (1975) and followed by a lot of similar observations in other hydrodynamic systems, nonlinear electronic, acoustic or optical devices, chemical reactions, etc. The observation of the period doubling transition scenario to chaos in a flow driven by thermal convection by Maurer and Libchaber (1979) opened the way to test quantitative predictions made using renormalisation group techniques on the logistic map (Feigenbaum 1978, Tresser & Coullet 1978). The observed quantitative agreement (Libchaber et al. 1982) clearly showed that the flow of a viscous fluid can display a transition to an erratic temporal behavior that is qualitatively and quantitatively described by a small number of interacting modes or degrees of freedom. Other transition scenarios to chaos were predicted, such as intermittency (Manneville & Pomeau 1979) or transitions from quasiperiodic states that were also found in good quantitative agreement with experimental observations in many different systems. The general picture that emerged from these experimental and theoretical studies is that there exist universal scenarios of transition to chaos. Similarly to phase transitions, there are quantitative characteristics of these scenarios that do not depend on the particular system under consideration.

However, in contrast to what has been believed for a while by some people, these achievements did not help to solve the problem of fluid turbulence. It has been realized that one usually needs to quench a lot of degrees of freedom of a fluid flow in order to observe chaotic regimes in agreement with the behaviors predicted for low-dimensional dynamical systems. For instance, the convection experiments mentioned above were performed in containers of small aspect ratio, i.e., involving a small number of convective rolls. In the case of large aspect ratio containers, no simple scenario of transition to chaos was identified (Ahlers & Behringer 1978). A more flexible way to control the number of degrees of freedom of a convective flow has been used by applying an external horizontal magnetic field to an electrically conducting fluid. This inhibits three-dimensional modes and makes the flow two-dimensional without variation of the velocity along the magnetic field axis. It has been shown that transitions to chaos in agreement with the behaviors predicted by low-dimensional dynamical systems were observed only for large enough values of the applied magnetic field (Fauve et al. 1984). Thus, although dynamical system theory provided new concepts in physics including the one of deterministic chaos, and triggered a new way of approaching nonlinear phenomena, it has been so far of little help to understand turbulent flows which involve the interaction of many different scales. Kolmogorov type qualitative arguments or a stochastic description are more efficient tools in these cases. This has given the impression that despite its success, deterministic chaos is restricted to a class of rather academic flows in which many degrees of freedom have been quenched.

On the other hand, it can be argued that the dynamics of some large scale modes in a strongly turbulent flow could display low-dimensional dynamics if they are weakly coupled to the turbulent background. This is the case in many geophysical and astrophysical flows where an oscillatory or a weakly chaotic behavior can occur in flows at huge Reynolds numbers. Some climatic phenomena indeed display a characteristic feature of low-dimensional chaos: well defined patterns occur within a random temporal behavior. Examples are atmospheric blockings that can affect the climate on a time scale of several days (Ghil & Childress 1987) or El Nino events that occur every few years (Vallis 1986). The qualitative features of these phenomena have been often modeled using a few coupled variables such as mean temperature, wind or current. This truncation is sometimes justified by scale separation (for instance, the dynamics of the ocean is much slower than the one of the atmosphere), but it is often arbitrary since many phenomena involve a continuum of scales without any clear gap among them. Another example of nearly periodic behavior superimposed on turbulence is provided by the solar cycle: it has been observed since several centuries that well-defined spatio-temporal patterns, dark spots on the solar surface that appear at mid-latitudes and migrate toward the equator, involve a 22 year period. It has been found later that these spots correspond to large values of the magnetic field generated by turbulent convection in the sun through the dynamo effect. The number of sun spots follows the cycle of the solar magnetic field that reverses roughly every 11 years. The amplitude of this oscillation varies on much longer time scales, and long periods with a very low solar activity (a small number of sun spots) randomly occurred in the past. Several low-dimensional models of the solar cycle, involving a few modes of the magnetic and velocity fields, have been proposed (see Section 5). It is very hard to justify this type of description; the level of turbulence is so high that both the velocity and the magnetic fields display a continuous range of scales and it looks rather arbitrary to write down a few equations for a small number of modes and forget about the effect of all the others.

It is known that the dynamics of a complex system is governed by a small number of relevant modes in the vicinity of bifurcations (see for instance, Arnold 1982). The amplitude of the unstable modes varying on long time scales compared to the stable ones, the later can be adiabatically eliminated, thus leading to a low-dimensional dynamical system. There exist various perturbation techniques that can be used to perform this elimination and find the differential equations governing the amplitudes of the unstable modes when the basic state of the system is stationary or time periodic. However, almost nothing is known when an instability occurs from a turbulent regime. It seems even difficult to give a proper definition of such a situation in hydrodynamic turbulence. The dynamo effect, i.e., the generation of a magnetic field by the flow of an electrically conducting fluid, provides a very interesting situation in this respect. It is an instability process that can occur in a liquid metal only when the kinetic Reynolds number of the flow is very large (see Section 2). The instability threshold can be easily defined. Although the experiments involve some cost and technical difficulties, once the dynamo regime is reached, the dynamics of the magnetic field can be easily measured. We will show here that even when the magnetic field is generated by a strongly turbulent flow that involves fluctuations as large as the mean flow, it displays low-dimensional dynamics. Thus, fluid dynamos provide an example in which a few modes are governed by a low-dimensional dynamical system although they are coupled to a strongly turbulent background.

This paper is organized as follows: definitions and elementary facts about fluid dynamos are shortly recalled in Section 2. In Section 3, the first experimental observations made in Karlsruhe and Riga are described. The results of the VKS experiment are reported in Section 4. Section 5 provides a short review of the phenomenon of magnetic reversals and of various related models. A model for the dynamics of the magnetic field observed in the VKS experiment is presented is Section 6. It is illustrated using direct numerical simulations in Section 7. Finally, a minimal model for field reversals is presented in Section 8. Some other systems displaying a low-dimensional large scale dynamics on a turbulent background are mentioned in the conclusion.

2. The dynamo effect

It is now believed that magnetic fields of planets and stars are generated by the motion of electrically conducting fluids through the dynamo process. This has been first proposed by Larmor (Larmor, 1919) for the magnetic field of the sun. Assuming the existence of an initial perturbation of magnetic field, he observed that "internal motion induces an electric field acting on the moving matter: and if any conducting path around the solar axis happens to be open, an electric current will flow round it, which may in turn increase the inducing magnetic field. In this way it is possible for the internal cyclic motion to act after the manner of the cycle of a self-exciting dynamo, and maintain a permanent magnetic field from insignificant beginnings, at the expense of some of the energy of the internal circulation" (for reviews of the subject, see for instance Moffatt (1978), Zeldovich et al. (1983), Busse (1977), Roberts (1994), Fauve & Pétrélis (2003)).

Maxwell's equations together with Ohm's law give the governing equation of the magnetic field, $\mathbf{B}(\mathbf{r}, t)$. In the approximation of magnetohydrodynamics (MHD), it takes the form

$$\frac{\partial \mathbf{B}}{\partial t} = \nabla \times (\mathbf{V} \times \mathbf{B}) + \frac{1}{\mu_0 \sigma} \nabla^2 \mathbf{B}, \tag{1}$$

where μ_0 is the magnetic permeability of vacuum and σ is the electrical conductivity. The last term on the right hand side of (1) represents ohmic dissipation, and the first one, electromagnetic induction due to the velocity field $\mathbf{V}(\mathbf{r}, t)$. $B = 0$ is an obvious solution of (1), and for $V = 0$, any perturbation of $\mathbf{B}(\mathbf{r}, t)$ (respectively of the current density $\mathbf{j}(\mathbf{r}, t)$) decays to zero due to ohmic diffusion. $B = 0$ can be an unstable solution if the induction term compensates ohmic dissipation. The ratio of these two terms defines the magnetic Reynolds number, $R_m = \mu_0 \sigma V L$, where V is the typical velocity amplitude and L the typical length scale of the flow. If $\mathbf{V}(\mathbf{r},t)$ has an appropriate geometry, perturbations of magnetic field grow when R_m becomes larger than a critical value R_m^c (in the range 10–1000 for most studied examples). Magnetic energy is generated from part of the mechanical work used to drive the flow.

In order to describe the saturation of the magnetic field above the dynamo threshold R_m^c, we need to take into account its back reaction on the velocity field. $\mathbf{V}(\mathbf{r},t)$ is governed by the Navier–Stokes equation

$$\frac{\partial \mathbf{V}}{\partial t} + (\mathbf{V}\cdot\nabla)\mathbf{V} = -\nabla\left(\frac{p}{\rho} + \frac{B^2}{2\mu_0\rho}\right) + \nu\nabla^2\mathbf{V} + \frac{1}{\mu_0\rho}(\mathbf{B}\cdot\nabla)\mathbf{B}, \qquad (2)$$

that we have restricted to the case of an incompressible flow ($\nabla\cdot\mathbf{V} = 0$). ν is the kinematic viscosity and ρ is the fluid density. In the MHD approximation, the Lorentz force, $\mathbf{j}\times\mathbf{B}$, can be split into the two terms involving $\mathbf{B}$ in (2). If the modification of the flow under the action of the growing magnetic field weakens the dynamo capability of the flow, the dynamo bifurcation is supercritical, i.e., the magnetic field grows continuously from zero when R_m is increased above R_m^c.

Thus, the minimum set of parameters involved in a fluid dynamo consists of the size of the flow domain, L, the typical fluid velocity, V, the density, ρ, the kinematic viscosity, ν, the magnetic permeability of vacuum, μ_0, and the fluid electrical conductivity, σ. For most astrophysical objects, the global rotation rate, Ω, also plays an important role and the Coriolis force $-2\Omega\times\mathbf{V}$ should be taken into account when the Navier–Stokes equation (2) is written in the rotating frame. Three independent dimensionless parameters thus govern the problem. We can choose the magnetic Reynolds number, R_m, the magnetic Prandtl number, P_m, and the Rossby number Ro

$$R_m = \mu_0\sigma VL, \quad P_m = \mu_0\sigma\nu, \quad Ro = \frac{V}{L\Omega}. \qquad (3)$$

Then, dimensional analysis implies that we have for the dynamo threshold

$$R_m^c = f(P_m, Ro), \qquad (4)$$

and for the mean magnetic energy generated above the dynamo threshold

$$\langle B^2\rangle = \mu_0\rho V^2\, g(R_m, P_m, Ro). \qquad (5)$$

f and g are arbitrary functions at this stage. Their dependence on P_m (or equivalently on the kinetic Reynolds number, $Re = VL/\nu$) and on Ro can be related to the effect of flow characteristics (in particular turbulence) on the dynamo threshold and saturation. In many realistic situations, more parameters should be taken into account. For instance, f and g also depend on the choice of boundary conditions (for instance their electrical conductivity or their magnetic permeability, etc).

For planets and stars as well as for all liquid metals in the laboratory, the magnetic Prandtl number is very small, $P_m < 10^{-5}$. Magnetic field self-generation can be obtained only for large enough values of R_m for which Joule dissipation can be overcome (for most known fluid dynamos, the dynamo threshold R_{mc} is roughly in the range 10–1000). Therefore, the kinetic Reynolds number, $Re = R_m/P_m$, is very large and the flow is strongly turbulent. This is of course the case of planets and stars which involve huge values of Re but is also true for dynamo experiments with liquid metals for which $Re > 10^5$. Direct numerical simulations are only possible for values of P_m orders of magnitude larger that the realistic ones for the sun, the Earth or laboratory experiments. First because it is not possible to handle a too large difference between the time scale of diffusion of the magnetic field and the one of momentum; second, a small P_m dynamo occurs for large Re and requires the resolution of the small spatial scales generated by turbulence. Strongly developed turbulence has also some cost for the experimentalists. Indeed, the power needed to drive a turbulent flow scales like $P \propto \rho L^2 V^3$ and we have

$$R_m \propto \mu_0 \sigma \left(\frac{PL}{\rho}\right)^{1/3}. \tag{6}$$

This formula has simple consequences: first, taking liquid sodium (the liquid metal with the highest electric conductivity), $\mu_0 \sigma \approx 10 \ m^{-2} s$, $\rho \approx 10^3 \ kg\, m^{-3}$, and with a typical length-scale $L \approx 1\text{m}$, we get $P \approx R_m^3$; thus a mechanical power larger than 100 kW is needed to reach a dynamo threshold of the order of 50. Second, it appears unlikely to ever operate experimental dynamos at R_m large compared with R_{mc}. Indeed, it costs 8 times more power to reach $2R_{mc}$ than to reach the dynamo threshold. In conclusion, most experimental dynamos should have the following characteristics:

- they bifurcate from a strongly turbulent flow regime,
- they operate in the vicinity of their bifurcation threshold.

Although the values of R_m and P_m that can be obtained in laboratory experiments using liquid sodium are not too far from the ones of the Earth core, it would be very difficult to perform experiments with large R_m at Ro significantly smaller than unity whereas we have $Ro \approx 10^{-6}$ for the Earth core. The comparison is of course also difficult in the case of the sun: although Ro is of order one for the solar convection zone, R_m is more than six orders of magnitude larger than in any laboratory experiment. As said above, the situation is worse when direct numerical simulations are considered. We thus cannot claim that cosmic magnetic fields can be reproduced at the laboratory scale except if we can show that the dynamics of the magnetic field weakly depends on some dimensionless parameters.

As already mentioned, laboratory dynamos operate in the vicinity of the instability threshold but at very high values of the Reynolds number. This give rise to a very interesting example of instability that differs in many respects from usual hydrodynamical instabilities. The dynamo bifurcation occurs from a base state which is fully turbulent. This may play a role on various aspects of the dynamo process and it raises several questions.

- What is the effect of turbulent velocity fluctuations on the dynamo onset? Can they change the nature of the bifurcation? Will they favor or inhibit

dynamo action? In other words, what is the behavior of f when $P_m \to 0$? In the absence of global rotation, $Ro^{-1} = 0$, is f constant with respect to P_m in this limit, thus giving R_m^c = constant, or does the threshold continuously increases when $P_m \to 0$? (Fauve & Pétrélis, 2007).
- Above onset, at which amplitude does the magnetic field saturate? In the absence of global rotation and for small P_m, do we have $\langle B^2 \rangle \propto \mu_0 \rho V^2 g(R_m)$, i.e., $\langle B^2 \rangle \propto [\rho/(\mu_0 \sigma^2 L^2)] g(R_m)$ close to threshold (Pétrélis & Fauve 2001)?
- Is there a parameter range for which we get energy equipartition, $\langle B^2 \rangle / \mu_0 \propto \rho V^2$? What is the behavior of the ohmic to viscous dissipation ratio? (Fauve & Pétrélis 2007).
- What is the effect of global rotation on the dynamo threshold and saturation?
- What is the effect of turbulent fluctuations on the bifurcation? Is $g(R_m) \propto R_m - R_m^c$ as for a usual supercritical bifurcation close to threshold, or should we expect a behaviour involving an anomalous exponent (Pétrélis et al. 2007)?
- What is the effect of turbulent fluctuations on the dynamics of the magnetic field? What are the statistical properties of the fluctuations of the magnetic field?

We will discuss these problems in connection with existing laboratory dynamo experiments. The first ones, performed in Karlsruhe and Riga, have been designed by taking into account the mean flow alone. Large scale turbulent fluctuations have been inhibited as much as possible by a proper choice of boundary conditions. On the contrary, the VKS experiment has been first motivated by the study of the possible effects of turbulence on the dynamo instability.

3. The Karlsruhe and Riga experiments

The first homogeneous fluid dynamos have been operated in liquid sodium in Karlsruhe (Stieglitz and Müller, 2001) using a flow in an array of pipes set-up in order to mimic a spatially periodic flow proposed by G.O. Roberts (1972), and in Riga (Gailitis et al., 2001) using a Ponomarenko-type flow (Ponomarenko, 1973). We first recall the flow geometries and briefly review the results obtained by both groups.

3.1. The Karlsruhe experiment

The experiment in Karlsruhe, Germany, was motivated by a kinematic dynamo model developed by G.O. Roberts (Roberts, 1972) who showed that various periodic flows can generate a magnetic field at large scale compared to the flow spatial periodicity. One of the cellular flows he considered is a periodic array of vortices with the same helicity. Flows with such topology drive an α-effect that can lead to dynamo action. This mechanism is quite efficient at self-generation (in the sense of generating a magnetic field at a low magnetic Reynolds number based on the wavelength of the flow).

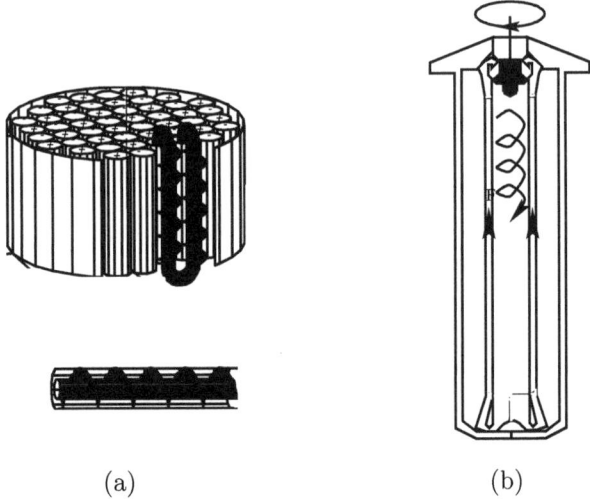

FIGURE 1. Schematics for the experiments from Karlsruhe (a) and Riga (b) which show how helical flow is forced by guiding the sodium through steel channels. From [88] with permission from © 2001 The American Physical Society.

A dynamo based on this mechanism was constructed and run successfully in Karlsruhe. A sketch of the experiment is shown in Figure 1a. The flow is located in a cylindrical vessel of width 1.85 m and height $H = 0.7$ m. It contains 52 elementary cells placed on a square lattice. Each cell is made of two coaxial pipes: an helical baffle drives the helical flow in the outer cylindrical shell whereas the flow in the inner shell is axial. In two neighbouring cells, the velocities are opposite such that the helicity has the same sign in all the cells. Although the volume is finite instead of the infinite extension assumed by G.O. Roberts, the dynamo capability of the flow is not strongly affected in the limit of scale separation, i.e., when the size L of the full volume is large compared to the wavelength l of the flow (Busse *et al.*, 1996). In this limit, the relevant magnetic Reynolds number involves the geometrical mean of the two scales as a length-scale, and the geometrical mean of axial and azimuthal velocities as a velocity scale. However, it can be shown using simple arguments that it is not efficient to increase too much the scale separation if one wants to minimise the power needed to reach the dynamo threshold (Fauve and Pétrélis, 2003). The flow is driven by three electromagnetic pumps and the axial and azimuthal velocities are independently controlled. The liquid sodium temperature is maintained fixed by three steam-evaporation heat exchangers. Measurements of the magnetic field were made both locally with Hall-probes and globally using wire coils. Pressure drops in the pipe and local velocity measurements were also performed.

When the flow rates are large enough, a magnetic field is generated by dynamo action. The bifurcation is stationary and the magnetic field displays fluctuations caused by the small scale turbulent velocity field (see Figure 2). This generation comes at a cost in the power necessary to drive the flow and the pressure drop increases.

Due to the Earth's magnetic field, the bifurcation is imperfect but both branches of the bifurcation can be reached by applying an external magnetic field, as displayed in Figure 2. Among others, the experimentalists performed careful studies of the dependence of the dynamo threshold on the axial and helical flow rates and on the electrical conductivity that can be varied by changing the temperature. They also considered the effect of flow modulation on the dynamo threshold and studied the amplitude and the geometry of the magnetic field in the supercritical regime.

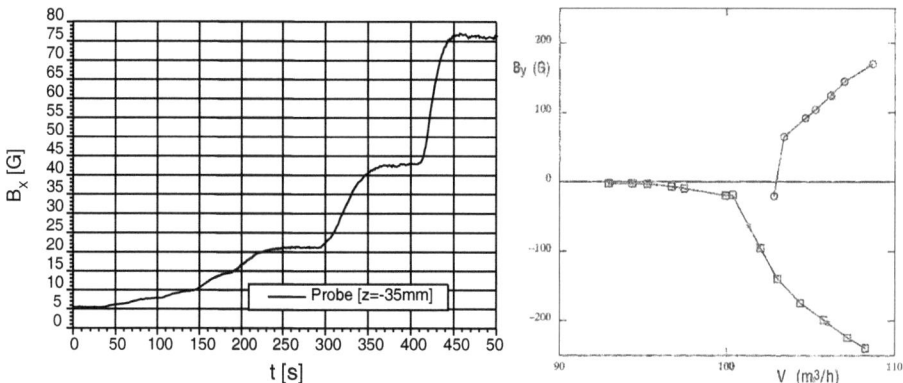

FIGURE 2. (Left) Time recording of one component of the magnetic field in the Karlsruhe experiment. The amplitude of the magnetic field increases after each increasing step of the flow rate of liquid sodium. Small fluctuations are visible once the magnetic field has saturated at a constant mean value. (Right) The magnetic field amplitude increases above the critical flow rate. Another branch of self-generation can be reached only by imposing an initial field. This other branch is disconnected from the main branch. The imperfection of the bifurcation has been ascribed to the Earth's field. From [89] with permission from © 2002 Institute of Physics, University of Latvia.

3.2. The Riga experiment

The experiment carried out by Gailitis *et al.* (2001) has been motivated by one of the simplest examples of a homogeneous dynamo found by Ponomarenko (1973). A conducting cylinder of radius R, embedded in an infinite static medium of the same conductivity with which it is in perfect electrical contact, is in solid body rotation at angular velocity Ω, and in translation along its axis at speed V. In an unbounded domain, this helical motion generates a travelling wave magnetic field. This Hopf bifurcation occurs for a minimum critical magnetic Reynolds number $R_{mc} = \mu_0 \sigma R \sqrt{(R\omega)^2 + V^2} = 17.7$ for an optimum Rossby number $Ro = V/(R\Omega) = 1.3$. We note that the maximum dynamo capability of the flow (R_{mc} minimum) is obtained when the azimuthal and axial velocities are of the same order of magnitude ($Ro \sim 1$). This trend is often observed with more complex flows for which the maximum dynamo capability is obtained when the poloidal and toroidal flow components are comparable.

The experiment set up by the Riga group is sketched in Figure 1b. Their flow is driven by a single propellor, generating helical flow down a central cylindrical cavity. The return flow is in an annulus surrounding this central flow. The geometry of the apparatus as well as mean flow velocity profiles have been optimized in order to decrease the dynamo threshold. In particular, it has been found that adding an outer cylindrical region with liquid sodium at rest significantly decreases R_{mc}. This can be understood if the axial mean flow as well as the rotation rate of the azimuthal mean flow are nearly constant except in boundary layers close to the inner cylinder. Then, the induction equation being invariant under transformation to a rotating reference frame and under Galilean transformations, the presence of some electrical conductor at rest is essential as it is in the case of the Ponomarenko dynamo. The three cylindrical chambers are separated by thin stainless steel walls, which were wetted to allow currents to flow through them.

Figure 3 displays the growth and saturation of a time periodic magnetic field at high enough rotation rate. The nature of the bifurcation as well as dynamo growth rates have been found in good agreement with kinematic theory (Gailitis et al., 2002) that predicts a Hopf bifurcation of convective nature at onset. In addition, the Riga group has made detailed observations of the magnetic field saturation value and the power dissipation needed to drive the flow. These measurements give indications of the effect of Lorentz forces in the flow in order to reach the saturated state. It has been found that one effect of the Lorentz force is to drive the liquid sodium in the outer cylinder in global rotation, thus decreasing the effective azimuthal velocity of the inner flow and therefore its dynamo capability. Dynamo generation does also correspond to an increase in the required mechanical power. However, a puzzling result is displayed in Figure 3: the amplitude of the magnetic field for supercritical rotation rates does not seem to show the universal $\sqrt{R_m - R_{mc}}$ law. In addition, the form of the law seems to depend on the location of the measurement point. This is to some extent due to the absence of temperature control in the Riga experiment. Variations in temperature modify the fluid parameters (electrical conductivity, viscosity and density) and this should be taken into account by plotting the results in dimensionless form (Fauve & Lathrop, 2005).

3.3. What have been learnt from the Karlsruhe and Riga experiments

Although there were no doubts about self-generation of magnetic fields by Roberts' or Ponomarenko-type laminar flows, these experiments have displayed several interesting features:

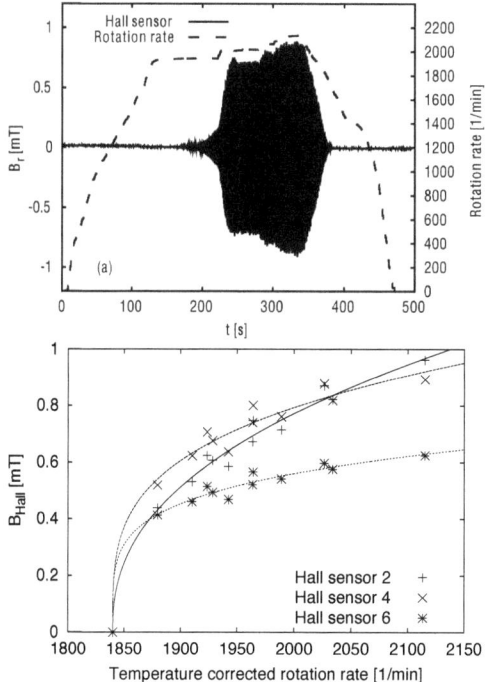

FIGURE 3. (Top) Time recording of the magnetic field from the Riga experiment. The dashed line gives the value of the rotation rate. The amplitude of roughly sinusoidal oscillations (not visible with the resolution of the picture) increases and then saturates when the rotation rate is increased above threshold (about 1850 rpm). The amplitude of saturation increases when the rotation rate is increased further. The dynamo switches off when the rotation rate is decreased below threshold. (Bottom) Magnetic field amplitude as the rotation rate is raised above the critical rotation rate. From [88] with permission from © 2001 The American Physical Society.

- the observed thresholds are in rather good agreement with theoretical predictions (Busse et al. 1996, Rädler et al., 1998, Gailitis et al., 2002) made by considering only the laminar mean flow and neglecting the small-scale turbulent fluctuations that are present in both experiments.
- The nature of the dynamo bifurcation, stationary for the Karlsruhe experiment or oscillatory (Hopf) in the Riga experiment, is also in agreement with laminar models.
- On the contrary, the saturation level of the magnetic field, due to the back reaction of the Lorentz force on the flow, cannot be predicted with a laminar flow model and different scaling laws exist in the supercritical dynamo regime

depending on the magnitude of the Reynolds number (Pétrélis and Fauve, 2001).
- Although secondary instabilities generating large scale dynamics of the magnetic field have not been observed in the Karlsruhe and Riga experiments, small scale turbulent fluctuations of the magnetic field are well developed.

These observations raise the following questions:
- What is the effect of turbulence, or of the magnitude of the Reynolds number, on the dynamo threshold R_{mc}? Is it possible to observe how R_{mc} depends on P_m for a dynamo generated by a strongly turbulent flow (by changing P_m in experiments with a given flow at different temperatures for instance)?
- What is the mechanism responsible for magnetic field fluctuations in the vicinity of the dynamo threshold: an on-off intermittency effect (Sweet et al., 2001) or the advection of the mean magnetic field by the turbulent flow?
- What is the mechanism for field reversals? Is it possible to observe them in laboratory experiments?

4. The VKS experiment

4.1. A bifurcation from a strongly turbulent flow

Using the Reynolds decomposition, we can write for a turbulent velocity field

$$\mathbf{V}(\mathbf{r},\mathbf{t}) = \overline{\mathbf{V}}(\mathbf{r}) + \tilde{\mathbf{v}}(\mathbf{r},\mathbf{t}), \tag{7}$$

where $\overline{\mathbf{V}}(\mathbf{r})$ is the mean flow and $\tilde{\mathbf{v}}(\mathbf{r},\mathbf{t})$ are the turbulent fluctuations. The overbar stands for a temporal average in experiments. Thus, both the mean flow $\overline{\mathbf{V}}(\mathbf{r})$ and the fluctuations $\tilde{\mathbf{v}}(\mathbf{r},\mathbf{t})$ are involved in the induction term of (1) and one has to understand their respective effects on the dynamo process. The Karlsruhe and Riga experiments have been designed by geometrically constraining a mean flow $\overline{\mathbf{V}}(\mathbf{r})$ known for its efficient dynamo action, the G.O. Roberts' flow (respectively the Ponomarenko flow) for the Karlsruhe (respectively Riga) experiment. Turbulent fluctuations, roughly an order of magnitude smaller than the mean flow, have been discarded, and the experimentally observed dynamo threshold as well as the geometry of the mean magnetic field, have been found in good agreement with these predictions, based only the mean flow.

As explained in the introduction, one of my early motivations for dynamo experiments have been the study of a system that displays a bifurcation from a strongly turbulent regime. Thus, I chose to try to generate a dynamo using a von Karman swirling flow, i.e., the flow generated in a cylinder by the motion of two coaxial rotating discs (Zandbergen & Dijkstra 1987). When the discs are operated in counter-rotation, these flows display various qualities of interest for a potential dynamo: a strong differential rotation and the lack of planar symmetry which are key ingredients for a closed loop induction by ω and α effects (Moffatt 1978). As shown by measurements of pressure fluctuations, large vorticity concentrations are produced (Fauve et al. 1993, Abry et al. 1994) which may also act in favour of the amplification of the magnetic field if the classical analogy between vorticity and magnetic field production is to be believed. The choice of VK flows was thus motivated by the hope that the above features will make possible the generation a magnetic field by a strongly turbulent flow, with fluctuations as large, or even larger than the mean flow.

4.2. The VKS experimental set-up

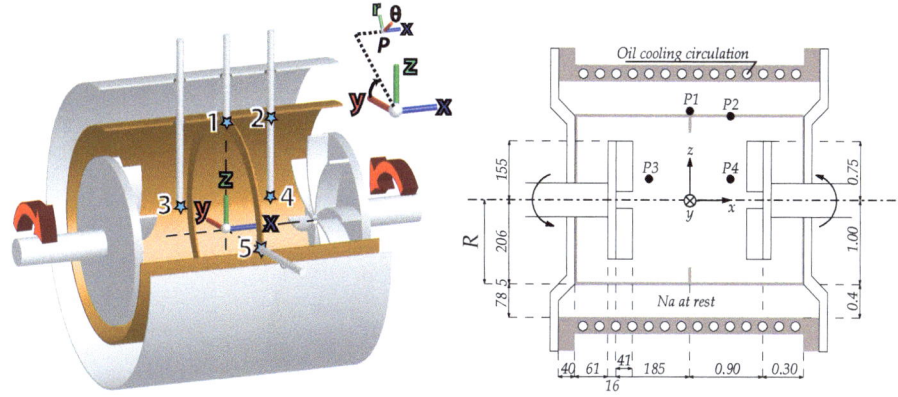

FIGURE 4. Experimental setup. From [58] with permission from © 2009 AIP Publishing LLC.

The VKS acronym stands for "von Kármán Sodium". The experiments has been developed in a collaborative work involving several french institutions: Direction des Sciences de la Matière of CEA, ENS-Lyon and ENS-Paris and have been realized in CEA/Cadarache-DEN/DTN. The VKS2 experiment is an evolution of a first design, VKS1 (Bourgoin et al. 2002, Pétrélis et al. 2003) which did not show any dynamo action. A sketch of the VKS2 set-up is displayed in Figure 4. The VK flow is generated in the inner cylinder of radius 206 mm and length 524 mm by two counter rotating discs of radius 154 mm and 371 mm apart. The disks are made of soft iron and are fitted with 8 curved blades of height $h = 41.2$ mm. An annulus of inner diameter 175 mm and thickness 5 mm is attached along the inner cylinder in the mid-plane between the disks.

This last configuration enabled the observation of a dynamo field as shown in Figure 5 (Monchaux et al. 2007). As the rotation rate of the discs is increased from 10 to 22 Hz, one observes at the location (1) (see Figure 4) the growth of a magnetic field: the azimuthal component acquires a nonzero average value of order 40 Gauss with relatively strong fluctuations. The two other components display small average values but fluctuate with rms values of order 5 gauss. Even though the fluctuation level is much higher than in the Karlsruhe or Riga experiments, we call this dynamo stationary in the sense that it is not displaying any kind of time-periodicity or reversals.

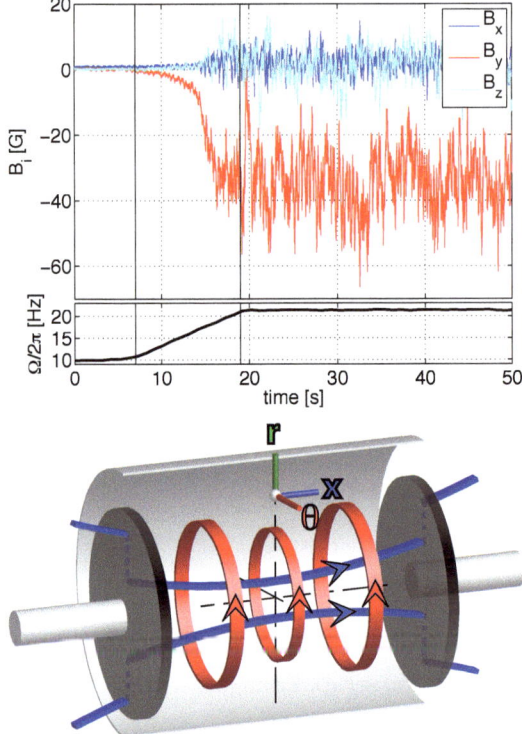

FIGURE 5. Three components of the magnetic field generated by dynamo action measured at point $P1$. Top: Growth of the magnetic field as the impellers' rotation rate F is increased from 10 to 22 Hz. Bottom: Sketch of the geometry of the mean magnetic field; toroidal component of the magnetic field (red), poloidal component (blue). From [57] & [72] with permission from © 2007, 2008 The American Physical Society.

Measurements of the magnetic field are fairly well resolved in the radial direction but have been performed only at a few axial and azimuthal locations (see Figure 4 left). Thus, although we cannot record higher-order modes, we observe that the mean magnetic field involves a leading order dipolar component with its axis along the rotation axis, $\mathbf{B_P}$, together with a related azimuthal component $\mathbf{B}_\theta$ (see Figure 5 right).

The amplitude of the magnetic field as a function of the magnetic Reynolds number is displayed in Figure 6. R_m has been defined as $R_m = K\mu_0\sigma R^2\Omega$ where R is the radius of the cylinder and Ω the rotation rate of the discs. $K = 0.6$ is a numerical coefficient relating ΩR to the maximum velocity in the flow. With this definition the critical magnetic Reynolds number is close to 31 as can be seen from Figure 6 (left) when the discs are rotated such that the leading edge of the

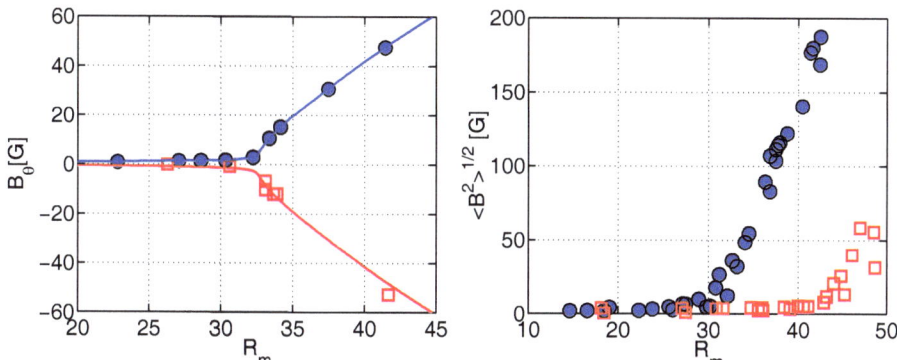

FIGURE 6. Left: Two independent realizations at same frequency above threshold showing opposite field polarities. Right: Magnetic field amplitude $\langle B^2 \rangle^{1/2} = \sqrt{B_x^2 + B_y^2 + B_z^2}$ at $P3$. Impellers are counterrotating at equal rotation rates, in the positive direction shown in Figure 4 (closed blue circles) or in the opposite direction, *i.e.*, with the blades on the impellers moving in a "scooping" or negative direction (open red squares). From [58] with permission from © 2009 AIP Publishing LLC.

curved blades is the convex one. The threshold is higher for the other direction (see Figure 6 right). Note that the two polarities of the magnetic field can be observed.

4.3. A possible dynamo mechanism for the VK flow

The mean VK flow has the following characteristics: the fluid is ejected radially outward by the discs; this drives an axial flow toward the discs along their axis and a recirculation in the opposite direction along the cylinder lateral boundary. In the case of counter-rotating impellers, the presence of a strong axial shear of azimuthal velocity in the mid-plane between the impellers generates a high level of turbulent fluctuations, roughly of the same order as the mean flow. It is thus unlikely that the fluctuations $\tilde{\mathbf{v}}$ can be neglected compared to $\overline{\mathbf{V}}$ in (1). It has been indeed observed that when the discs counter-rotate with the same frequency, a mean magnetic field is generated with a dominant axial dipolar component. Such an axisymmetric mean field cannot be generated by the mean flow alone, $\overline{\mathbf{V}}(r,x)$, that would give a non axisymmetric magnetic field according to Cowling theorem (Moffatt 1978). Non axisymmetric fluctuations $\tilde{\mathbf{v}}(r,\theta,x)$ thus play an essential role. As explained by Pétrélis *et al.* (2007), a possible mechanism is of $\alpha - \omega$ type, the α-effect being related to the helical motion of the radially expelled fluid between two successive blades of the impellers, and the ω-effect resulting from differential rotation due to counter-rotation of the impellers. This has been modeled using mean field MHD with an ad hoc α-effect related to this helical motion (Laguerre *et al.* 2009). The $\alpha - \omega$ mechanism has been illustrated without using mean field MHD by Gissinger (2009). When only the mean field velocity is taken into account in a numerical simulation of the induction equation, the generated magnetic field is an equatorial dipole as displayed in Figure 7 (left) and observed earlier in several numerical works (Marié et al. 2003, Bourgoin et al. 2004, Ravelet et al. 2005, Stefani et al. 2006). When a non axisymmetric velocity component that mimics helical flow along the blades is taken into account, it is found that an axial dipole becomes the preferred growing mode (see Figure 7 right) in agreement with the VKS experiment. Thus, the VKS dynamo is not generated by the mean flow alone in contrast to Karlsruhe and Riga experiments, and non-axisymmetric fluctuations play an essential role in the dynamo process.

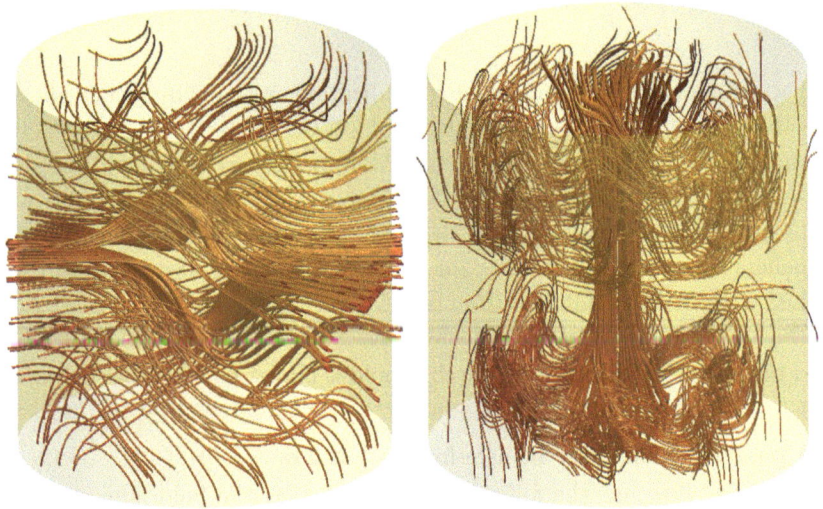

FIGURE 7. Geometry of the dynamo mode obtained by kinematic numerical simulations: (Left) an equatorial dipole is generated by the mean flow alone. (Right) When a non axisymmetric velocity component is taken into account, an axial dipole is generated. From [32] with permission from © 2009 IOP Publishing.

Another very important experimental fact is that the VKS dynamo has been observed so far only when impellers made of soft iron have been used. More precisely, one impeller at least should be made of soft iron and, it should be the rotating one if one of the impellers is at rest. It has been shown that magnetic boundary conditions corresponding to the high permeability limit significantly decrease the dynamo threshold (Gissinger et al. 2008, Gissinger 2009). However, it has been also claimed that other mechanisms could play a role: coupling with the magnetization inside the discs (Pétrélis et al. 2007), a possible additional source of ω-effect (Verhille et al. 2010) and the effect of a spatially periodic magnetic permeability along the azimuthal direction related to the blades (Giesecke et al. 2010). The later is the only one that has been simulated and found to decrease the dynamo threshold further. However, recent experiments have shown that a spatially periodic magnetic permeability alone is not enough to generate a dynamo, thus these simulations deserve to be checked.

The essential role of ferromagnetic blades (in addition to ferromagnetic discs) can be also explained using the following simple argument: any azimuthal magnetic field is refracted when it hits the blades and channeled inside each blade. It is conveyed along the blades and should emerge near the center with an axial component in order to satisfy $\nabla \cdot \mathbf{B} = 0$. According to this simple mechanism, ferromagnetic blades convert some azimuthal field component into the poloidal one and thus provide some additional contribution to the α-effect.

4.4. Dynamics of the magnetic field in the VKS experiment

As said above, the magnetic field generated by impellers counter-rotating at the same speed is statistically stationary. No secondary bifurcation is observed up to the maximum possible speed allowed by the available motor power. The dynamics are much richer when the impellers are rotated at different speed as displayed by the parameter space (see Figure 8).

Different dynamical regimes are observed when, starting from impellers rotating at 22 Hz, the frequency of an impeller, say F_2, is decreased, F_1 being kept

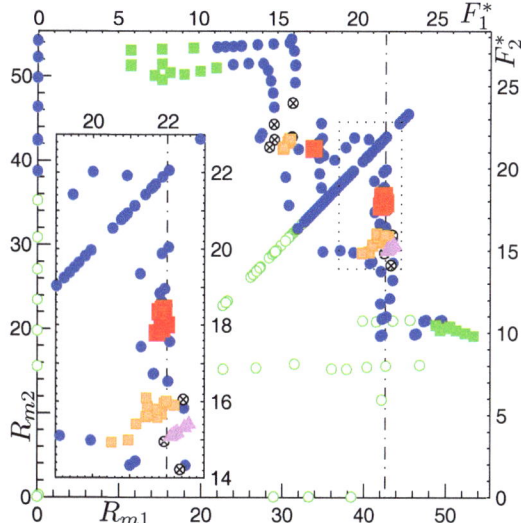

FIGURE 8. Dynamo regimes observed when the rotation frequencies F_1 and F_2 of the impellers are varied: no dynamo (green ○; $B \lesssim 10$G for more than 180 s), statistically stationary dynamos (blue ●); oscillatory dynamos (green squares); limit cycles (red squares), magnetic reversals (orange squares), bursts (purple triangles) and transient magnetic extinctions (⊗). From [72] with permission from © 2008 The American Physical Society.

constant. As said above, we first observe a statistically stationary dynamo regime with a dominant azimuthal mean field close to the flow periphery (see Figure 9, top left). This corresponds to the trace labelled 22–22 in the (B_r, B_θ) plane of Figure 9 (middle). As the frequency of the slower impeller is decreased, we obtain other stationary dynamo regimes for which the radial component of the mean field increases and then becomes larger than the azimuthal one (22–20 and 22–19). When we tune the impeller frequencies to 22 and 18.5 Hz respectively, a global bifurcation to a limit cycle occurs. We observe that the trajectory of this limit cycle goes through the location of the previous fixed points related to the stationary regimes. Direct time-recordings of the magnetic field, measured at the periphery of the flow in the mid-plane between the two impellers, are displayed in Figure 9 (bottom). We propose to ascribe the strong radial component (in green) that switches between ± 25 G to a quadrupolar mode (see Figure 9, top right). Its interaction with the dipolar mode that is the dominant one for exact counter-rotation, gives rise to the observed relaxation dynamics. The relaxation oscillation is observed in a rather narrow range of impeller frequency F_2 (less than 1 Hz). When the frequency of the slowest impeller in decreased further, statistically stationary regimes are recovered (22–18 to 22–16.5 Hz in Figure 9, middle). They also correspond to fixed points located on the trajectory of the limit cycle, except for the case 22–16.5 Hz that separates from it.

When the rotation frequency of the slowest impeller is decreased further, new dynamical regimes occur. One of them consists in field reversals (Berhanu *et al.* 2007). The three components of the magnetic field reverse at random intervals (Figure 10, left). The average length of phases with given polarity is two orders of magnitude larger than the duration of a reversal that corresponds to an ohmic diffusion time scale ($\tau_\sigma = \mu_0 \sigma L^2 \sim 1$ s on the scale L of the experiment). We emphasize that the trajectories connecting the $\mathbf{B}$ and $-\mathbf{B}$ states are robust despite the strong fluctuations of the flow. This is displayed in Figure 10 (right): the time evolution of reversals from up to down states can be superimposed by shifting the origin of time such that $\mathbf{B}(t=0) = 0$ for each reversal. Down-up reversals are superimposed in a similar way on up-down ones by plotting $-\mathbf{B}$ instead of $\mathbf{B}$. The time evolution averaged on 12 successive reversals can be represented as the trajectory of the system in phase space using a plot of $[B_\theta(t), B_\theta(t - \delta t)]$ where $\delta t = 1$ s $\sim \tau_\sigma$ (see Figure 11). For each reversal, the field first decays exponentially with a rate 0.8 s^{-1}. The system then moves on a faster time scale to reach the state with opposite polarity after displaying an overshoot in the direct time recording (Figure 10, right). In phase space, this is related to the fact that the trajectory has to circle around each fixed point in order to reach it. The trajectory in phase space is amazingly robust despite strong velocity fluctuations. These fluctuations put an upper bound on the duration of phases with a given polarity. However, this does not suppress the scale separation between the length of the phases with given polarity and the duration of a reversal. In that sense, turbulent fluctuations have a weak effect on the large scale dynamics of the magnetic field.

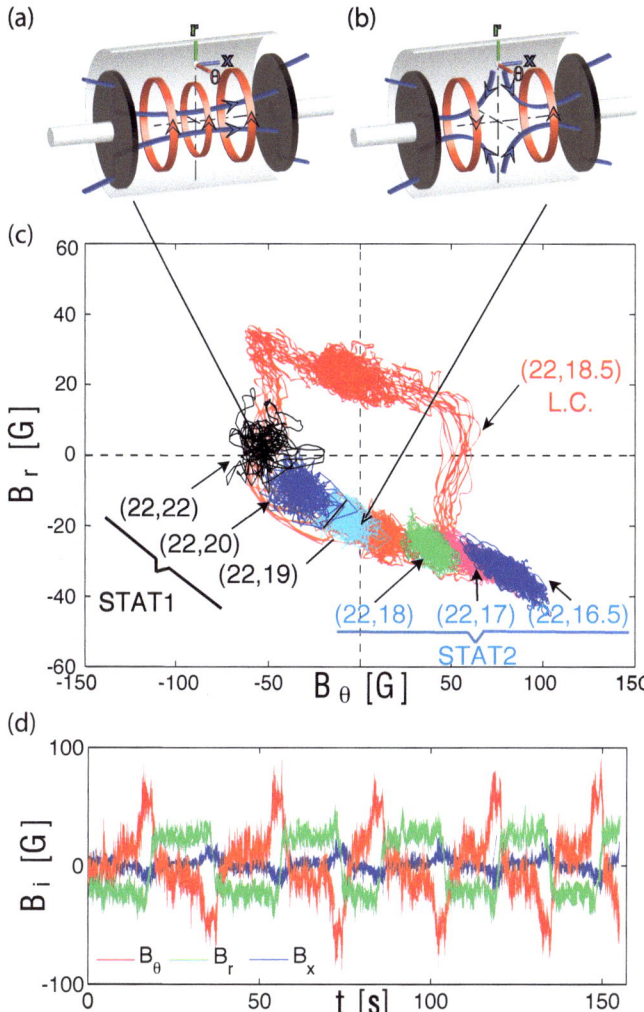

FIGURE 9. Sketch of the axial dipolar (a) and quadrupolar (b) magnetic modes. (c) location of the different states in the (B_r, B_θ) plane: fixed points corresponding to the stationary regimes for frequencies (F_1, F_2); limit cycle (L.C.) observed for impellers counterrotating at different frequencies $(22, 18.5)$Hz (red). The magnetic field is time averaged over 1 s to remove high frequency fluctuations caused by the turbulent velocity fluctuations. (d) time recording of the components of the magnetic field for frequencies $(22, 18.5)$Hz. From [58] with permission from © 2009 AIP Publishing LLC.

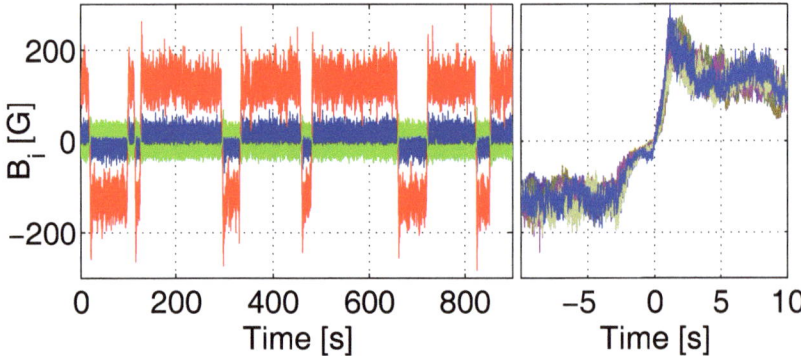

FIGURE 10. Reversals of the magnetic field generated by driving the flow with counterrotating impellers at frequencies $F_1 = 16$ Hz and $F_2 = 22$ Hz. Left: Time recording of the three magnetic field components at $P2$: axial (x) in blue, azimuthal (y) in red and radial (z) in green. Right: Superimposition of the azimuthal component for successive reversals from negative to positive polarity together with successive reversals from positive to negative polarity with the transformation $B \to -B$. For each of them the origin of time has been shifted such that it corresponds to $B = 0$. From [5] with permission from © 2009 AIP Publishing LLC.

When the magnetic field amplitude starts to decrease, either a reversal occurs, or the magnetic field first decays and then grows again with its direction unchanged. Similar sequences, called excursions, are also observed in recordings of the Earth's magnetic field. The phase space representation of Figure 11 is very appropriate to display them. One observes that the trajectory of an excursion is first similar to the one of reversals during their slow phase.

Other dynamical regimes are displayed in Figure 12. First, it is shown that there is a continuous transition from random reversals to noisy periodic behavior (top left and right) without any modification of the mechanical driving of the flow. Only the sodium temperature, and thus the Reynolds numbers are varied. Since the kinetic Reynolds number of the flow is very large, it is unlikely that its variation strongly affects the large scale flow. Observing nearly periodic oscillations rules out the naive picture in which reversals would only result from turbulent fluctuations driving the system away from a metastable state. For rotation frequencies 22–15 Hz, the magnetic field displays intermittent bursts (Figure 12, bottom right). The most probable value of the azimuthal field is roughly 20 G but bursts up to more than 100 G are observed such that the probability density function of the field has an exponential tail (not shown). For rotation frequencies 21–15 Hz, the same type of dynamics occur, but in a symmetric fashion, both positive and negative values of the field being observed (bottom left).

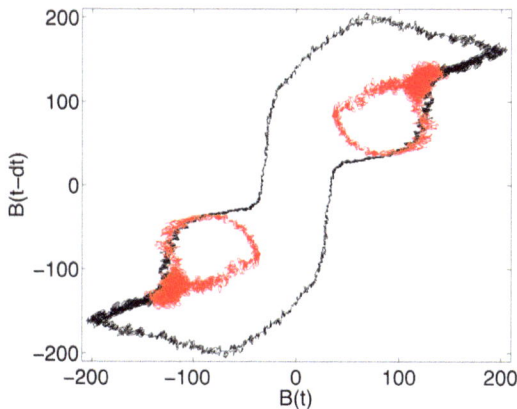

FIGURE 11. Plot of a planar cut of the phase space $[B_\theta(t), B_\theta(t-\delta t)]$ where $\delta t = 1$ s. The thick dots correspond to the two polarities of the magnetic field in its nearly stationary phases. The trajectories of reversals are represented in black. One excursion (together with the symmetric one) is represented in red. From [8], courtesy of E. Dormy, with permission from © 2007 IOP Publishing.

The different dynamical regimes of the magnetic field generated by the VKS flow display several characteristic features of low-dimensional dynamical systems. These dynamics will be understood below as the ones resulting from the competition between a few nearly critical modes. We emphasize that what is remarkable in these experiments is the robustness of these low-dimensional dynamical features that are not smeared out despite large turbulent fluctuations of the flow that generates the dynamo field.

5. Low-dimensional models of field reversals

It has been known since the work of Brunhes (1906) that Earth's magnetic field remains roughly parallel to the same direction, almost its rotation axis, for long durations (100000 years or much longer) but from time to time, it flips with the poles reversing sign. Polarity reversals are also observed for the magnetic field of the sun. Both its large scale dipolar component as well as the intense concentrations of magnetic field observed at smaller scales in the sun spots reverse sign roughly every 11 years.

Flows in the interiors of planets or stars have huge kinetic Reynolds numbers. For instance, $Re \sim 10^9$ in the Earth's liquid core or $Re \sim 10^{15}$ in the convective zone of the Sun. These flows being strongly turbulent, we would expect them to advect and distort the magnetic field lines in a very complicated way both in space and time. Thus, it is puzzling that the generated magnetic fields display a large

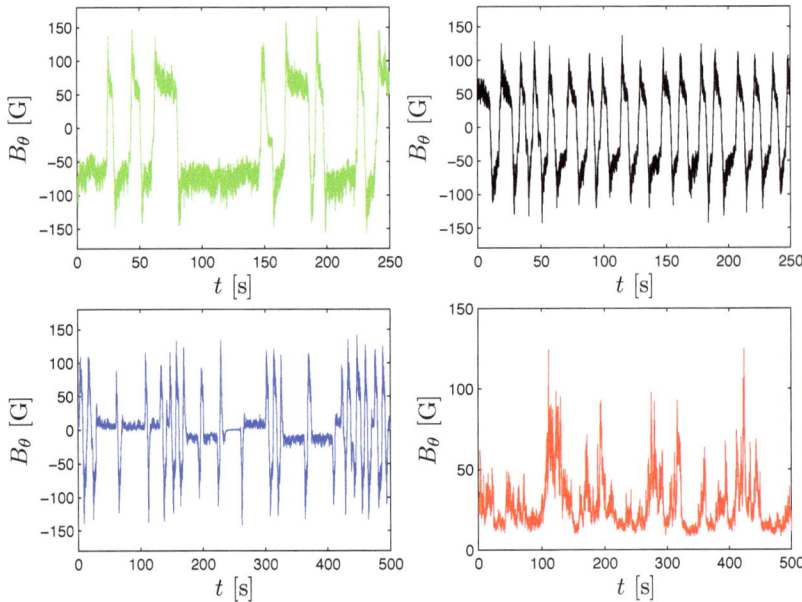

FIGURE 12. Top: Time recordings of the azimuthal component of the magnetic field observed for impellers rotating at (22–16) Hz. The sodium temperature is 131° C (left) and 147° C (right). Bottom: Time recordings of the azimuthal component of the magnetic field observed for impellers rotating at (22–15) Hz (left), (21–15) Hz (right). From [72] with permission from © 2008 The American Physical Society.

scale coherent component with rather simple dynamics: a dominant dipolar component, roughly aligned with the rotation axis for Earth and an oscillatory magnetic field with well-characterized spatial features for the sun (Zeldovich *et al.* 1983).

As just described, qualitatively similar dynamics have been observed in the VKS experiment. We will first shortly review the different models that have been proposed for the dynamics of the magnetic fields of the Earth and the Sun. Then, we will present in Section 6 a simple model of the dynamics observed in the VKS experiment.

5.1. Disc dynamos and truncations of the MHD equations

The first simple models of field reversals involved couple rotor disc dynamos (Rikitake 1958, Allan 1962, Cook and Roberts 1970) or even a Bullard disc dynamo when a shunt is added (Malkus 1972, Robbins 1977). The equations for the currents are of same type as Lorenz model (Lorenz 1963). When the two solutions $\pm I_0$ related one to the other by the $B \to -B$ symmetry are unstable and chaotic regimes occur, these systems stay for a while in the vicinity of one solution and then flips to the neighborhood of the other. These transitions occur in a random fashion and this can be considered as reversal dynamics. However, both the shape of the transitions displayed by direct recordings or in the phase space as well as their statistical properties differ from the experimental observations of field reversals and from paleomagnetic records. In addition, equations governing disc dynamos strongly differ from full MHD equations and cannot be obtained from them in any consistent approximation. It is however possible to relate Rikitake equations with a simple model of an $\alpha - \omega$ dynamo (Moffatt 1978). It has been also shown by Nozières (1978) that equations similar to the ones of disc dynamos can be obtained by truncating the full MHD equations. Keeping two magnetic modes of the diffusion operator and one velocity mode, equations similar (but not identical) to Rikitake (1958) are found. Nozières then describe reversals as a relaxation limit cycle between two quasi-stationary states related by the $B \to -B$ symmetry.

The main problem with truncated systems is that they usually describe dynamics that do not persist when higher modes are taken into account. The most famous example is the Lorenz attractor (Lorenz 1963) obtained from a drastic truncation of a Rayleigh–Bénard convection problem. Although its discovery has been one of the major steps of dissipative dynamical system theory and triggered a lot of studies, its chaotic dynamics does not subsist when higher modes of the convection problem are kept. The dynamo problem is even more sensitive to truncation. The growth of the magnetic field itself can result from a truncation of the velocity field even when a fair enough number of modes are kept whereas no dynamo exists when the resolution is good enough. It seems therefore unlikely that truncated systems involving a few coupled velocity and magnetic modes would correctly describe reversals of the magnetic field.

5.2. Normal forms

A different class of models, also involving a few coupled differential equations, is based on the assumption that several magnetic eigenmodes are competing above the dynamo threshold. These models have been mostly used in the context of low-order stellar dynamo models and more particularly to describe the solar cycle and its slow modulation. Tobias *et al.* (1995) take into account two magnetic modes (a poloidal and a toroidal one) undergoing a Hopf bifurcation in the framework of Parker's model (Parker 1955). They assume that the velocity field generating the magnetic field is close to a saddle-node bifurcation and couple the marginal velocity mode to the magnetic modes in order to obtain a third-order system which displays periodic, quasiperiodic and chaotic behaviours when the system parameters are varied. Wilmot-Smith *et al.* (2005) obtain similar results but with a coupling term that does not break the $B \to -B$ symmetry.

Knobloch & Landsberg (1996) consider a different model that does not involve marginal velocity modes but two magnetic modes, a dipolar and a quadrupolar one, both generated through a Hopf bifurcation. Taking into account 1 : 1 resonant coupling terms, they find aperiodic regimes that can also represent the modulation of the cyclic activity of the solar magnetic field. Finally, Knobloch *et al.* (1998)

assume the existence of two velocity modes, symmetric (respectively antisymmetric) with respect to the equatorial plane, and couple them to the dipolar and quadrupolar magnetic modes of the previous model. They show that two different types of modulation of the cyclic activity can be described.

In the framework of normal forms, it has been proposed to relate reversals to trajectories close to heteroclinic cycles that connect unstable fixed points $\pm\mathbf{B}$ (Armbruster et al. 2001, Chossat & Armbruster 2003). Heteroclinic cycles provide a simple framework to describe separation of time scales between quasi-steady states with a given polarity, related to the slowing down of the system in the vicinity of a saddle point, and rapid reversal events. Heteroclinic cycles are generally structurally unstable except in the presence of symmetries that lead to invariant subspaces of the dynamical system. Melbourne et al. (2001) try to describe the dynamics of Earth's magnetic field by writing amplitude equations for an equatorial dipole coupled to axial dipole and quadrupole. This model has heteroclinic cycles but no connection of states with opposite polarities except when additional coupling terms that break the symmetries are taken into account. Strictly speaking, a stable heteroclinic cycle connecting $\pm\mathbf{B}$ cannot describe reversals because the period of the trajectory in phase space goes to infinity as the trajectory is attracted on the cycle. However, an arbitrary amount of noise is enough to kick the system away from the saddle points and to generate random reversals with a finite mean period (Stone & Holmes 1990).

5.3. Metastable states in the presence of external noise

Other models rely on external noise in a stronger way. They start from a dipolar magnetic mode with amplitude $D(t)$ that bifurcates supercritically and model the effect of hydrodynamic turbulence through random fluctuations of the coefficients of the dynamical system governing D and the amplitudes of the stable modes in the vicinity of the bifurcation threshold. Fluctuations only in the amplitude equation, $\dot{D} = \mu D - D^3$, i.e., a growth rate μ that involves a noisy component, does not lead to reversals between the two stationary solutions $D = \pm\sqrt{\mu}$. However, taking into account that D is coupled with the amplitudes of the stable modes which are also excited by fluctuations, can lead to reversals (Schmitt et al. 2001). D behaves as the position of a strongly damped particle driven by random noise in a two-well potential. The crucial role of damped modes has been emphasized further by Hoyng & Duistermaat (2004). The reversals are triggered by large fluctuations of damped modes driven by noise. These modes act on D as an effective additive noise.

Recent numerical simulations have modelled hydrodynamic fluctuations with a noisy α-effect (Giesecke et al. 2005, Stefani & Gerbeth 2005, Stefani et al. 2007). The deterministic part of this model can generate periodic relaxation oscillations with the system slowing down in the vicinity of two states with opposite polarities $\pm\mathbf{B}$. In this respect, it belongs to the class of systems described by Nozières (1978). The addition of external noise is thus crucial to generate random reversals. It is likely that the phenomenology of this model is related to the proximity of a codimension-two point that results from two interacting modes with different radial structures.

5.4. Hydrodynamic mechanisms and direct numerical simulations

The above descriptions of reversals assume the existence of some large scale dominant modes of the magnetic field. The random dynamics of reversals are either of deterministic nature (low-dimensional chaos) or result from the addition of external noise that describes hydrodynamic fluctuations.

A different approach, initiated by Parker (1969), consists in trying to identify the nature of the fluctuations of the velocity field that is required to generate a reversal. In the case of Earth, it is believed that the magnetic field is generated through an α-ω mechanism, ω being related to differential rotation and α resulting from the existence of a mean number of cyclonic convective cells in Earth's core, that fluctuate both in number and position. When strong enough, these fluctuations can reverse the magnetic field (Parker 1969, Levy 1972).

Another mechanism has been also proposed by Parker (1979). It follows from the observation by Roberts (1972) that a meridional circulation favors stationary dipolar α-ω dynamos in spherical geometries. Parker (1979) suggested that if the meridional circulation is altered for a while, an oscillatory magnetic mode may become dominant and generates a reversal of the magnetic field. It has been claimed later that this mechanism can be also suggested from palaeomagnetic data (McFadden & Merrill 1995). Numerical simulations of the MHD equations in a rotating sphere have displayed this in a clear-cut way: it has been shown by Sarson & Jones (1999) and Sarson (2000) that the random emission of poleward light plumes, or "buoyancy surge", generates fluctuations of the meridional flow that can trigger a reversal. They also found that this mechanism is not affected much by the back reaction of the magnetic field on the flow and does result from the proximity in parameter space of stationary and time periodic dynamo modes, depending on the intensity of the meridional flow. A process also related to convective plumes has been observed by Wicht & Olson (2004). They found that a magnetic field with an opposite polarity is produced locally in the convective plumes and that the transport of this reversed flux can generate a reversal. They also showed that the observed reversals are almost unchanged when the Lorentz force is removed from the numerical code. Other advection processes of the magnetic field by the flow have been studied in detail by Aubert et al. (2008). It should be noted that all these numerical simulations have been performed with large values of P_m ($1 < P_m < 20$). Local modifications of the magnetic field by the flow are likely to play a less important role for small values of P_m because they are strongly damped by ohmic diffusion.

Since 1995 (Glatzmaier & Roberts 1995), a lot of three-dimensional numerical simulations of the MHD equations in a rotating sphere have been able to simulate a self-consistent magnetic field that displays reversals (see the reviews by Dormy et al. 2000, Roberts & Galtzmaier 2000). However, it has been emphasized that most relevant dimensionless parameters that can be achieved in direct simulations are orders of magnitude away from their value in Earth's core or laboratory experiments. Even in the limited range accessible to direct simulations, it has been shown that the geometry of the generated magnetic field and the properties of field reversals can strongly depend on the values of the relevant dimensionless numbers (Kutzner & Christensen 2002, Busse & Simitev 2006). Thus, one may conclude as in Coe et al. (2000) that "each reversal in the simulations has its own unique character, which can differ greatly in various aspects from others". However, we emphasize that a lot of these numerical simulations also display similar properties at a more global level, if one considers the how the symmetries of the flow and the magnetic field evolve during a reversal. We will discuss this aspect in the next section.

6. A simple model for the dynamics observed in the VKS experiment

The most striking feature of the VKS experiment is that time dependent magnetic fields are generated only when the impellers rotate at different frequencies (Berhanu et al. 2007, Ravelet et al. 2008). We have shown in Pétrélis & Fauve (2008) that this is related to the additional invariance under $\mathcal{R}_\pi$ when $F_1 = F_2$ (rotation of an angle π along any axis in the mid-plane). We indeed expect that in that case, the modes involved in the dynamics are either symmetric or antisymmetric. Such modes are displayed in Figure 13. A dipolar mode is changed to its opposite by $\mathcal{R}_\pi$, whereas a quadrupolar mode is unchanged.

We assume that the magnetic field is the sum of a dipolar component with an amplitude D and a quadrupolar one, Q. We define $A = D + iQ$ and we assume that an expansion in power of A and its complex conjugate $\bar{A}$ is pertinent close to threshold in order to obtain an evolution equation for both modes. Taking into

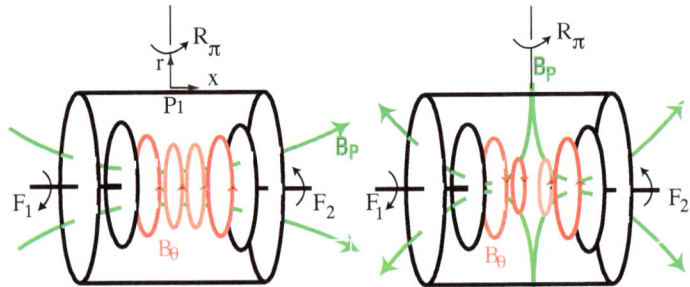

FIGURE 13. Possible eigenmodes of the VKS experiment. The two discs counter-rotate with frequency F_1 and F_2. Left: Magnetic dipolar mode. Right: Magnetic quadrupolar mode. Poloidal (green) and toroidal (red) components are sketched.

account the invariance $\mathbf{B} \to -\mathbf{B}$, i.e., $A \to -A$, we obtain

$$\dot{A} = \mu A + \nu \bar{A} + \beta_1 A^3 + \beta_2 A^2 \bar{A} + \beta_3 A \bar{A}^2 + \beta_4 \bar{A}^3 \,, \tag{8}$$

where we limit the expansion to the lowest-order nonlinearities. In the general case, the coefficients are complex and depend on the experimental parameters.

Symmetry of the experiment with respect to $\mathcal{R}_\pi$ when the discs exactly counter-rotate, amounts to constraints on the coefficients. Applying this transformation to the magnetic modes, changes D into $-D$ and Q into Q, thus $A \to -\bar{A}$. We conclude that, in the case of exact counter-rotation, all the coefficients are real. When the frequency difference $f = F_1 - F_2$ is increased from zero, we obtain that the real parts of the coefficients are even and the imaginary parts are odd functions of f. When the coefficients are real, the growth rate of the dipolar component is $\mu_r + \nu_r$ and that of the quadrupolar component is $\mu_r - \nu_r$. The dipole being observed for exact counter-rotation implies that $\nu_r > 0$ for $f = 0$. By increasing f, we expect that ν_r changes sign and favors the quadrupolar mode according to the experimental results (see Figure 9). We will explain in the next section how modifying the parameters of (8) leads to bifurcation to time dependent solutions.

6.1. A mechanism for oscillations and reversals

As shown in Pétrélis & Fauve (2008), the planar system (8) explains the dynamical regimes observed so far in the VKS experiment (Ravelet et al. 2008). It is invariant under the transformation $\mathbf{B} \to -\mathbf{B}$. Thus, in the case of counter-rotating impellers, $F_1 = F_2$, it has two stable dipolar solutions $\pm D$ and two unstable quadrupolar solutions $\pm Q$. When the frequency difference f is increased, these solutions become more and more mixed due to the increase of the strength of the coupling terms between dipolar and quadrupolar modes. Dipolar (respectively quadrupolar) solutions get a quadrupolar (respectively dipolar) component and give rise to the stable solutions $\pm B_s$ (respectively unstable solutions $\pm B_u$) displayed in Figure 14. When f is increased further, a saddle-node bifurcation occurs, i.e., the stable and unstable solutions collide by pairs and disappear. This generates a limit cycle that connects the collision point with its opposite. This result can be understood as follows: the solution $B = 0$ is unstable with respect to the two different fixed points, and their opposite. It is an unstable point, whereas one of the two bifurcating solutions is a stable point, a node, and the other is a saddle. If the saddle and the node collide, say at B_c, what happens to initial conditions located close to these points? They cannot be attracted by $B = 0$ which is unstable and they cannot reach other fixed points since they just disappeared. Therefore the trajectories describe a cycle. The associated orbit contains $B = 0$ since, for a planar problem, in any orbit, there is a fixed point. Suppose that the orbit created from B_c is different from the one created by $-B_c$. These orbits being images by the transformation $\mathbf{B} \to -\mathbf{B}$, they must intersect at some point. Of course, this is not possible for a planar system because it would violate the uniqueness of the solutions. Therefore, there is only one cycle that connects points close to B_c and $-B_c$.

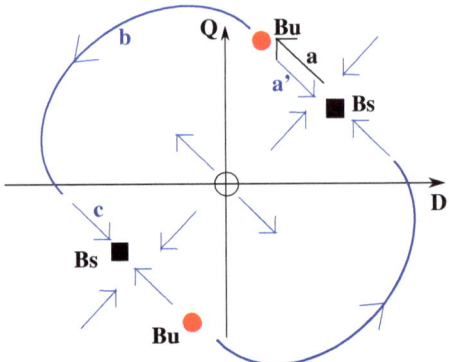

FIGURE 14. A generic saddle-node bifurcation in a system with the $\mathbf{B} \to -\mathbf{B}$ invariance: below threshold, fluctuations can drive the system against its deterministic dynamics (phase a). If the effect of fluctuations is large enough, this generates a reversal (phases b and c). Otherwise, an excursion occurs (phase a'). From [68] with permission from © 2008 IOP Publishing.

This provides an elementary mechanism for field reversals in the vicinity of a saddle-node bifurcation. First, in the absence of fluctuations, the limit cycle generated at the saddle-node bifurcation connects $\pm B_c$. This corresponds to periodic reversals. Slightly above the bifurcation threshold, the system spends most of the time close to the two states of opposite polarity $\pm B_c$. Second, in the presence of fluctuations, random reversals can be obtained slightly below the saddle-node bifurcation. B_u being very close to B_s, even a fluctuation of small intensity can drive the system to B_u from which it can be attracted by $-B_s$, thus generating a reversal.

The effect of turbulent fluctuations on the dynamics of the two magnetic modes governed by (8) can be easily modeled by adding some noisy component to the coefficients (Pétrélis & Fauve 2008). Random reversals are displayed in Figure 15 (top left). The system spends most of the time close to the stable fixed points $\pm B_s$. We observe in Figure 15 (top right) that a reversal consists of two phases. In the first phase, the system evolves from the stable point B_s to the unstable point B_u (in the phase space sketched in Figure 14). The deterministic part of the dynamics acts against this evolution and the fluctuations are the motor of the dynamics. That phase is thus slow. In the second phase, the system evolves from B_u to $-B_s$, the deterministic part of the dynamics drives the system and this phase is faster.

The behaviour of the system close to B_s depends on the local flow. Close to the saddle-node bifurcation, the position of B_s and B_u defines the slow direction of the dynamics. If a component of B_u is smaller than the corresponding one of B_s, that component displays an overshoot at the end of a reversal. In the opposite case, that component will increase at the beginning of a reversal. For instance, in the phase space sketched in Figure 14, the component D decreases at the end of a reversal and the signal displays an overshoot. The component Q increases just before a reversal.

For some fluctuations, the second phase does not connect B_u to $-B_s$ but to B_s. It is an aborted reversal or an excursion in the context of the Earth dynamo. Note that during the initial phase, a reversal and an excursion are identical. In the second phase, the approaches to the stationary phase differ because the trajectory that links B_u and B_s is different form the trajectory that links B_u and $-B_s$. In particular, if the reversals display an overshoot this will not be the case of the excursion (see Figure 15 top right) and the sketch of the cycle in Figure 14).

Finally, it is illustrated in Figure 15 (bottom left and right), that the other dynamical regimes of the VKS experiment, such as symmetric or asymmetric intermittent bursts, can be described with the same model (Pétrélis & Fauve 2008).

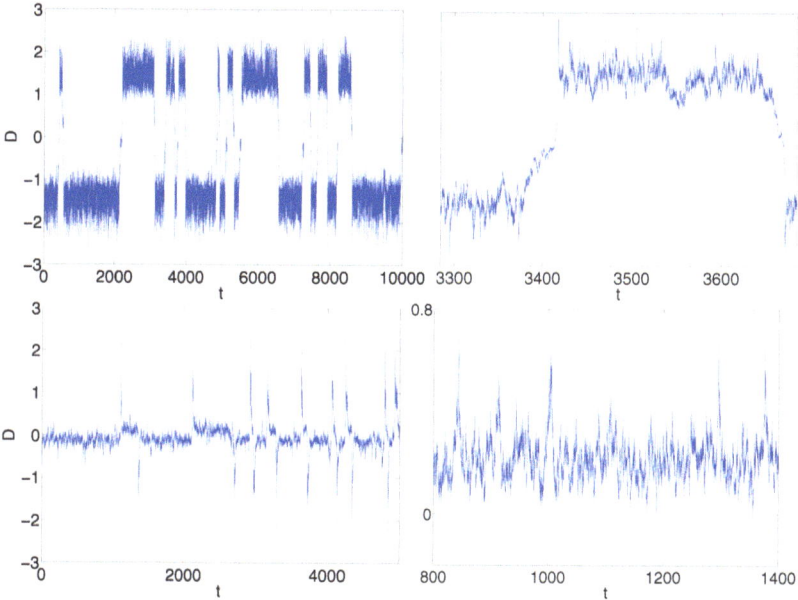

FIGURE 15. Time recordings obtained from equation (8) displaying different dynamics of the magnetic field, as observed in the VKS experiment: reversals, symmetric bursts and asymmetric bursts. From [68] with permission from © 2008 IOP Publishing.

6.2. A simple model for Earth's magnetic field reversals

The above model of reversals of magnetic field in the vicinity of a saddle-node bifurcation in a system with the invariance $\mathbf{B} \to -\mathbf{B}$ explains many intriguing features of the reversals of Earth magnetic field (Pétrélis et al. 2009). The most significant output is that the mechanism predicts specific characteristics of the field obtained from paleomagnetic records (Valet et al. 2005), in particular their asymmetry: the Earth's dipole decays on a slower time scale than it recovers after a reversal. In addition, it displays an overshoot that immediately follows the reversals. Other characteristic features such as excursions as well as the existence of superchrons are understood in the same framework.

Although the symmetries of the flow in the Earth's core strongly differ from the ones of the VKS experiment, dipolar and quadrupolar modes can be defined with respect to equatorial symmetry such that model (8) can be transposed for Earth's magnetic field. From an analysis of paleomagnetic data, it has been proposed that reversals involve an interaction between dipolar and quadrupolar modes (McFadden et al. 1991). We thus obtain an interesting prediction about the liquid core in that case: if reversals involve a coupling of the Earth's dipole with a quadrupolar mode, then this requires that the flow in the core has broken mirror symmetry. In contrast, another scenario has been proposed in which the Earth's dipole is coupled to an octupole, i.e., another mode with a dipolar symmetry. This does not require additional constraint on the flow in the core in the framework of our model. In any case, the existence of two coupled modes allows the system to evolve along a path that avoids $\mathbf{B} = \mathbf{0}$. In physical space, this means that the total magnetic field does not vanish during a reversal but that its spatial structure changes.

7. Different morphologies for field reversals

We have shown in the previous section that an efficient way to reverse an axial dipolar field is to couple it with another mode. In the case of axisymmetric mean fields, an axial quadrupole is a natural choice. However, one can imagine that it can be also possible to involve a non-axisymmetric mode in the dynamics of reversals. In that case, the leading order choice would be an equatorial dipole. This type of scenario has been recently observed in numerical simulations of a flow driven by counter-rotating propellers in a spherical domain. This geometry displays many similarities with the one of the VKS experiment (in both cases, a cylindrical symmetry is related to the rotation axis). It corresponds to the Madison experiment. Although no dynamo has been observed yet, numerical simulations have been performed (Bayliss et al. 2007, Gissinger et al. 2008, 2010). In the case of counter-rotating propellers, an equatorial dipole is observed when the kinetic Reynolds number is small (Gissinger et al. 2008). However, for moderate kinetic Reynolds numbers, $Re \sim 300$, the flow involves fluctuations that drives an axial dipole first (see Figure 16).

As for the VKS experiment, this axial dipole displays reversals only when the $\mathcal{R}_\pi$ symmetry is broken by rotating the propellers at different speeds. In the simulations, this is achieved by multiplying the forcing by a parameter C, with $C = 1$ for the lower hemisphere but can be different from one for the upper

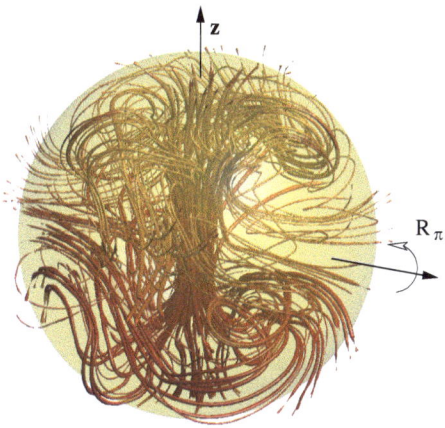

FIGURE 16. Magnetic field lines obtained with a symmetric forcing ($C = 1$) for $R_m = 300$ and $P_m = 1$. Note that the field involves a dipolar component with its axis aligned with the axis z of rotation of the propellers. From [33] with permission from © 2010 IOP Publishing.

one. Time recordings of some components of the magnetic field are displayed in Figure 17 for $R_m = 300$, $P_m = 1$ and $C = 2$. We observe that the axial dipolar component (in black) randomly reverses sign. The phases with given polarity are an order of magnitude longer than the duration of a reversal that corresponds to an ohmic diffusion time. The magnetic field strongly fluctuates during these phases because of hydrodynamic fluctuations. It also displays excursions. All of these

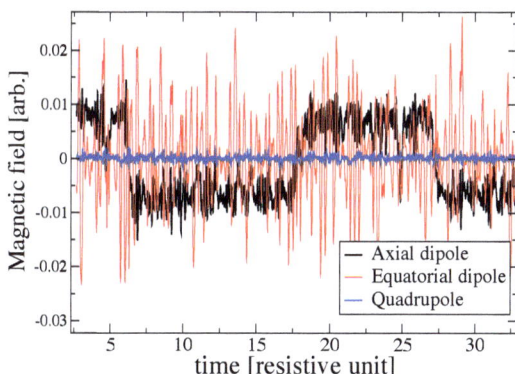

FIGURE 17. Time recording of the axial dipolar magnetic mode (in black), the axial quadrupolar mode (in blue) and the equatorial dipole (in red) for $R_m = 300$, $P_m = 1$ and $C = 2$. From [33] with permission from © 2010 IOP Publishing.

features are also observed in the VKS experiment. However, the simulation for $P_m = 1$ also displays strong differences with the VKS experiment. The equatorial dipole is the mode with the largest fluctuations whereas the axial quadrupolar component is an order of magnitude smaller than the dipolar modes. In addition, it does not seem to be coupled to the axial dipolar component.

We now turn to simulations using only smaller values of P_m (values comparable to the ones of the VKS experiment are out of reach in direct numerical simulations). The time evolution of the magnetic modes for $R_m = 165$, $P_m = 0.5$ and $C = 1.5$ is represented on Figure 18 (left). It differs significantly from the

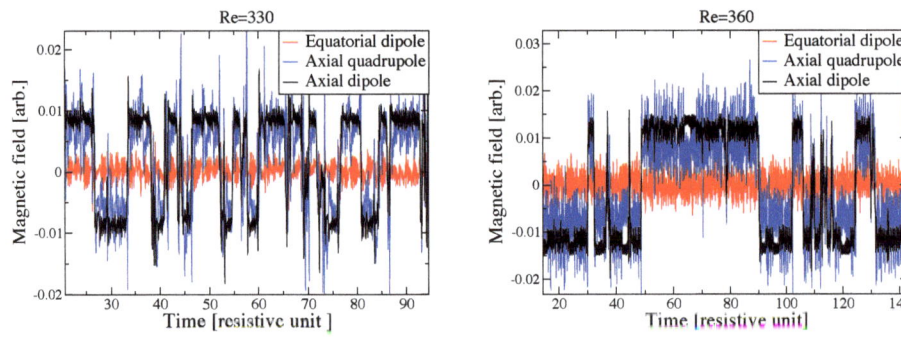

FIGURE 18. Time recordings of the axial dipole (black), the axial quadrupole (blue) and the equatorial dipole (red). Left: $R_m = 165$, $P_m = 0.5$ and $C = 1.5$. Right: $R_m = 180$, $P_m = 0.5$ and $C = 2$. From [33] with permission from © 2010 IOP Publishing.

previous case ($P_m = 1$). First of all, the quadrupole is now a significant part of the field, and reverses together with the axial dipole. The equatorial dipole remains comparatively very weak and unessential to the dynamics. One can argue that R_m has also been modified when changing P_m from 1 to 0.5. However, for $P_m = 1$, we have observed the same dynamics of reversals when R_m has been decreased down to $R_m = 220$ below which reversals are not observed any more.

The high amount of fluctuations observed in these signals is related to hydrodynamic fluctuations. One could be tempted to speculate that a higher degree of hydrodynamic fluctuations necessarily yields a larger reversal rate. Such is in fact not the case. A more sensible approach could be to try to relate the rate of reversals to the amount of fluctuations of the magnetic modes in a phase with given polarity. Increasing R_m from 165 to 180 does yield larger fluctuations as shown in Figure 18 (right). However the reversal rate is in fact lowered because C was modified to $C = 2$. This clearly shows that the asymmetry parameter C plays an important role in addition to the fluctuations of the magnetic field. For $P_m = 0.5$, reversals occur only in a restricted region, $1.1 < C < 2.5$, which is also a feature of the VKS experiment. The reversal rate strongly depends on the value of C with respect to these borders, in good agreement with the model presented in Pétrélis & Fauve (2008). Thus, the transition from a stationary regime to a reversing one is not generated by an increase of hydrodynamic fluctuations.

We have thus shown that different types of random reversals of a dipolar magnetic field can be obtained by varying the magnetic Prandtl number in a rather small range around $P_m = 1$. This may be of interest for simulations of the magnetic field of the Earth that have been mostly restricted to values of P_m larger than one. We have observed that axisymmetric dipolar and quadrupolar modes decouple from the other magnetic modes while getting coupled together when P_m is decreased. Although we do not claim to have reached an asymptotic low P_m regime which is out of reach of the present computing power, we observe that dominant axial dipole and quadrupole are also observed in the VKS experiment for which $P_m \sim 10^{-5}$.

8. A minimal model for field reversals

These direct numerical simulations illustrate the role of the magnetic Prandtl number (or possibly of the distance to the dynamo threshold) in the dynamics of reversals. We now write the simplest dynamical system that involves the three modes that look important in the low P_m simulations: the dipole D, the quadrupole Q, and the zonal velocity mode V that breaks the $\mathcal{R}_\pi$ symmetry. These modes transform as $D \to -D$, $Q \to Q$ and $V \to -V$ under the $\mathcal{R}_\pi$ symmetry. Keeping nonlinear terms up to quadratic order, we get

$$\dot{D} = \mu D - VQ, \tag{9}$$

$$\dot{Q} = -\nu Q + VD, \tag{10}$$

$$\dot{V} = \Gamma - V + QD. \tag{11}$$

A non zero value of Γ is related to a forcing that breaks the $\mathcal{R}_\pi$ symmetry, i.e., propellers rotating at different speeds.

The dynamical system (9)–(11) with $\Gamma = 0$ occurs in different hydrodynamic problems and has been analyzed in detail (Hughes & Proctor 1990). The relative signs of the coefficients of the nonlinear terms have been taken such that the solutions do not diverge when $\mu > 0$ and $\nu < 0$. Their modulus can be taken equal to one by appropriate scalings of the amplitudes. The velocity mode is linearly damped and its coefficient can be taken equal to -1 by an appropriate choice of the time scale. Note that similar equations were obtained with a drastic truncation of the linear modes of MHD equations (Nozières 1978). However, in that context μ should be negative and the damping of the velocity mode was discarded, thus strongly modifying the dynamics.

This system displays reversals of the magnetic modes D and Q for a wide range of parameters. A time recording is shown in Figure 6. The mechanism for these reversals results from the interaction of the modes D and Q coupled by the broken $\mathcal{R}_\pi$ symmetry when $V \neq 0$. It is thus similar to the one described in

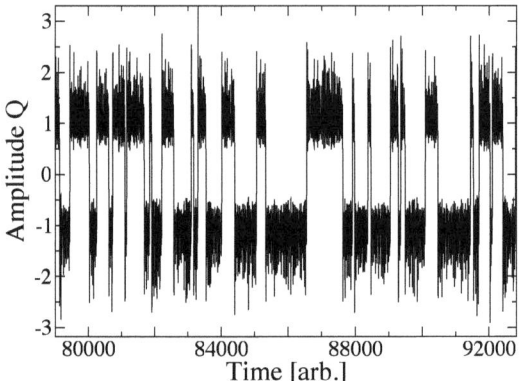

FIGURE 19. Numerical integration of the amplitude equations (9)–(11). Time recording of the amplitude of the quadrupolar mode for $\mu = 0.119$, $\nu = 0.1$ and $\Gamma = 0.9$. From [33] with permission from © 2010 IOP Publishing.

Section 6 but also involves an important difference: keeping the damped velocity mode into the system generates chaotic fluctuations. It is thus not necessary to add external noise to obtain random reversals. This system is fully deterministic as opposed to the one of Pétrélis and Fauve (2008). The phase space displayed in Figure 20 (left) shows the existence of chaotic attractors in the vicinity of the $\pm\mathbf{B}$ quasi-stationary states. When these symmetric attractors are disjoint, the magnetic field fluctuates in the vicinity of one of the two states $\pm\mathbf{B}$ and the dynamo

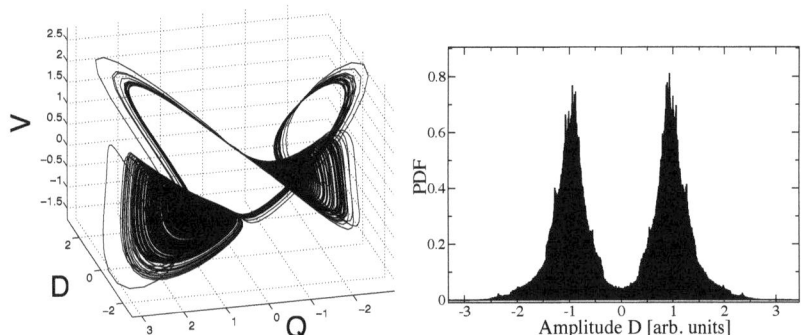

FIGURE 20. Numerical integration of the amplitude equations (9)–(11). Left: Three-dimensional phase space. Right: Probability density function of D ($\mu = 0.119$, $\nu = 0.1$ and $\Gamma = 0.9$). From [33] with permission from © 2010 IOP Publishing.

is statistically stationary. When μ is varied, these two attractors can get connected through a crisis mechanism, thus generating a regime with random reversals.

We do not claim that this minimal low-order system fully describes the direct simulations presented here. For instance, in the case of exact counter-rotation ($C = 1$, i.e., $\Gamma = 0$), equations (9)–(11) do not have a stable stationary state with a dominant axial dipole. The different solutions obtained when μ is increased cannot capture all the dynamo regimes of the VKS experiment or of the direct simulations when R_m is increased away from the threshold. Taking into account cubic nonlinearities provides a better description of the numerical results for $P_m = 0.5$. However, this three mode system with only quadratic nonlinearities involves the basic ingredients of the reversals observed in the present numerical simulations for low enough values of the magnetic Prandtl number. As recalled in Section 5, geomagnetic reversals have been modeled since a long time using low-dimensional dynamical systems or equations involving a noisy forcing. The above model (9)–(11) does not rely on an external noise source to generate random reversals. Compared to previous deterministic models, it displays dynamical and statistical properties that are much closer to the ones of our direct simulations at low P_m or of the VKS experiment. For instance, the direct recordings of D or Q do not involve the growing oscillations characteristic of reversals displayed by the Rikitake or Lorenz systems but absent in dynamo experiments or in direct simulations. Correspondingly, the probability density function of D displayed in Figure 20 (right) is also much closer to the one obtained in experiments or direct simulations than the one of previous deterministic models.

9. Conclusion

We have studied dynamical regimes that can arise when two axisymmetric magnetic eigenmodes are coupled. Symmetry considerations allow to identify properties of the magnetic modes and, in some cases, put constraints on the coupling between the modes. We have shown that when a discrete symmetry is broken by the flow that generates the magnetic field, the coupling between an odd and an even magnetic mode (with respect to the symmetry) can generate a bifurcation from a stationary state to a periodic state. This behaviour is generic when a saddle-node bifurcation occurs in a system that is invariant under $\mathbf{B} \to -\mathbf{B}$. Close to the bifurcation threshold, fluctuations drive the system into a state of random reversals that connect a solution B_s to its opposite $-B_s$. This scenario provides a simple explanation for many features of the dynamics of the magnetic field observed in the VKS experiment: alternation of stationary and time dependent regimes when a control parameter is varied, continuous transition from random reversals to time periodic ones, characteristic shapes of the time recordings of reversals versus excursions.

Although the discrete symmetry involved for the flow in the Earth core is different from the one of the VKS experiment, a similar analysis can be performed for the geodynamo (Pétrélis *et al.* 2009).

More generally, our scenario can be applied to purely hydrodynamic systems. Cellular flows driven by thermal convection (Krishnamurti & Howard 1981) or by volumic forces (Sommeria 1986) display a transition for which a large scale circulation is generated on a smaller scale turbulent background. This large scale flow can display random reversals, very similar to the ones observed for the magnetic field. A model analogue to the present one, can explain how this large scale field can reverse without the need of a very energetic turbulent fluctuation acting coherently in the whole flow volume.

Acknowledgment

I thank the members of the VKS team with whom the experimental data reported in Section 4 have been obtained. The model of Section 6 has been worked out with F. Pétrélis and its application to the geodynamo resulted from a collaboration we had with E. Dormy and J.P. Valet. Finally, the simulations presented in Section 7 have been performed by E. Dormy and C. Gissinger.

References

[1] P. Abry, S. Fauve, P. Flandrin, and C. Laroche, Analysis of pressure fluctuations in swirling turbulent flows. J. Physique II **4**, 725–733 (1994).

[2] G. Ahlers & R.P. Behringer, Evolution of turbulence from the Rayleigh–Bénard instability, Phys. Rev. Lett. **40**, 712–716 (1978).

[3] D.W. Allan, *On the behavior of systems of coupled dynamos.* Proc. Camb. Phil. Soc. **58**, 671–693 (1962).

[4] J. Aubert, J. Aurnou & J. Wicht, The magnetic structure of convection-driven numerical dynamos. Geophys. J. Int. **172**, 945–956 (2008).

[5] S. Aumaître, *et al.*, The VKS experiment: Turbulent dynamical dynamos. Phys. Fluids **21**, 035108 (2009).

[6] D. Armbruster, P. Chossat & I. Oprea, *Structurally stable heteroclinic cycles and the dynamo dynamics*, In *Dynamo and Dynamics, a Mathematical Challenge* (eds. Chossat P., Armbruster D. & Oprea I) (2001), pp. 313–322, Nato Science Series II, vol 26, Kluwer Academic Publishers.

[7] V. Arnold, *Geometrical Methods in the Theory of Ordinary Differential Equations*, Springer-Verlag (1982).

[8] M. Berhanu, *et al.*, Magnetic field reversals in an experimental turbulent dynamo. Europhys. Lett. **77**, 59001 (2007).

[9] M. Bourgoin, *et al.*, Magnetohydrodynamics measurements in the von Kármán sodium experiment. Phys. Fluids **14**, 3046–3058 (2002).

[10] M. Bourgoin, P. Odier, J.F. Pinton, and Y. Ricard, An iterative study of time independent induction effects in magnetohydrodynamics. *Phys. Fluids* **16**, 2529–2547 (2004).

[11] B. Brunhes, Recherches sur la direction d'aimantation des roches volcaniques. J. de Phys. Théor. App. **5**, 705–724 (1906).

[12] F.H. Busse, *Mathematical problems of dynamo theory*, In *Applications of bifurcation theory* (1977) pp. 175–202, Academic Press.

[13] F.H. Busse, U. Müller, R. Stieglitz & A. Tilgner, A two-scale homogeneous dynamo, and extended analytical model and an experimental demonstration under development. Magnetohydrodynamics **32**, 235–248 (1996).

[14] F.H. Busse & R. Simitev, Parameter dependences of convection driven dynamos in rotating spherical fluid shells. GAFD **100**, 341–361 (2006).

[15] M. Ghil & S. Childress, Topics in geophysical fluid dynamics: atmospheric dynamics, dynamo theory, and climate dynamics. Appl. Math. Sci. **60** (1987), New York: Springer Verlag.

[16] P. Chossat & D. Armbruster, Dynamics of polar reversals in spherical dynamos. Proc. Roy. Soc. Lond. A **459**, 577–596 (2003).

[17] R.S. Coe, L. Hongre & G.A. Glatzmaier, An examination of simulated geomagnetic reversal from a paleomagnetic perspective. Phil. Transact. Royal Soc. A **358**, 1141–1170 (2000).

[18] A.E. Cook & P.H. Roberts, The Rikitake two-disc dynamo system. Proc. Camb. Phil. Soc. **68**, 547–569 (1970).

[19] E. Dormy, J.-P. Valet & V. Courtillot, Numerical models of the geodynamo and observational constraints. Geochem. Geophys. Geosyst. **1**, 2000GC000062 (2000).

[20] S. Fauve, C. Laroche, and B. Castaing, Pressure fluctuations in swirling turbulent flows. J. Physique II **3**, 271–278 (1993).

[21] S. Fauve, C. Laroche, A. Libchaber & B. Perrin, Chaotic Phases and Magnetic Order in a Convective Fluid. Phys. Rev. Lett. **52**, 1774–1777 (1984).

[22] S. Fauve & D.P. Lathrop, *Laboratory Experiments on Liquid Metal Dynamos and Liquid Metal MHD Turbulence*. In Fluid Dynamics and Dynamos in Astrophysics and Geophysics, (ed. Soward, A. *et al.*, (2003), pp. 393–425.

[23] S. Fauve & F. Pétrélis, *The dynamo effect*. In Peyresq Lectures on Nonlinear Phenomena, vol. II (2003), pp. 1–64 (ed. Sepulchre, J.-A.) Singapore: World Scientific.

[24] S. Fauve & F. Pétrélis, Scaling laws of turbulent dynamos. C. R. Physique **8**, 87–92 (2007).

[25] M.J. Feigenbaum, Quantitative universality for a class of nonlinear transformations. J. Stat. Phys. **19**, 25–52 (1978).

[26] A. Gailitis, O. Lielausis, E. Platacis, S. Dement'ev, A. Cifersons, G. Gerbeth, T. Gundrum, F. Stefani, M. Christen and G. Will, Magnetic field saturation in the Riga dynamo experiment. Phys. Rev. Lett. **86**, 3024–3027 (2001).

[27] A. Gailitis, O. Lielausis, E. Platacis, E. Dement'ev, A. Cifersons, G. Gerbeth, T. Gundrum, F. Stefani, M., Christen, and G. Will, Dynamo experiments at the Riga sodium facility. Magnetohydrodynamics **38**, 5–14 (2002).

[28] A. Giesecke, G. Rüdiger & D. Elstner, Oscillating α^2-dynamos and the reversal phenomenon of the global geodynamo. Astron. Nach. **326**, 693–700 (2005).

[29] A. Giesecke, F. Stefani & G. Gerbeth, Role of Soft-Iron Impellers on the Mode Selection in the von Kármán–Sodium Dynamo Experiment. Phys. Rev. Lett. **104**, 044503 (2010).

[30] C. Gissinger, A. Iskakov, S. Fauve & E. Dormy, Effect of magnetic boundary conditions on the dynamo threshold of von Karman swirling flows. Europhys. Lett. **82**, 29001 (2008).

[31] C. Gissinger, E. Dormy & S. Fauve, Bypassing Cowling's theorem in axisymmetric fluid dynamos. Phys. Rev. Lett. **101**, 144502 (2008).

[32] C.J.P. Gissinger, A numerical model of the VKS experiment. Europhys. Lett. **87**, 39002 (2009).

[33] C. Gissinger, E. Dormy & S. Fauve, Morphology of field reversals in turbulent dynamos. Europhys. Lett. **90**, 49001 (2010).

[34] G.A. Glatzmaier & P.H. Roberts, A three-dimensional self-consistent computer simulation of a geomagnetic field reversal. Nature **377**, 203–209 (1995).

[35] J.P. Gollub & H.L. Swinney, Onset of Turbulence in a Rotating Fluid. Phys. Rev. Lett. **35**, 927–930 (1975).

[36] P. Hoyng & J.J. Duistermaat, Geomagnetic reversals and the stochastic exit problem. Europhys. Lett. **68** (2), 177–183 (2004).

[37] D. Hughes & M.R.E. Proctor, A low-order model for the shear instability of convection: chaos and the effect of noise. Nonlinearity **3**, 127–153 (1990).

[38] E. Knobloch & A.S. Landsberg, A new model for the solar cycle. Mon. Not. R. Astron. Soc. **278**, 294–302 (1996).

[39] E. Knobloch, S.M. Tobias & N.O. Weiss, Modulation and symmetry changes in stellar dynamos. Mon. Not. R. Astron. Soc. **297**, 1123–1138 (1998).

[40] R. Krishnamurti & L.N. Howard, Large-scale flow generation in turbulent convection. Proc. Natl. Sci. USA **78**, 1981–1985 (1981).

[41] C. Kutzner & U.R. Christensen, From stable dipole towards reversing numerical dynamos. Physics of the Earth and Planetary Interior **131**, 29–45 (2002).

[42] R. Laguerre, *et al.*, Impact of Impellers on the Axisymmetric Magnetic Mode in the VKS2 Dynamo Experiment. Phys. Rev. Lett. **101**, 104501 and 219902 (2008).

[43] L.D. Landau, and E.M. Lifshitz, *Fluid Mechanics*. Oxford: Pergamon (1959).

[44] J. Larmor, *How could a rotating body such as the sun become a magnet?*. Rep. 87th Meeting Brit. Assoc. Adv. Sci. Bornemouth, Sept. 9–13, 1919 pp. 159–160, London:John Murray.

[45] E.H. Levy, Kinematic reversal schemes for the geomagnetic dipole. Astophys. J. **171**, 635–642 (1972).

[46] A. Libchaber, C. Laroche & S. Fauve, Period doubling cascade in mercury, a quantitative measurement. J. Physique Lettres **43**, 211–216 (1982).

[47] B. Liu & J. Zhang, Self-Induced Cyclic Reorganization of Free Bodies through Thermal Convection. Phys. Rev. Lett. **100**, 244501 (2008).

[48] E. Lorenz, Deterministic non periodic flow. Journal of the Atmospheric Sciences **20**, 130–141 (1963).

[49] W.V.R. Malkus, Reversing Bullard's dynamo. EOS Tran. Am. Geophys. Union **53**, 617 (1972).

[50] P. Manneville & Y. Pomeau, Intermittency and the Lorenz model. Phys. Lett. **A 75**, 1–2 (1979).

[51] L. Marié, J. Burguete, F. Daviaud, and J. Léorat, Numerical study of homogeneous dynamo based on experimental von Kármán type flows. Eur. Phys. J. B **33**, 469–485 (2003).

[52] J. Maurer & A. Libchaber, Rayleigh–Bénard experiment in liquid helium; frequency locking and the onset of turbulence. J. Physique Lettres **40**, 419–423 (1979).

[53] P.L. McFadden, R.T. Merril & M.W. McElhinny, Dipole/quadrupole family modeling of paleosecular variations. J. Geophys. Research **93**, 11583–11588 (1991).

[54] P.L. McFadden & R.T. Merrill, Fundamental transitions in the deodynamo as suggested by palaeomagnetic data. Physics of the Earth and Planetary Interior **91**, 253–260 (1995).

[55] I. Melbourne, M.R.E. Proctor & A.M. Rucklidge, *A heteroclinic model of geodynamo reversals and excursions*. In Dynamo and Dynamics, a Mathematical Challenge (eds. Chossat P., Armbruster D. & Oprea I.), pp. 363–370, Nato Science Series II, vol. 26 (2001), Kluwer Academic Publishers.

[56] H.K. Moffatt, *Magnetic Field Generation in Electrically Conducting Fluids*. Cambridge: Cambridge University Press (1978).

[57] R. Monchaux, et al., Generation of a magnetic field by dynamo action in a turbulent flow of liquid sodium. Phys. Rev. Lett. **98**, 044502 (2007).

[58] R. Monchaux, et al., *The von Kármán Sodium experiment: Turbulent dynamical dynamos*. Phys. Fluids **21**, 035108 (2009).

[59] N. Nishikawa & K. Kusano, Simulation study of the symmetry-breaking instability and the dipole field reversal in a rotating spherical shell dynamo. Physics of Plasma **15**, 082903 (2008).

[60] P. Nozières, Reversals of the Earth's magnetic field: an attempt at a relaxation model. Physics of the Earth and Planetary Interior **17**, 55–74 (1978).

[61] D. Sweet, E. Ott, J. Finn, T.M. Antonsen & D.A. Lathrop, Blowout bifurcations and the onset of magnetic activity in turbulent dynamos. Phys. Rev. E **63**, 066211 (2001).

[62] E.N. Parker, Hydromagnetic dynamo models. Astophys. J. **122**, 293–314 (1955).

[63] E.N. Parker, The occasional reversal of the geomagnetic field. Astophys. J. **158**, 815–827 (1969).

[64] E.N. Parker, *Cosmical magnetic Fields*. Oxford: Clarendon Press (1979).

[65] F. Pétrélis & S. Fauve, Saturation of the magnetic field above the dynamo threshold. Eur. Phys. J. B **22**, 273–276 (2001).

[66] F. Pétrélis, M. Bourgoin, L. Marié, J. Burgete, A. Chiffaudel, F. Daviaud, S. Fauve, P. Odier, and J.F. Pinton, Nonlinear magnetic induction by helical motion in a liquid sodium turbulent flow. Phys. Rev. Lett. **90**, 174501 (2003).

[67] F. Pétrélis, E. Dormy, J.-P. Valet & S. Fauve, Simple mechanism for the reversals of Earth magnetic field. Phys. Rev. Lett. **102**, 144503 (2009).

[68] F. Pétrélis & S. Fauve, Chaotic dynamics of the magnetic field generated by dynamo action in a turbulent flow. J. Phys.: Condens. Matter **20**, 494203 (2008).

[69] F. Pétrélis, N. Mordant & S. Fauve, On the magnetic fields generated by experimental dynamos. G. A. F. D. **101**, 289–323 (2007).

[70] Yu.B. Ponomarenko, Theory of the hydromagnetic generator. Appl. Mech. Tech. Phys. **14**, 775–778 (1973).

[71] F. Ravelet, A. Chiffaudel, F. Daviaud, and J. Léorat, Toward an experimental von Kármán dynamo: Numerical studies for an optimized design. Phys. Fluids **17**, 117104 (2005).

[72] F. Ravelet, et al., Chaotic dynamos generated by a turbulent flow of liquid sodium. Phys. Rev. Lett. **101**, 074502 (2008).

[73] T. Rikitake, Oscillations of a system of disc dynamos. Proc. Camb. Phil. Soc. **54**, 89–105 (1958).

[74] G.O. Roberts, Dynamo action of fluid motions with two-dimensional periodicity. Phil. Trans. Roy. Soc. London A **271**, 411–454 (1972).

[75] P.H. Roberts, Kinematic dynamo models. Phil. Trans. Roy. Soc. London A **272**, 663–698 (1972).

[76] P.H. Roberts, *Dynamo theory*. Irreversible phenomena an dynamical systems analysis in geosciences, (eds. Nicolis C. & Nicolis G.), Reidel Publishing Company (1987).

[77] P.H. Roberts, *Fundamentals of dynamo theory*. In Lectures on solar and planetary dynamos, chap. 1, pp. 1–57, eds. M.R.E. Proctor & A.D. Gilbert, Cambridge University Press (1994).

[78] P.H. Roberts & G.A. Galtzmaier, Geodynamo theory and simulations. Rev. Mod. Phys. **72**, 1081–1123 (2000).

[79] K.A. Robbins, A new approach to subcritical instability and turbulent transitions in a simple dynamo. Math. Proc. Camb. Phil. Soc. **82**, 309–325 (1977).

[80] D. Ruelle & F. Takens, On the nature of turbulence. Commun. Math Phys. **20**, 167–192 (1971).

[81] G.R. Sarson, Reversal models from dynamo calculations. Phil. Trans. R. Soc. Lond. A **358**, 921–942 (2000).

[82] G.R. Sarson & C.A. Jones, A convection driven geodynamo reversal model. Physics of the Earth and Planetary Interior **111**, 3–20 (1999).

[83] D. Schmitt, M.A.J.H. Ossendrijver & P. Hoyng, Magnetic field reversals and secular variation in a bistable dynamo model. Physics of the Earth and Planetary Interior **125**, 119–124 (2001).

[84] J. Sommeria, Experimental study of the two-dimensional inverse energy cascade in a square box. J. Fluid Mech. **170**, 139–168 (1986).

[85] F. Stefani & G. Gerbeth, Asymmetric polarity reversals, bimodal field distribution and coherence resonance in a spherically symmetric mean-field dynamo model. Phys. Rev. Lett. **94**, 184506 (2005).

[86] F. Stefani, M. Xu, G. Gerbeth, F. Ravelet, A. Chiffaudel, F. Daviaud, and J. Léorat, Ambivalent effects of added layers on steady kinematic dynamos in cylindrical geometry: application to the VKS experiment. Eur. J. Mech. B **25**, 894 (2006).

[87] F. Stefani, M. Xu, L. Sorriso-Valvo, G. Gerbeth & U. Günther, Oscillation or rotation: a comparison of two simple reversal models. Geophysical and Astrophysical Fluid Dynamics **101**, 227–248 (2007).

[88] R. Stieglitz & U. Müller, Experimental demonstration of a homogeneous two-scale dynamo. Phys. Fluids **13**, 561–564 (2001).

[89] R. Stieglitz & U. Müller, Experimental demonstration of a homogeneous two-scale dynamo. Magnetohydrodynamics **38**, 27–34 (2002).

[90] E. Stone & P. Holmes, Random Perturbations of Heteroclinic Attractors. SIAM J. Appl. Math. **50**, 726–743 (1990).

[91] S.M. Tobias, N.O. Weiss & V. Kirk, Chaotically modulated stellar dynamos. Mon. Not. R. Astron. Soc. **273**, 1150–1166 (1995).

[92] C. Tresser & P. Coullet, Itèrations d'endomorphismes et groupe de renormalisation. C. R. Acad. Sci. Paris **A 287**, 577–580 (1978).

[93] J.-P. Valet, L. Meynadier & Y. Guyodo, Geomagnetic field strength and reversal rate over the past 2 Million years. Nature **435**, 802–805 (2005).

[94] G.K. Vallis, El Nino: A Chaotic Dynamical System? Science **232**, 243–245 (1986).

[95] G. Verhille, et al., Induction in a von Kármán flow driven by ferromagnetic impellers. New Journal of Physics **12**, 033006 (2010).

[96] J. Wicht & P. Olson, A detailed study of the polarity reversal mechanism in a numerical dynamo model. Geochemistry, Geophysics, Geosystems **5**, Q03GH10 (2004).

[97] A.L. Wilmot-Smith, P.C.H. Martens, D. Nandy, E.R. Priest & S.M. Tobias, Low order stellar dynamo models. Mon. Not. R. Astron. Soc. **363**, 1167–1172 (2005).

[98] P.J. Zandbergen & D. Dijkstra, von Kármán swirling flows. Annu. Rev. Fluid Mech. **19**, 465–491 (1987).

[99] Ya.B. Zeldovich, A.A. Ruzmaikin & D.D. Sokoloff, *Magnetic fields in astrophysics.* New York: Gordon and Breach (1983).

Stéphan Fauve
École Normale Supérieure
24, rue Lhomond
F-75005 Paris, France
e-mail: `fauve@lps.ens.fr`

Discrete Graphs – A Paradigm Model for Quantum Chaos

Uzy Smilansky

Abstract. The research in Quantum Chaos attempts to uncover the fingerprints of classical chaotic dynamics in the corresponding quantum description. To get to the roots of this problem, various simplified models were proposed and used. Here a very simple model of a random walker on large d-regular graphs, and its quantum analogue are proposed as a paradigm which shares many salient features with realistic models – namely the affinity of the spectral statistics with random matrix theory, the role of cycles and their statistics, and percolation of level sets of the eigenvectors. These concepts will be explained and reviewed with reference to the original publications for further details.

1. Introduction

The question "What is Quantum Chaos?" seems to be best replied by an operational answer: The research in Quantum Chaos attempts to uncover the fingerprints of classical chaotic dynamics in the corresponding quantum description. In particular, since many attributes of classical (Hamiltonian) chaos are universal, quantum chaos attempts to focus on analogous, but not necessarily the same universal features. One of the most fruitful approaches in the development of classical chaos was to focus on toy models, which are simple enough to allow detailed analysis, without losing the essential dynamical complexity which marks chaotic systems. This same route was followed in Quantum Chaos – some of the toy models were constructed by quantizing the classical toy models. Other models went further and devised leaner and leaner problems which allowed deeper penetration into the essential ingredients of Quantum Chaos. In the present talk I shall describe a recent model – the Laplacian on d-regular graphs – which would seem at first sight as pushing the strife for simplicity *ad absurdum*. However, this is not the case, and by presenting the model, I shall explain by analogy the corresponding concepts which prevail in quantum chaos. It will also give me a chance to comment on the fruitful interaction of quantum chaos research with other fields of Mathematical Physics.

This review is based on the work performed during the past three years by Yehonatan Elon, Amit Godel, Idan Oren and the author. The original papers and thesis [1, 2, 3, 4, 5, 6, 7, 8] contain much more data, results and proofs, and the interested reader is referred there for the information which is lacking here. Our attention to d-regular graphs was brought by the pioneering work of Jakobson et al. [9] which will be discussed at length in the sequel. Previous attempts to relate quantum chaos to combinatorial graphs (not necessarily d-regular) are described in [10].

1.1. The Model: d-regular graphs

Deferring formal definitions to a later stage, the objects of this study are the d-regular graphs on V vertices. These are graphs where each vertex is connected to precisely d other vertices (no loops and no parallel edges). They have very attractive properties which made them very popular in various fields, ranging from computer science to number theory (see [11] for a review). Their most prominent feature is that they are **expanders** – a concept which can be explained through their geometrical and dynamical properties.

From the geometrical point of view, an expanding graph has the property that the volume of any ball is proportional to that of its boundary. Thus, when V increases, the growing rate of the graph is exponential, hence the name "expander".

From the dynamical point of view, a random walker on the graph would visit the vertices with uniform probability exponentially fast. The mixing rate is completely determined by the spectrum of the graph Laplacian to be defined below. It can be easily shown that for connected (non bipartite) d-regular graphs the equilibrium state is reached exponentially fast and at a rate which is independent of V.

The spectral theory of graphs shows that both features can be traced to the same root – namely, the expansion parameter and the mixing rate are related through an inequality attributed to Cheeger [12, 13, 14]. This will not be discussed here any further.

Besides of being expanders, d-regular graphs have another important property of a more local nature:
The local tree structure: Inside a ball of radius $\log_{d-1} V$ about most of its vertices, the local structure of a d-regular graph is identical to that of a d-regular tree. Stated differently, loops of lengths shorter than $\log_{d-1} V$ are rare on the graph.

The spectrum of the d-regular graphs and the corresponding eigenvectors in the limit $V \to \infty$ share many properties with their counterparts in quantum chaos. To introduce them systematically a few concepts and facts should be defined and stated.

1.2. Definitions and facts

A graph G is a set $\mathcal{V}$ of vertices connected by a set $\mathcal{E}$ of edges. The number of vertices is denoted by $V = |\mathcal{V}|$ and the number of edges is $E = |\mathcal{E}|$. A *simple graph* is a graph where an edge cannot connect a vertex to itself, nor can it connect already connected vertices. The connectivity of the graph is specified by the $V \times V$ adjacency (connectivity) matrix A,

$$A_{i,j} = \begin{cases} 1 & (i,j) \text{ connected,} \\ 0 & (i,j) \text{ not connected.} \end{cases} \quad (1)$$

We shall deal with *d-regular graphs* where each vertex is connected to exactly d vertices, d stands for the *degree*.

The ensemble of all simple, d-regular graphs with V vertices will be denoted by $\mathcal{G}_{V,d}$. The cardinality of $\mathcal{G}_{V,d}$ increases faster than exponentially in V. Therefore the density of disconnected graphs in $\mathcal{G}_{V,d}$ vanishes in the limit of large V. Averaging over $\mathcal{G}_{V,d}$ will be carried out with uniform probability and will be denoted by $\langle \cdots \rangle$.

While the vertices can be considered as a discrete version of classical *configuration space* for the dynamics on the graph, the *directed edges*, $e = (i,j)$ provide a description of the graph which is the discrete analogue of *phase-space*. It is convenient to associate with each directed edge $e = (j,i)$ its *origin* $o(e) = i$ and *terminus* $t(e) = j$ so that e points from the vertex i to the vertex j. The edge e' follows e if $t(e) = o(e')$. The reverse edge of e will be denoted by $\hat{e} = (i,j)$. The phase-space connectivity of the graph can be specified by the $2E \times 2E$ matrix B:

$$B_{e',e} = \begin{cases} 1 & o(e') = t(e), \\ 0 & o(e') \neq t(e). \end{cases} \quad (2)$$

Classical trajectories are replaced here by walks on the graph. A *walk* of length t from the vertex x to the vertex y on the graph is a sequence of successively connected vertices $x = v_1, v_2, \ldots, v_t = y$. Alternatively, it is a sequence of $t-1$ directed edges $e_1, \ldots, e_{t-1}$ with $o(e_i) = v_i$, $t(e_i) = v_{i+1}$, $o(e_1) = x$, $t(e_{v-1}) = y$. A *closed walk* is a walk with $x = y$. The number of walks of length t between x and y equals $(A^t)_{y,x}$, or equivalently $\sum_{e,e'} (B^{t-1})_{e',e} \delta_{o(e),x} \delta_{t(e')=y}$.

In the sequel walks without back-scatter in which $e_{i+1} \neq \hat{e}_i$, , $1 \leq i \leq t-2$ will play an important rôle. We shall refer to them as *nb-walks* for short. To study nb-walks it is useful to define

$$J_{e,e'} = \delta_{\hat{e},e'}, \quad (3)$$

which singles out edges connected by backscatter. The matrix

$$Y = B - J \quad (4)$$

connects bonds which are not reversed. Thus, e.g., the number of t-periodic nb-walks is $\text{tr} Y^t$.

The probability that a long random walk visits any given vertex is completely controlled by the spectrum of the discrete Laplacian on the graph which is defined as

$$\Delta \equiv -A + dI^{(V)}, \quad (5)$$

where $I^{(V)}$ is the unit matrix in V dimensions. It is convenient to study the spectrum of A which differs from the Laplacian by a change of sign and a constant shift. The highest eigenvalue of A is d corresponding to a uniform distribution on the graph vertices. This implies that for d-regular graphs the limit distribution for a random walker is uniform (if the graph is connected and not bipartite). The distance between d and the next eigenvalue (the spectral gap $s(G)$) determines the speed at which a random walker covers the graph uniformly. For (V,d) regular graphs it is known [15] that the spectral gap is given asymptotically by

$$s(G) = d - 2\sqrt{d-1} + o\left(\frac{d}{\log V}\right) \qquad (6)$$

with probability which converges to 1 exponentially fast in V. (Here and in the sequel, logarithms are calculated in base $d-1$.) This result complements a previously established bound, [16]

$$s(G) < d - 2\sqrt{d-1} + \frac{2}{\log V}, \qquad (7)$$

which implies that in the limit $V \to \infty$, the spectral gap cannot exceed $d - 2\sqrt{d-1}$. Hence, the limiting maximal mixing rate for regular graphs is attained asymptotically for almost any (V,d) graph. In other words, the d-regular graphs are optimal mixers. This strong mixing property is typical of hard classical chaos and it justifies the use of this model as a paradigm of quantum chaos as defined in the introduction.

Adjacency matrices of random d-regular graphs, have some remarkable spectral properties, which can be studied using the *spectral density* $\rho(\mu)$ and the *spectral counting function* $N(\mu)$ defined as:

$$\rho(\mu) \equiv \frac{1}{V} \sum_{\mu_a \in \sigma(A)} \delta(\mu - \mu_a) \ ; \ N(\mu) = V \int^{\mu} d\rho(\mu) = \sum_{\mu_a \in \sigma(A)} \Theta(\mu - \mu_a). \qquad (8)$$

Here, μ_a are the adjacency eigenvalues corresponding to eigenvectors $f^{(a)} \in \mathbb{R}^V$.

An important discovery which marked the starting point of the study of spectral statistics for d-regular graphs was the derivation of the mean (in $\mathcal{G}(n,d)$) of the spectral density by Kesten [17] and McKay [18]:

$$\rho_{KM}(\mu) = \lim_{V \to \infty} \langle \rho(\mu) \rangle = \begin{cases} \frac{d}{2\pi} \frac{\sqrt{4(d-1)-\mu^2}}{d^2 - \mu^2} & \text{for } |\mu| \leq 2\sqrt{d-1}, \\ 0 & \text{for } |\mu| > 2\sqrt{d-1}. \end{cases} \qquad (9)$$

The proof of this result relies on the local tree property of random d-regular graphs, namely, that almost surely every subgraph of diameter less than $\log V$ is a tree. Counting periodic orbits on the tree can be done explicitly, and using the close relations between these numbers and the spectrum, one obtains (17). Note that the Kesten–Mckay density approaches Wigner's semi-circle law for large d.

A finite number of eigenvalues outside the support of ρ_{KM} cannot be excluded. Graphs for which the entire spectrum of A (except from the largest eigenvalue) lies within the support $[-2\sqrt{d-1}, 2\sqrt{d-1}]$, are called Ramanujan (for a review, see, e.g., [19] and references cited therein). They are known to exist for some particular values of d, but much of their properties and their distribution are not known.

Finally, we shall consider the spectral properties of the "magnetic" Laplacian [20] which is defined in terms of the *magnetic adjacency matrix*

$$M_{i,j} = A_{i,j} e^{i\phi_{i,j}}, \qquad (10)$$

where the phases $\phi_{i,j}$ attached to the edges (i,j) play the rôle of "magnetic fluxes". The magnetic ensemble $\mathcal{G}^{(M)}(V,d)$ consists of $\mathcal{G}(V,d)$ augmented with independently and uniformly distributed phases $\phi_{i,j}$. In the following we shall use the superscript M consistently when referring to the magnetic ensemble. M is complex Hermitian, and therefore the evolution it induces breaks time reversal symmetry. The largest eigenvalue of M may be different than d, but the asymptotic spectral density approaches the Kesten–McKay distribution.

2. Spectral statistics and trace formulae

One of the central themes in quantum chaos was the study of the statistics of spectral fluctuations. Analyzing the numerically computed spectral sequences of quantum billiards, led Bohigas, Giannoni and Schmit [21] to propose their conjecture: The spectral fluctuations for quantum chaotic systems follow the predictions of the Gaussian ensembles of Random Matrix Theory (RMT), while for integrable systems, they are Poissonian [22]. In the next paragraphs we shall review the numerical evidence which illustrates the excellent agreement between the spectral statistics of the adjacency matrices of d-regular graphs (and their magnetic analogues) and the predictions of RMT [9, 6]. While describing these results a few concepts from RMT will be introduced [23].

2.1. Some numerical evidence and a few more definitions

For reasons which will become clear as the theory unfolds, it is advantageous to map the spectrum from the real line to the unit circle,

$$\phi_j = \arccos \frac{\mu_j}{2\sqrt{d-1}} \quad ; \quad 0 \leq \phi_j \leq \pi. \qquad (11)$$

This change of variables is allowed since in the limit of large V, only a fraction of order $1/V$ of the spectrum is outside the support of the Kesten–McKay distribution $[-2\sqrt{d-1}, 2\sqrt{d-1}]$ [25].

The mean (Kesten–McKay) spectral density on the circle is not uniform,

$$\begin{aligned} \rho_{KM}(\phi) &= \frac{2(d-1)}{\pi d} \frac{\sin^2 \phi}{1 - \frac{4(d-1)}{d^2}\cos^2 \phi} \\ N_{KM}(\phi) &= V\frac{d}{2\pi}\left(\phi - \frac{d-2}{d}\arctan\left(\frac{d}{d-2}\tan\phi\right)\right). \end{aligned} \qquad (12)$$

Following the standard methods of spectral statistics, one introduces a new variable θ, which is uniformly distributed on the unit circle. This "unfolding" procedure is explicitly given by

$$\theta_j = \frac{2\pi}{V} N_{KM}(\phi_j). \qquad (13)$$

The nearest spacing distribution defined as

$$P(s) = \lim_{V \to \infty} \frac{1}{V} \left\langle \sum_{j=1}^{V} \delta\left(s - \frac{V}{2\pi}(\theta_j - \theta_{j-1})\right) \right\rangle \qquad (14)$$

is often used to test the agreement with the predictions of RMT (this was also the test conducted in [9]). In this definition of the nearest spacing distribution, θ_0 coincides with θ_V, since the phases lie on the unit circle. In Figure 1 we show

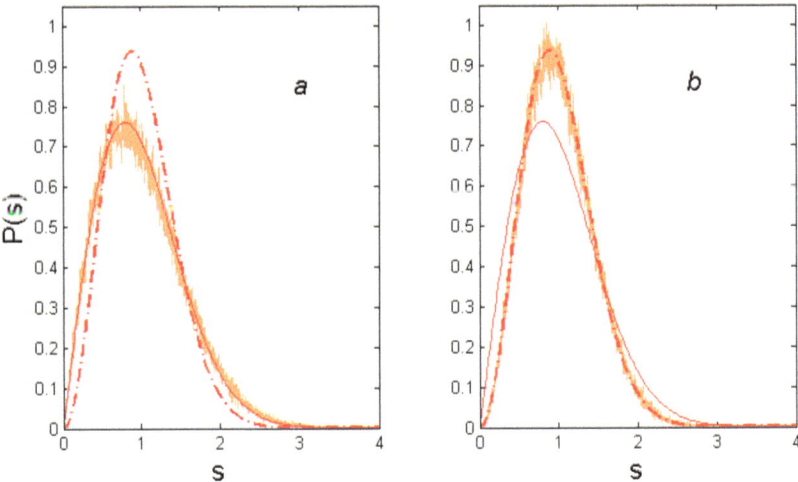

FIGURE 1. Nearest level spacings for: (a) Graphs possessing time reversal symmetry. (b) Magnetic graphs. Both figures are accompanied with the RMT predictions: Solid line – COE, dashed line – CUE.

numerical simulations obtained by averaging over 1000 randomly generated 3-regular graphs on 1000 vertices and their "magnetic" counterparts, together with the predictions of RMT for the COE and the CUE ensembles [24], respectively. The agreement is quite impressive.

Another quantity which is often used for the same purpose is the spectral form-factor,

$$K_V(t) = \frac{1}{V} \left\langle \left| \sum_{j=1}^{V} e^{it\theta_j} \right|^2 \right\rangle. \qquad (15)$$

The form-factor is the Fourier transform of the spectral two point correlation function and it plays a very important rôle in the understanding of the relation between RMT and the quantum spectra of classically chaotic systems [23, 28].

In RMT the form factor displays scaling: $\lim_{V \to \infty} K_V(t = \tau V) = K(\tau)$. The explicit limiting expressions for the COE and CUE ensembles are [24]:

$$K_{COE}(\tau) = \begin{cases} 2\tau - \tau \log(2\tau + 1), & \text{for } \tau < 1, \\ 2 - \tau \log \frac{2\tau+1}{2\tau-1}, & \text{for } \tau > 1. \end{cases} \quad (16)$$

$$K_{CUE}(\tau) = \begin{cases} \tau, & \text{for } \tau < 1, \\ 1, & \text{for } \tau > 1. \end{cases} \quad (17)$$

The numerical data used to compute the nearest neighbor spacing distribution $P(s)$, was used to calculate the corresponding form factors for the non-magnetic and the magnetic graphs, as shown in Figure 2. The agreement between the numerical results and the RMT predictions is apparent.

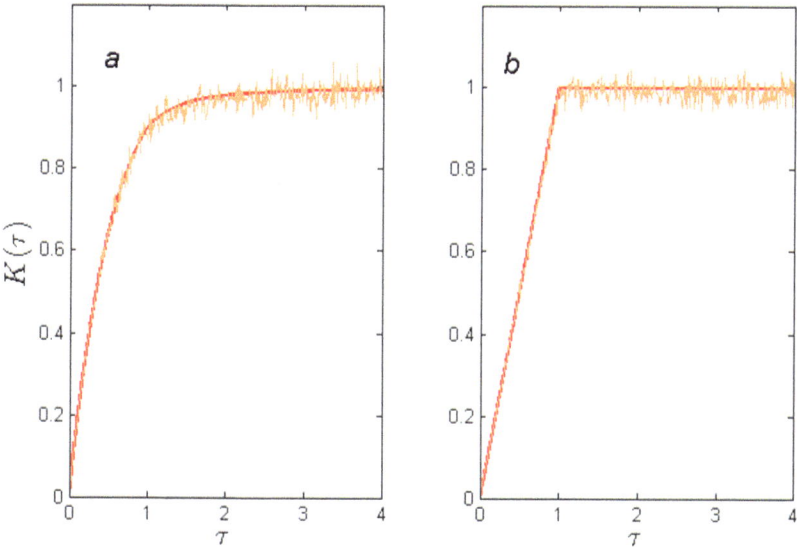

FIGURE 2. The form factor $K(\tau)$ (unfolded spectrum) for (a) 3-regular graphs numerical *vs.* the COE prediction. (b) 3-regular magnetic graphs *vs.* the CUE prediction.

The above comparisons between the predictions of RMT and the spectral statistics of the eigenvalues of d-regular graphs was based on the unfolding of the phases ϕ_j into the uniformly distributed phases θ_j. As will become clear in the next sections, it is more natural to study here the fluctuations in the original spectrum and in particular the form factor

$$\widetilde{K}_V(t) = \frac{1}{V} \left\langle \left| \sum_{j=1}^{V} e^{it\phi_j} \right|^2 \right\rangle. \tag{18}$$

The transformation between the two spectra was effected by (13) which is one-to-one and its inverse is defined:

$$\phi = S(\theta) \doteq N_{KM}^{-1}\left(V\frac{\theta}{2\pi}\right). \tag{19}$$

This relationship enables us to express $\widetilde{K}_V(t)$ in terms of $K_V(t)$. In particular, if $K_V(t)$ scales by introducing $\tau = \frac{t}{V}$ then,

$$\widetilde{K}\left(\tau = \frac{t}{V}\right) = \frac{1}{\pi} \int_0^\pi d\theta K\left(\tau S'(\theta)\right). \tag{20}$$

The derivation of this identity is straightforward. In the limit $\tau \to 0$, (20) reduces to

$$\widetilde{K}\left(\tau = \frac{t}{V}\right) \approx \frac{1}{2} K(\tau). \tag{21}$$

Figure 3 shows $\widetilde{K}(\tau = \frac{t}{V}) = \widetilde{K}_V(t)$ computed by assuming that its unfolded analogue takes the RMT form (16) or (17), and it is compared with the numerical data for graphs with $d = 10$. It is not a surprise that this way of comparing

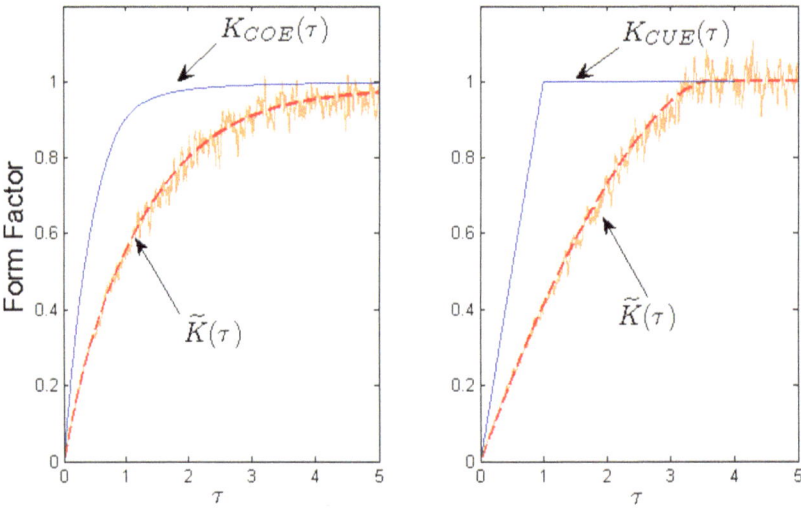

FIGURE 3. The form factor $\widetilde{K}(\tau)$ (original spectrum) for 10-regular graphs. The numerical results are presented vs. the expression (20) assuming RMT in the dashed line, and $K(\tau)$ (16), (17) in the solid line.

between the predictions of RMT and the data, shows the same agreement as the one observed previously.

2.2. Trace formulae

The theoretical attempts to justify the Bohigas–Gianonni–Schmit conjecture were all based on *trace formulae* which establish a link between the quantum, spectral information, and the periodic manifolds in the phase-space of the underlying classical dynamics. The dichotomy observed in the quantum spectra is due to the intrinsic difference between the periodic manifolds in the corresponding classical dynamics: In chaotic systems the manifolds are discrete, unstable periodic orbits, while for integrable dynamics they correspond to periodic tori where the dynamics is marginally stable. Trace formulae take the general form [27]:

$$\rho(E) \approx \rho_{\text{smooth}}(E) + \Im \left\{ \sum_p A_p e^{i(S_p(E)/\hbar + \frac{\pi}{2}\nu_p)} \right\}, \qquad (22)$$

where E is the spectral parameter (Energy). ρ_{smooth} is a smooth function of E which provides the asymptotic behavior of $\rho(E)$. The sum is responsible to the oscillatory or fluctuating part of the spectral density. It goes over contributions from the periodic classical manifolds. The leading terms in the amplitudes A_p depend on the stability of the periodic orbits for chaotic billiards and on the phase space volume occupied by the corresponding torus for integrable cases. $S_p(E)$ are the action integrals along the periodic orbits and ν_p are the Maslov index which have a clear geometric meaning related to the orbit. The function which was most frequently studied in quantum chaos is the spectral 2-points correlation function:

$$R(\epsilon) = \frac{1}{\Delta} \int_{E_0-\Delta/2}^{E_0+\Delta/2} \tilde{\rho}(E+\epsilon/2)\tilde{\rho}((E-\epsilon/2)\mathrm{d}E, \quad \tilde{\rho}(E) = \rho(E) - \rho_{\text{smooth}}(E), \quad (23)$$

or the *form-factor* which is its Fourier transform. Substituting the trace formula (22) in (23), one could express the spectral correlations in terms of correlations between classical actions [29]. The short time limit of the form factor predicted by RMT can be recovered by assuming that the actions of different periodic orbits are not correlated (the "diagonal" approximation [28]). Further work elucidated the dynamical origin of the periodic orbit correlations [30]. This opened the way for a complete reconstruction of the RMT expression for the form factor based on the classical information and general symmetry considerations [31].

Following the quantum chaos example, the connection between spectral statistics and periodic orbit theory will be demonstrated by using trace formulae for d-regular graphs.

We shall start by considering the spectrum of the adjacency matrices A. The trace formula can be obtained by making use of a well-known identity [32, 33] which relates the characteristic polynomials and the spectra of the *vertex* adjacency matrix A and the *edge* nb-adjacency matrix Y (4):

$$\det(I^{(2E)} - sY) = (1-s^2)^{E-V} \det(I^{(V)}(1+(d-1)s^2) - sA). \qquad (24)$$

For Ramanujan graphs, the spectrum consists of d and $|\mu_k| \le 2\sqrt{d-1}$, $k = 1, \ldots, (V-1)$. Transforming the μ_k to the unit circle as in (11), the spectrum of Y reads

$$\sigma(Y) = \left\{(d-1), 1, +1 \times (E-V), -1 \times (E-V), \right. \\ \left. (\sqrt{d-1}\, e^{i\phi_k}, \sqrt{d-1}\, e^{-i\phi_k}, k=1,\ldots,(V-1))\right\}. \quad (25)$$

(The restriction to Ramanujan graphs can be removed [5] but this will not be discussed here.) For large t, the number of t-periodic nb-walks given by $\text{tr}Y^t$ is dominated by the largest eigenvalue, so that asymptotically $\text{tr}Y^t \sim (d-1)^t$. It is advantageous to introduce

$$y_t^{(A)} = \frac{1}{V} \frac{\text{tr}Y^t - (d-1)^t}{(\sqrt{d-1})^t} \quad (26)$$

which is the properly regularized number of t-periodic nb-walks. (The superscript (A) indicates reference to the spectrum of the adjacency matrix A.) The explicit expressions for the eigenvalues of Y are used now to write

$$y_t^{(A)} = \frac{1}{V}\left(\frac{1}{d-1}\right)^{\frac{t}{2}} + \frac{d-2}{2}\left(\frac{1}{d-1}\right)^{\frac{t}{2}}(1+(-1)^t) + \frac{2}{V}\sum_{k=1}^{V-1}\cos(t\phi_k). \quad (27)$$

Multiplying by $e^{it\phi}$ and summing one gets

$$\rho(\phi) = \frac{2(d-1)}{\pi d}\frac{\sin^2\phi}{1 - \frac{4(d-1)}{d^2}\cos^2\phi} + \frac{1}{\pi}\text{Re}\sum_{t=3}^{\infty}y_t^{(A)}e^{it\phi} + \mathcal{O}\left(\frac{1}{V}\right). \quad (28)$$

The first term is the smooth part, which is just the Kesten–McKay density. It is known from combinatorial graph theory that the counting statistics of t-periodic nb-walks with $t < \log V$ is Poissonian, with $\langle \text{tr}Y^t\rangle = (d-1)^t$. Thus, the mean value of y_t vanishes as $\mathcal{O}\left(\frac{1}{V}\right)$. Hence, as expected,

$$\lim_{V\to\infty}\langle\rho(\mu)\rangle = \rho_{KM}(\mu). \quad (29)$$

The $\mathcal{O}\left(\frac{1}{V}\right)$ correction to the smooth part is known explicitly but will not be quoted here. The trace formula (28) was derived previously by P. Mněv [35] in an entirely different way.

Equation (28) is the desired form of the trace formula. The summation extends over the contribution of t-periodic nb-periodic walks, and the only information required for the spectral density are the $y_t^{(A)}$ – the normalized number of the periodic nb-walks. Note in passing that nb-walks appear in various contexts in combinatorial graph theory as in, e.g., [34]. Further comments on the trace formula are deferred to the last paragraphs in this chapter.

The trace formula for the spectrum of M has the same form. The only difference comes by replacing $y_t^{(A)}$ by:

$$y_t^{(M)} = \frac{1}{V}\frac{\sum_\alpha e^{i\chi_\alpha}}{(\sqrt{(d-1)})^t}, \quad (30)$$

where the sum above is over all the nb t-periodic walks, and χ_α is the total phase (net magnetic flux) accumulated along the t-periodic walk. For finite V, $\left\langle y_t^{(M)} \right\rangle \neq 0$ since $\chi_\alpha = 0$ for nb-periodic walks where each bond is traversed equal number of times in the two directions. However, in the limit of large V, the number of such walks is small and therefore $\left\langle y_t^{(M)} \right\rangle \to 0$.

2.3. A periodic orbit expression for the spectral form-factor

Here and in the sequel, whenever the expression applies to both the A and M spectra, the superscripts (A) or (M) are deleted. Writing $\tilde{\rho}(\phi) = \rho(\phi) - \rho_{KM}(\phi)$ we get from the trace formula (28)

$$y_t = 2\int_0^\pi \cos(t\phi)\tilde{\rho}(\phi)d\phi \ . \tag{31}$$

Thus,

$$\langle y_t^2 \rangle = 4\int_0^\pi \int_0^\pi \cos(t\phi)\cos(t\psi)\langle\tilde{\rho}(\phi)\tilde{\rho}(\psi)\rangle\,d\phi d\psi \ . \tag{32}$$

Recalling (33)

$$\tilde{K}_V(s) \equiv 2V\int_0^\pi\int_0^\pi \cos(s\phi)\cos(s\psi)\langle\tilde{\rho}(\phi)\tilde{\rho}(\psi)\rangle d\phi d\psi \ , \tag{33}$$

and comparing (32) and (33) we get:

$$\tilde{K}_V(t) = \frac{V}{2}\langle y_t^2 \rangle \ . \tag{34}$$

So far the treatment of the two ensembles was carried on the same formal footing. We shall now address each ensemble separately.

2.3.1. The form factor for the $\mathcal{G}_{V,d}$ ensemble.
In graph theory it is customary to discuss the number of nb t-cycles on the graph $C_t = \frac{Y_t}{2t}$ (in this definition, one does not distinguish between cycles which are conjugate to each other by time reversal). From combinatorial graph theory it is known that on average $\langle C_t \rangle = \frac{(d-1)^t}{2t}$ [36, 37, 38, 39]. Hence $\tilde{K}$ can also be written as:

$$\tilde{K}_V^{(A)}(t) = \frac{t}{V} \cdot \frac{\left\langle (C_t - \langle C_t \rangle)^2 \right\rangle}{\langle C_t \rangle} \ . \tag{35}$$

The expression of the spectral form factor $\tilde{K}_V^{(A)}(t)$ in terms of combinatorial quantities is the main result of the present work. In particular, it shows that the form factor is the ratio between the variance of C_t – the number of nb t-cycles – and its mean. This relation is valid for all t in the limit $V \to \infty$.

For t satisfying $t < \log V$, it is known that asymptotically, for large V, the C_t's are distributed as independent Poisson variables. For a Poisson variable, the variance and mean are equal. Using (21) we get that for $\tau \ll 1$

$$K^{(A)}(\tau) = 2\tau \ . \tag{36}$$

This result coincides with the COE prediction (16). It provides the first rigorous support of the connection, established so far numerically, between RMT and the spectral statistics of graphs. It is analogous to Berry's "diagonal approximation" [28] in quantum chaos.

2.3.2. The form factor for the $\mathcal{G}_{V,d}^M$ ensemble. In the magnetic ensemble the matrices (10) are complex valued and Hermitian, which is tantamount to breaking time reversal symmetry. The relevant RMT ensemble in this case is the CUE.

In the ensuing derivation we shall take advantage of the statistical independence assumed for the magnetic phases which are uniformly distributed on the circle. Ensemble averaging will imply averaging over both the magnetic phases and the graphs.

Recall that $y_t^{(M)}$, in the case of magnetic graphs, was defined by (30):

$$y_t^{(M)} = \frac{1}{V} \frac{\sum_\alpha e^{i\chi_\alpha}}{(\sqrt{(d-1)})^t} .$$

$y_t^{(M)}$ is the sum of interfering phase factors contributed by the individual nb t-periodic walks on the graph. The phase factors of periodic walks which are related by time reversal are complex conjugated. Periodic walks which are self tracing (meaning that every bond on the cycle is traversed the same number of times in both directions), have no phase: $\chi_\alpha = 0$. Using standard arguments from combinatorial graph theory one can show that for $t < \log_{d-1} V$, self tracing nb t-periodic walks are rare. Moreover, the number of t-periodic walks which are repetitions of shorter periodic walks can also be neglected. Hence

$$y_t^{(M)} \approx \frac{1}{V} \frac{2t \sum'_\alpha \cos(\chi_\alpha)}{(\sqrt{(d-1)})^t} , \qquad (37)$$

where $\sum'$ includes summation over the nb t-cycles excluding self tracing and non-primitive cycles. The number of t-cycles on the graph is C_t, hence (37) has approximately C_t terms. From (37) and the definition of $y_t^{(M)}$, it is easily seen that

$$\left\langle \left(y_t^{(M)}\right)^2 \right\rangle = \frac{1}{V^2(d-1)^t} 4t^2 \left\langle \left(\sum'\cos(\chi_\alpha)\right)^2 \right\rangle .$$

Averaging over the independent magnetic phases we get that

$$\widetilde{K}_V^{(M)}(t) = \frac{V}{2} \left\langle \left(y_t^{(M)}\right)^2 \right\rangle \approx \frac{t}{2V} \equiv \frac{\tau}{2}. \qquad (38)$$

For $\tau \to 0$, and using (21) we get

$$K^{(M)}(\tau) = \tau , \qquad (39)$$

which agrees with the CUE prediction.

2.4. From RMT to combinatorial graph theory

We have shown above that by making use of the known asymptotic statistics of Y_t one can derive the leading term in the expansion of $K_V(t)$. It behaves as $g\tau$ where $g = 1, 2$ for the two graph ensembles, which is consistent with the predictions of RMT. Had we known more about the counting statistics, we could make further predictions and compare them to RMT results. However, to the best of our knowledge we have exhausted what is known from combinatorial graph theory, and the only way to proceed would be to take the reverse approach, and *assume* that the form factor for graphs is given by the predictions of RMT, and see what this implies for the counting statistics. Checking these predictions from the combinatorial point of view is beyond our scope. However, we shall show that they are accurately supported by the numerical simulations. In a way, this approach is similar to that of Keating and Snaith [40] who deduced the asymptotic behavior of the high moments of the Riemann ζ function on the critical line, by assuming that the Riemann zeros statistics follow the RMT predictions for the GUE ensemble.

The starting points for the discussion are the relations (34) and (20) which can be combined to give

$$\langle (y_t)^2 \rangle = \frac{2}{V}\widetilde{K}_V(t) = \frac{2}{V\pi}\int_0^\pi d\theta K\left(\tau S'(\theta)\right). \tag{40}$$

Our strategy here will be to use for the unfolded form factor the known expressions from RMT (16) and (17) and compute $\langle (y_t)^2 \rangle$. This will provide an expression for the combinatorial quantities defined for each of the graph ensembles, and expanding in τ we shall compute the leading correction to their known asymptotic values. The actual computations are somewhat cumbersome and will not be repeated here. Starting with the simpler CUE form we get for the $\mathcal{G}_M(V, d)$ assemblage,

$$\widetilde{K}^{(M)}(\tau) = \frac{\tau}{2} + f_1(d)\tau^{\frac{3}{2}} + \mathcal{O}(\tau^2) \tag{41}$$

where

$$f_1(d) = -\frac{1}{3\pi\sqrt{D}} \quad ; \quad D \equiv \frac{d(d-1)}{(d-2)^2}. \tag{42}$$

Thus, the difference $(\widetilde{K}^{(M)}(\tau) - \frac{\tau}{2})/f_1(d)$, should scale for small τ as $\tau^{\frac{3}{2}}$ for all values of d. This data collapse is shown in Figure 4.

For the $\mathcal{G}(V, d)$ assemblage:

$$\widetilde{K}^{(A)}(\tau) = \tau + f_2(d)\tau^{\frac{3}{2}} + \cdots \tag{43}$$

where

$$f_2(d) = \frac{1}{\sqrt{2D}}\left(\frac{2}{\pi}\cdot\operatorname{arccoth}(\sqrt{2}) - \frac{2\sqrt{2}}{3\pi} - 1\right). \tag{44}$$

Thus, the difference $(\widetilde{K}^{(A)}(\tau) - \tau)/f_2(d)$, at small τ, should scale as $\tau^{\frac{3}{2}}$ independently of d. This data collapse is shown in Figure 5.

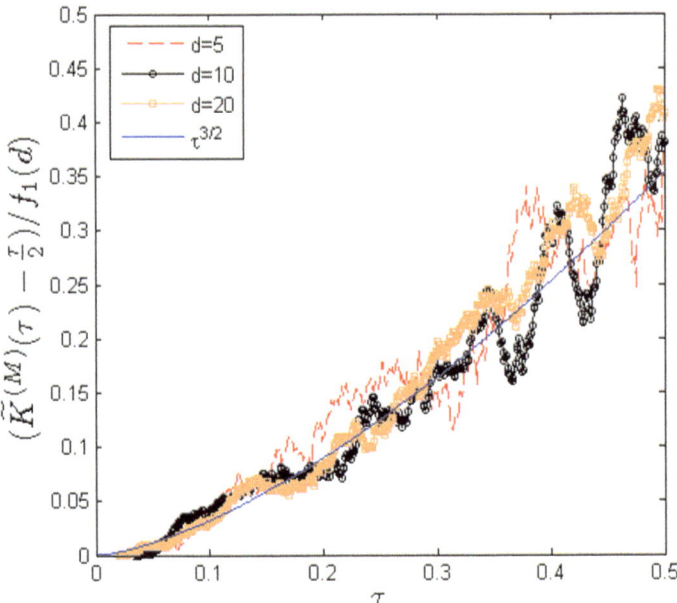

FIGURE 4. $(\widetilde{K}^{(M)}(\tau) - \frac{\tau}{2})/f_1(d)$ for various values of d vs. the curve $\tau^{\frac{3}{2}}$.

Using the above and (35) we can write,

$$\frac{\left\langle (C_t - \langle C_t \rangle)^2 \right\rangle}{\langle C_t \rangle} = \frac{1}{\tau\pi} \int_0^\pi d\theta K_{COE}\left(\tau S'(\theta)\right) \xrightarrow[\tau \to 0]{} 1 + f_2(d)\sqrt{\tau} + \cdots \quad (45)$$

If the C_t's were Poissonian random variables, the expansion above would terminate at 1. Since it does not, we must conclude that the C_t's are not Poissonian. The highest-order deviation comes from the next order term in the expansion which is proportional to $\tau^{\frac{1}{2}}$. The coefficient, $f_2(d)$, is explicitly calculated above.

We can examine the behavior at another domain of τ, namely $\tau > 1$. It can easily be shown that $S'(\theta) \geq \frac{d}{4(d-1)}$. Consequently, for $\tau > \frac{4(d-1)}{d}$, the argument of K in (20) is larger than one, and so

$$\lim_{\tau \to \infty} \widetilde{K}^{(A)}(\tau) = 1. \quad (46)$$

Combining this result with (35), provides the asymptotic of the variance-to-mean ratio:

$$\lim_{\tau \to \infty} \tau \frac{\text{var}(C_t)}{\langle C_t \rangle} = 1, \quad \text{for } V, t \to \infty; \quad \frac{t}{V} = \tau \quad (47)$$

This is a new interesting combinatorial result, since very little is known about the counting statistics of periodic orbits in the regime of $\tau > 1$.

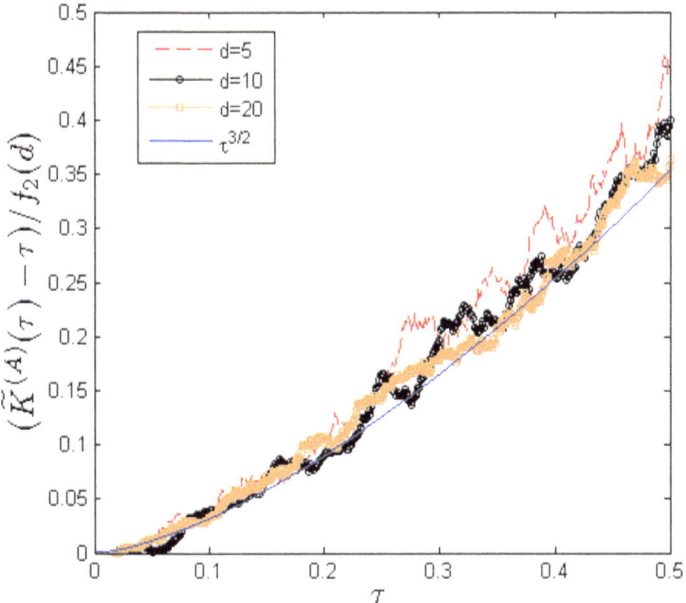

FIGURE 5. $(\widetilde{K}^{(A)}(\tau) - \tau)/f_2(d)$ for various values of d vs. the curve $\tau^{\frac{3}{2}}$.

2.5. Some comments on trace formulae

The trace formula quoted in (28) is but a single expression taken from a continuum of formulae, all of them are exact, pertain to the same spectral density, but make use of a different class of periodic orbits. For every real parameter $w \in [1, \frac{d-1}{2}]$, and $d \geq 3$ one can write a trace formula for the spectrum of the adjacency matrix. Writing the spectral density in terms of the spectral parameter μ the trace assumes the standard separation between the smooth and the oscillatory part:

$$\rho(\mu) = \rho_{\text{smooth}}(\mu; w) + \tilde{\rho}(\mu; w) + \mathcal{O}\left(\frac{1}{V}\right). \quad (48)$$

Here, however, the expression for the smooth part

$$\rho_{\text{smooth}}(\mu; w) = \frac{d/(2\pi)}{\sqrt{4w(d-w) - \mu^2}} \left(1 - \frac{(d-2w)(d-2)}{d^2 - \mu^2} + \frac{(w-1)^2(\mu^2 - 2w(d-w))}{w^2(d-w)^2}\right) \quad (49)$$

differs from the Kesten–McKay expression. In other words, when both sides of (48) are averaged over the ensemble, $\langle \tilde{\rho}(\mu, w) \rangle \neq 0$! The oscillatory part takes the form,

$$\tilde{\rho}(\mu; w) = \frac{1}{\pi} \mathcal{R}e \left(\sum_{t=3}^{\infty} \frac{y_t(w)}{\sqrt{4w(d-w) - \mu^2}} \exp^{it \frac{\mu}{2\sqrt{w(d-w)}}} \right). \quad (50)$$

The information about the periodic walks is stored in $y_t(w)$ (the analogue of y_t above (26)). Here, *all the t-periodic walks* are contributing, and every backscatter is endowed with a weight $(1-w)$. Denoting by $N(t;g)$ the number of t-periodic walks which backscatter exactly g times, $y_t(w)$ is

$$y_t(w) = \frac{1}{V} \frac{\sum_g N(t;g)(1-w)^g - (d-w)^t}{(\sqrt{w(d-w)})^t} . \tag{51}$$

Each of the two terms which contribute to $\rho(\mu)$ depend on w yet, their sum does not. The dependence of $\tilde{\rho}(\mu;w)$ on w comes from two sources – the weights of the backscattered orbits and the phase factor. Hence, the re-summation required to obtain (28) which corresponds to $w=1$ from (50) with $w \neq 1$ is far from being trivial. Exploratory studies [5] show how combinatorial expressions for counting certain families of periodic walks can be derived by studying the w dependence of the trace formula. Reviewing this study exceeds the scope of the present review, and it is mentioned only to indicate the potential wealth stored in this and other parametric dependent trace formulae. Another family of trace formula was written down in [4], but will not be discussed here.

3. Eigenfunctions

There are several interesting properties which mark the eigenfunctions of the Schrödinger operator in quantum chaotic systems. The most frequently studied models are quantum chaotic billiard domains $\Omega \in \mathbb{R}^2$, and their properties will serve as a bench-mark for the comparison with the properties of eigenvectors of the adjacency matrices A corresponding to d-regular graphs.

The most important features displayed by wave-functions of quantum chaotic billiards are:

- Almost all eigenfunctions ψ_n (ordered by increasing value of the spectral parameter and square normalized) for sufficiently large n, are distributed uniformly in the sense that given any smooth "observable" $A(x)$, $x \in \Omega$

$$\int_\Omega \psi_n^2(x) A(x) \mathrm{d}^2 x \to \int_\Omega A(x) \mathrm{d}^2 x \tag{52}$$

when $n \to \infty$ on all integer sequences from which sequences of zero measure might be excluded [41].
- On the atypical sequences, the eigenfunction might show high concentration near classically periodic orbits – a phenomenon known as scarring [42].
- In the limit of large n, the wave functions are conjectured to behave in probability as uniformly distributed Gaussian variables with the covariance

$$\langle \psi_n(x)\psi_n(y) \rangle = J_0(k_n|x-y|) . \tag{53}$$

Here, the triangular brackets stand for an average over a properly defined spectral interval about the mean eigenvalue k_n^2. The points x,y are in the billiard domain and sufficiently remoted from the boundaries, and $J_0(x)$ is

the Bessel function of order 0. While the statement that the distribution is Gaussian is not proven rigorously, it is supported by numerical simulations, the correlation function can be justified in the semi-classical limit. This property is sometimes referred to as the random wave model [43].
- The nodal domains (defined as the connected domains where $\psi_n(x)$ have a constant sign) in quantum chaotic billiards display two features – their normalized count $\frac{\nu_n}{n}$ narrowly distribute around a universal number. Moreover, they cover Ω in patterns typical to critical percolation [21, 46].

Back to d-regular graphs: The local tree property of d-regular graphs is at the basis of our present understanding of the statistical properties of the adjacency eigenvectors. Considering the infinite d-regular tree, Y. Elon [2] was recently able to properly construct a "random wave" model on the infinite d-regular tree whose Gaussian distribution can be rigorously assessed. The covariance for the Gaussian field is

$$\phi^{(\lambda)}(s) \doteq \langle f_i f_j \rangle = (d-1)^{-s/2} \left(\frac{d-1}{d} U_s \left(\frac{\lambda}{2\sqrt{d-1}} \right) - \frac{1}{d} U_{s-2} \left(\frac{\lambda}{2\sqrt{d-1}} \right) \right) \quad (54)$$
$$s = |i - j|.$$

Here, $f_i(\lambda)$ stands for the component of the adjacency eigenvector at the vertex i, with the normalization $\langle f_i^2 \rangle = 1$. The distance $s = |i - j|$ on the graph is the length of the minimal walk between the vertices i, j, $U_k(x)$ are the Chebyshev polynomials of the second kind, and the triangular brackets denote averaging over the random wave ensemble. Taking into account the asymptotic behavior of the Chebyshev polynomials, one can recognize (54) as the analogue of the correlation function of random waves for planar random waves (53).

Due to the local tree property, it is natural to expect, (but not yet rigorously proved), that in the limit $V \to \infty$ the properties derived for the tree graph, will also hold for $\mathcal{G}(V, d)$, for distances within the range $s \leq \log V$. This was tested numerically, and the (multivariate) Gaussian distribution for the components of the adjacency eigenvectors was checked within the numerical accuracy. Figures 6 and 7 summarize the numerical tests.

As in billiards, one can define nodal domains of the adjacency vectors as the connected sub-graphs where the eigenvector components have a constant sign. The number of nodal domains obeys the analogue of Courant theorem [44]. There is no known way to count nodal domains in billiards or even in arbitrary graphs. Y. Elon [1] proposed recently a method to compute the expected number of nodal domains in d-regular graphs as a function of the corresponding eignevalue. His argument will be sketched here.

For a graph G and an adjacency eigenvector $f(\lambda)$, the *induced nodal graph* $\tilde{G}_f$ is obtained by the deletion of edges, which connect vertices that possess opposite signs in f

$$\tilde{G}_f = (V, \tilde{E}_f) \quad , \quad \tilde{E}_f = \{(v_i, v_j) \in E | f_i f_j > 0\}. \quad (55)$$

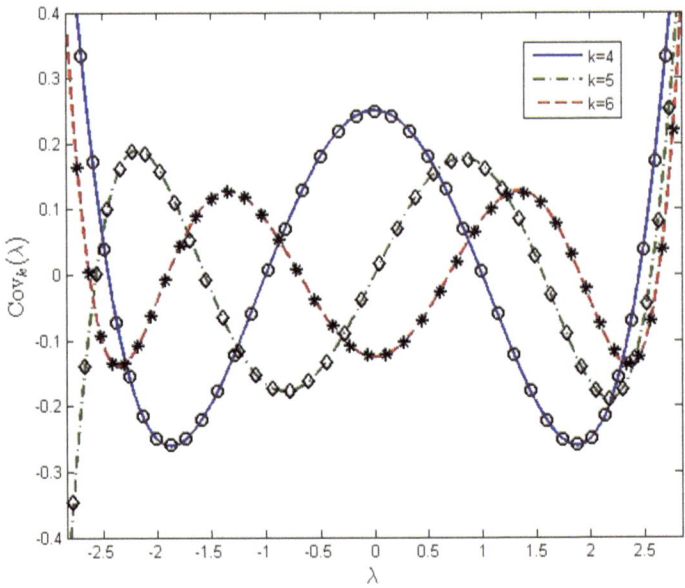

FIGURE 6. A comparison between (54) (marked by lines), and the numerical covariance, for a single realization of $G(4000, 3)$ (denoted by different symbols), where $\log_{d-1}(n) = 11.97$, for $k = 4, 5, 6$ (the local tree regime).

We also define

$$p_e(f(\lambda)) = \mathbb{P}(e \in \tilde{E}_f) \qquad (56)$$

to denote the probability that a random edge $e \in E$ survives the nodal trimming. According to the random waves hypothesis, $p_e(f(\lambda))$ should typically converge as $V \to \infty$ into

$$p_e(\lambda) = 2 \int_0^\infty \int_0^\infty d\mathbf{f} \frac{1}{2\pi\sqrt{|\mathcal{C}_\lambda|}} \exp\left(-\frac{1}{2}\langle \mathbf{f}, \mathcal{C}_\lambda^{-1} \mathbf{f}\rangle\right) = \frac{1}{2} + \frac{1}{\pi} \arcsin\left(\frac{\lambda}{d}\right) \qquad (57)$$

where $\mathbf{f} = (f_1, f_2)$ and the covariance operator elements are $(\mathcal{C}_\lambda)_{11} = (\mathcal{C}_\lambda)_{22} = \phi^{(\lambda)}(0) = 1$, $(\mathcal{C}_\lambda)_{12} = (\mathcal{C}_\lambda)_{21} = \phi^{(\lambda)}(1) = \lambda/d$. ($\phi^{(\lambda)}(s)$ is the covariance defined in (54)). This result which follows from the random waves conjecture was checked numerically and found to reproduce the simulations in a perfect way.

Since on average, the induced nodal graph $\tilde{G}_f$ posses $p_e(\lambda) \cdot |E| = \frac{nd}{2} p_e(\lambda)$ edges and V vertices, the expected nodal count may be bounded from below (for all of the eigenvectors but the first) by neglecting the cycles in $\tilde{G}$:

$$\mathbb{E}\left(\frac{\nu(\lambda)}{V}\right) \geq \max\left\{\frac{2}{V}, \left(1 - \frac{d}{2} p_e(\lambda)\right)\right\}. \qquad (58)$$

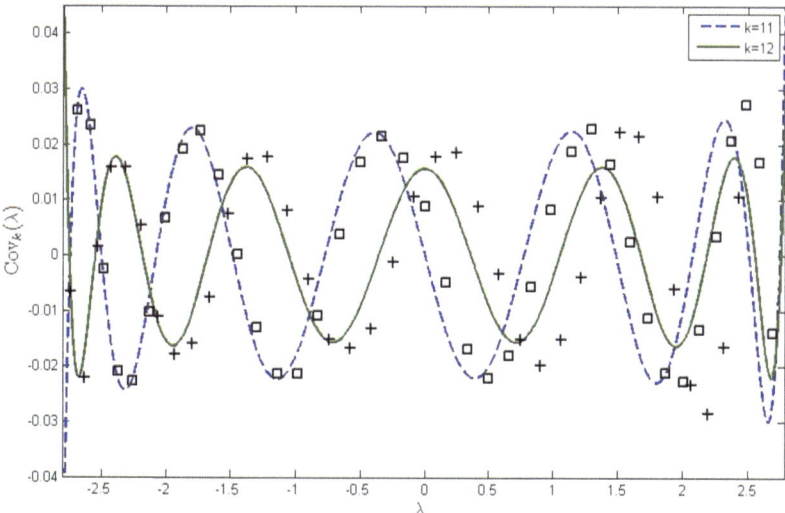

FIGURE 7. A comparison between (54) (marked by lines), and the numerical covariance, for $G(4000,3)$ and $k = 11, 12$ (beyond the tree regime).

While (58) lose its efficiency as d increases[1], for low values of d, this crude bound matches surprisingly well the observed nodal count as shown in Figure 8. By taking

---
[1] In fact, for $d > 7$, $\forall \lambda, 1 - dp_e(\lambda)/2 < 0$. Therefore tn that case the bound becomes trivial.

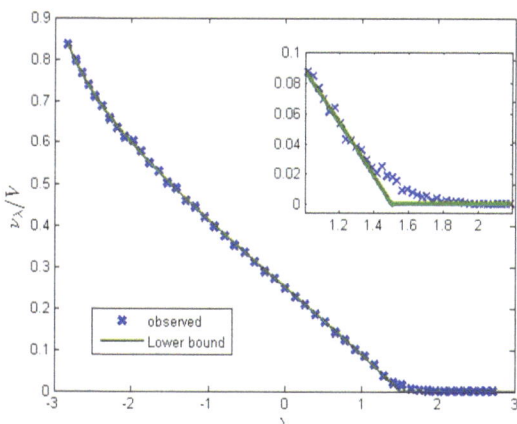

FIGURE 8. A comparison between the theoretical estimate for the nodal count and the observed count for a single realization of $G(4000, 3)$. The inset is a magnification of the spectral window near the bound's flexion – the only part of the spectrum, in which the observed count deviates considerably from the bound.

into account more correlations, Elon was able to improve the crude bound above [1]. Due to space limitation this will not be discussed here any further.

Percolation is an important topic of research in probabilistic graph theory [47]. This subject was recently extended [3] by defining percolation of level sets of the adjacency eigenvectors as explained below. The α level set of an eigenvector is defined as the subgraph where $\tilde{G}_\alpha = \{v_i \in G : f_i > \alpha\}$ (the normalization $\frac{1}{V}\sum_i f_i^2 = 1$ is assumed). Given an adjacency eigenvector (corresponding to an eigenvalue λ) one can study the α dependence of the ratio

$$\frac{|\tilde{G}_\alpha^{\max}|}{|\tilde{G}_\alpha|} \qquad (59)$$

where $\tilde{G}_\alpha^{\max}$ is the maximal connected component in $\tilde{G}_\alpha$. The α level set will be called percolating once $\tilde{G}_\alpha^{\max} \sim V$. A percolation transition occurs at $\alpha_c(\lambda)$ if, in the limit $V \to \infty$ and for almost all graphs in $\mathcal{G}(V,d)$, the ratio (59) is discontinuous at $\alpha = \alpha_c(\lambda)$. The transition for a single large graph is shown in Figure 9 for a few values of λ. Assuming the validity of the random wave conjecture, Y. Elon was able to compute the dependence of the critical level on both λ and d, and found a very good agreement with the numerical results, as shown in Figure 10.

The applicability of the random wave conjecture of the adjacency vectors is supported very robustly by the numerical and theoretical considerations. A rigorous validation is still lacking.

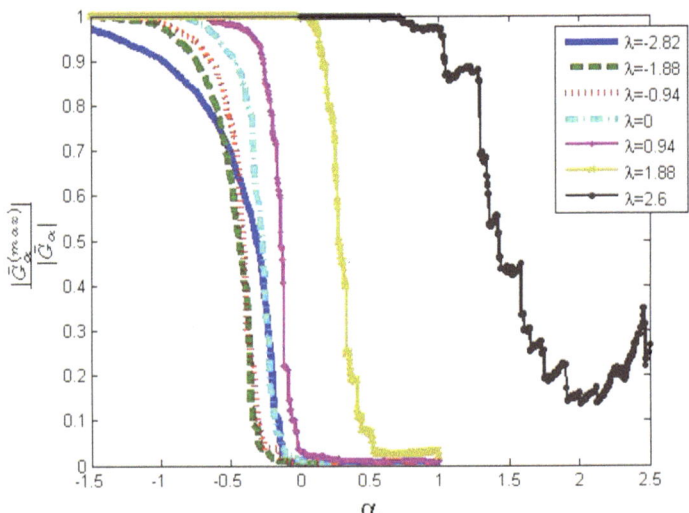

FIGURE 9. The ratio between the magnitude of the largest level set to the size of the induced graph $\tilde{G}_\alpha$ for a single realization of a $(4000, 3)$ graph. Each curve corresponds to one eigenvector, while α is varied along the curve.

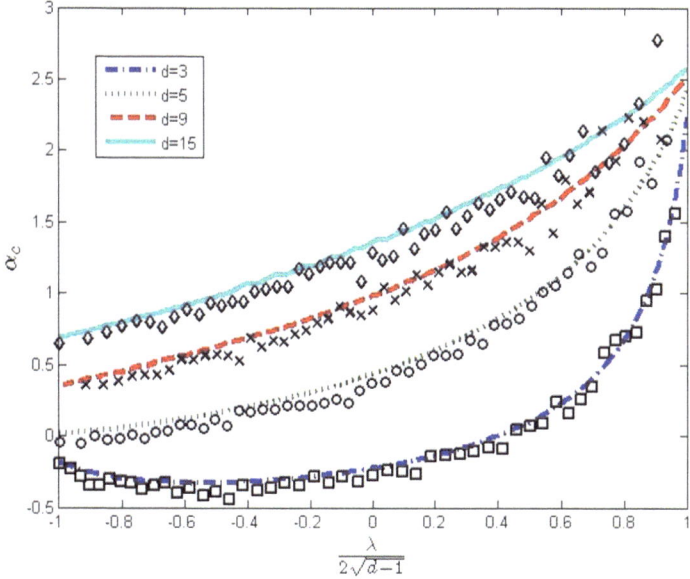

FIGURE 10. A comparison between the theoretical estimates of $\alpha_c(\lambda, d)$ based on the random wave model (lines) to the numerically computed $\alpha_c(\lambda, d)$ for (V, d) graphs (markers).

4. Scattering

Chaos is usually associated with bounded dynamical systems. However, classical chaos can be defined and studied also in open (scattering) systems, and its quantum analogue provided many interesting theoretical problems and practical applications [48, 49, 50]. In this section, it will be shown how a bound finite graph can be converted to a scattering system by attaching "leads" to infinity [7, 51]. An expression for the scattering matrix will be provided, which can be further studied in various contexts such as, e.g., resonance distribution and conductance fluctuations and correlations.

Consider a finite (not necessarily d-regular) graph $\mathcal{G}^{(0)}$. In what follows, all quantities belonging to $\mathcal{G}^{(0)}$ will be denoted by a superscript (0). In particular, $D_i^{(0)}$ stands for the degree of the vertex i in $\mathcal{G}^{(0)}$. Scattering is defined by attaching to any subset of the vertices lead graphs which are defined as follows: A lead graph l is a semi-infinite set of vertices $(l, 1), (l, 2), \ldots$ which are connected linearly. A vertex is identified by a double index (l, i), l denotes the lead, and i enumerates the vertex position on the lead. The lead connectivity (adjacency) matrix is $A_{(l,n),(l',n')}^{(\text{Lead})} = w\delta_{l,l'}\delta_{|n-n'|,1}$, $n, n' \in \mathbb{N}^+$, where w stands for the number of parallel bonds which connect successive vertices. (All quantities related to the leads will be denoted by the superscript (Lead).) The spectrum of the lead Laplacian $\Delta_l^{(\text{Lead})}$ and the corresponding eigenfunctions $\mathbf{f} = (f_{(l,1)}, f_{(l,2)}, \dots)^\top$ satisfy

$$(\Delta_l^{(\text{Lead})}\mathbf{f})_{(l,n)} = -w(f_{(l,n+1)} + f_{(l,n-1)}) + 2wf_{(l,n)} = \lambda f_{(l,n)} \quad \text{for } n > 1,$$
$$= -wf_{(l,2)} + wf_{(l,1)} = \lambda f_{(l,1)} \quad \text{for } n = 1. \tag{60}$$

The spectrum is continuous and supported on the spectral band $\lambda \in [0, 4w]$ (the conduction band). For any λ in the conduction band, there correspond two eigenfunctions which can be written as linear combinations of counter-propagating waves:

$$f_{(l,n)}^{(\pm)} = \xi_\pm^{n-1} \quad \text{where} \quad \xi_\pm = 1 - \frac{\lambda}{2w} \pm \sqrt{\left(1 - \frac{\lambda}{2w}\right)^2 - 1} = e^{\pm i\alpha(\lambda)}. \tag{61}$$

For $\lambda > 4w$, $|\xi_-| > |\xi_+|$. The reason for constructing the leads with w parallel bonds is because the conduction band can be made arbitrarily broad. In the present application, an appropriate choice of w would be of the order of the mean valency in the interior graph, so the spectrum of $\mathcal{G}^{(0)}$ falls well within the conduction band.

A function which satisfies the boundary condition at $n = 1$ (second line of (60)) is,

$$f_{(l,n)} = f_{(l,n)}^{(-)} + s_l(\lambda) f_{(l,n)}^{(+)}, \tag{62}$$

where

$$s_l(\lambda) = -\frac{1}{1 - \xi_-} \frac{\xi_+}{} = \xi_+, \quad |s_l(\lambda)| = 1 \text{ for } \lambda \in [0, 4w]. \tag{63}$$

The lead scattering amplitude $s_l(\lambda)$ provides the phase gained by scattering at the end of the lead (as long as λ is in the conduction band).

Returning now to the interior graph $\mathcal{G}^{(0)}$ it is converted into a scattering graph by attaching to its vertices semi-infinite leads. At most one lead can be attached to a vertex, but not all vertices should be connected to leads. Let $\mathcal{L}$ denote the set of leads, and $L = |\mathcal{L}|$. The connection of the leads to $\mathcal{G}^{(0)}$ is given by the $V^{(0)} \times L$ "wiring" matrix

$$W_{j,(l,1)} = \begin{cases} 1 & \text{if } j \in \mathcal{V}^{(0)} \text{ is connected to } l \in \mathcal{L} \\ 0 & \text{otherwise} \end{cases}. \tag{64}$$

The number of leads which emanate from the vertex i is either 0 or 1, and is denoted by $d_i = W_{i,(l,1)}$. Define also the diagonal matrix $\tilde{D} = \text{diag}(d_i)$ so that

$$WW^\top = \tilde{D} \quad ; \quad W^\top W = I^{(L)}, \tag{65}$$

where $I^{(L)}$ is the $L \times L$ unit matrix.

The scattering graph $\mathcal{G}$ is the union of $\mathcal{G}^{(0)}$ and the set of leads $\mathcal{L}$. Its vertex set is denoted by $\mathcal{V}$ and its adjacency matrix A for $\mathcal{G}$ is given by

$$\forall i, j \in \mathcal{V}, \; : \; A_{i,j} = \begin{cases} A_{i,j}^{(0)} & \text{if } i, j \in \mathcal{V}^{(0)} \\ A_{i=(l,i),j=(l,j)}^{(\text{Lead})} & \text{if } l \in \mathcal{L} \\ wW_{i,j=(l,1)} & \text{if } i \in \mathcal{V}^{(0)} \text{ and } l \in \mathcal{L} \end{cases}. \tag{66}$$

As is usually done in scattering theory, one attempts to find eigenfunctions $\mathbf{f}$ of the discrete Laplacian of the scattering graph, subject to the condition that on the leads $l = 1, \ldots, L$ the wave function consists of counter propagating waves:

$$f_{(l,n)} = a_l \xi_-^{n-1} + b_l \xi_+^{n-1}, \quad n \geq 1. \tag{67}$$

where a_l and b_l are the incoming and outgoing amplitudes. They are to be determined from the requirement that $\mathbf{f}$ is an eigenfunction of the scattering graph Laplacian. It will be shown below that this requirement suffices to provide a linear relationship between the incoming and outgoing amplitude. The $L \times L$ scattering matrix $S^{(\text{Lead})}(\lambda)$ is defined as the mapping from the incoming to the outgoing amplitudes:

$$\mathbf{b} = S^{(\text{Lead})}(\lambda) \mathbf{a}. \tag{68}$$

To compute $S^{(\text{Lead})}(\lambda)$, consider the action of the Laplacian on an eigenvector $\mathbf{f}$.

$$\forall i \in \mathcal{V}^{(0)} : (\Delta \mathbf{f})_i = -\sum_{j \in \mathcal{V}^{(0)}} A_{i,j}^{(0)} f_j - w \sum_{l \in \mathcal{L}} W_{i,(l,1)} f_{(l,1)} + (D_i^{(0)} + w d_i) f_i = \lambda f_i.$$

$$\forall l \in \mathcal{L} : (\Delta \mathbf{f})_{(l,1)} = -w \sum_{i \in \mathcal{V}^{(0)}} W_{(l,1),i}^\top f_i - w f_{(l,2)} + 2w f_{(l,1)} = \lambda f_{(l,1)}.$$

$$(\Delta \mathbf{f})_{(l,n)} = -w f_{(l,n+1)} - w f_{(l,n-1)} + 2w f_{(l,n)} = \lambda f_{(l,n)}. \tag{69}$$

The equations for $i \in \mathcal{V}^{(0)}$ (first line in (69) above), can be put in a concise form:

$$\left(\Delta^{(0)} + w\tilde{D} - \lambda I^{(\mathcal{V}^{(0)})}\right) \mathbf{f}^{(\mathcal{V}^{(0)})} = w W \mathbf{f}_1^{(L)}, \tag{70}$$

where $I^{(\mathcal{V}^{(0)})}$ is the unit matrix in $\mathcal{V}^{(0)}$ dimension, $\mathbf{f}^{(\mathcal{V}^{(0)})}$ is the restriction of $\mathbf{f}$ to the vertices of the interior graph $\mathcal{G}^{(0)}$ and $\mathbf{f}_1^{(L)}$ is the L-dimensional vector with components $f_{(l,1)}$, $l = 1, \ldots, L$. For λ away from the eigenvalues of $\Delta^{(0)} + w\tilde{D}$ the $\mathcal{V}^{(0)} \times \mathcal{V}^{(0)}$ matrix $R^{(0)}(\lambda)$ is defined as

$$R^{(0)}(\lambda) = \left(\Delta^{(0)} + w\tilde{D} - \lambda I^{(\mathcal{V}^{(0)})}\right)^{-1}. \tag{71}$$

Thus,

$$\mathbf{f}^{(\mathcal{V}^{(0)})} = w R^{(0)} W \mathbf{f}_1^{(L)}. \tag{72}$$

Substituting in the second set of equations in (69) and using (67),

$$\left(-w^2 W^\top R^{(0)}(\lambda) W + (2w - \lambda) I^{(L)}\right)(\mathbf{a} + \mathbf{b}) = w(\mathbf{a} \, \xi_- + \mathbf{b} \, \xi_+). \tag{73}$$

This can be easily brought into the form (68). Using the fact that $2 - \lambda/w = \xi_- + \xi_+$ we get

$$S^{(\text{Lead})}(\lambda) = -\left(w W^\top R^{(0)}(\lambda) W - \xi_- I^{(L)}\right)^{-1} \left(w W^\top R^{(0)}(\lambda) W - \xi_+ I^{(L)}\right). \tag{74}$$

This is the desired form of the scattering matrix. It has a few important properties.

i. As long as λ is in the conduction band, ξ_- and ξ_+ are complex conjugate and unitary. Since $W^\top R^{(0)}(\lambda) W$ is a symmetric real matrix, $S^{(\text{Lead})}(\lambda)^\top =$

$S^{(\text{Lead})}(\lambda)$ and $S^{(\text{Lead})}(\lambda)S^{(\text{Lead})}(\lambda)^\dagger = I^{(L)}$, that is, $S^{(\text{Lead})}(\lambda)$ is a symmetric and unitary matrix.

ii. Once λ is outside of the conduction band, the $f_{(l,n)}^{(\pm)}$ are exponentially increasing or decreasing solutions – they are the analogues of the evanescent waves encountered in the study of wave-guides. One of the reason for the introduction of the w parallel bonds in the leads was to broaden the conduction band and avoid the spectral domain of evanescent waves. However, for the sake of completeness one observes that the scattering matrix as defined above can be analytically continued outside of the conduction band by using (61) which is valid for any λ. The $S^{(\text{Lead})}(\lambda)$ matrix outside the conduction band loses its physical interpretation, and it remains symmetric but is not any more unitary. However, it is a well-defined object, and can be used in the sequel for any real or complex λ. In the limit $\lambda \to \infty$, $S^{(\text{Lead})}(\lambda) \to \xi_+ W^\top R^{(0)}(\lambda) W \approx \left(\frac{w}{\lambda}\right)^2 I^{(L)}$.

iii. At the edges of the conduction band $\xi_\pm(\lambda = 0) = 1$; $\xi_\pm(\lambda = 4w) = -1$. Substituting in (74) one finds that at the band edges, $S^{(\text{Lead})} = -I^{(L)}$.

iv. The matrix $R^{(0)}(\lambda)$ is well defined for λ away from the spectrum of $\Delta^{(0)} + w\tilde{D}$. Approaching these values does not cause any problem in the definition of $S^{(\text{Lead})}(\lambda)$ since there $S^{(\text{Lead})} = -I^{(L)}$. However, for sufficiently large w the singularities of $R^{(0)}(\lambda)$ can be separated away from the domain where the spectrum of $\mathcal{G}^{(0)}$ is supported.

v. The resonances are defined as the poles of the scattering matrix in the complex λ plane. They are the solution of the equation

$$z_{\text{res}}(\lambda) = \det\left(wW^\top R^{(0)}(\lambda)W - \xi_- I^{(L)}\right) = 0 . \qquad (75)$$

The point $\lambda = 0$ is not a pole since $S^{(\text{Lead})}(\lambda = 0) = -I^{(L)}$.

vi. Finally, it might be instructive to note that the matrix $R^{(0)}(\lambda)$ is closely related to the discrete analogue of the Dirichlet to Neumann map. This can be deduced from the following construction: add to each vertex $i \in \mathcal{V}^{(0)}$ a new auxiliary vertex $\tilde{i}$ connected exclusively to i. (Here we use $w = 1$ to make the analogy clearer.) Write the discrete Laplacian for the new graph, and solve $(\Delta - \lambda I)\tilde{\mathbf{f}} = 0$, where $\tilde{\mathbf{f}}$ is a $2V^{(0)}$-dimensional vector, the first V entries correspond to the original vertices, and the last V entries correspond to the auxiliary vertices: $\tilde{\mathbf{f}} = (\mathbf{f}, \mathbf{g})^\top$. Assuming that the values $g_{\tilde{i}}$ on the auxiliary vertices are given, the entries in $\mathbf{f}$ can be expressed as $\mathbf{f} = R(\lambda)\mathbf{g}$, where $R(\lambda)$ as defined in (71). To emphasize the connection to the Dirichlet to Neumann map, define $\psi = \frac{1}{2}(\mathbf{g} + \mathbf{f})$ (the "boundary function") and $\partial\psi = (\mathbf{g} - \mathbf{f})$ (the "normal derivative") then,

$$\partial\psi = M(\lambda)\psi \;\; ; \;\; M(\lambda) = 2(I^{(V^{(0)})} + R(\lambda))^{-1}(I^{(V^{(0)})} - R(\lambda)) . \qquad (76)$$

The Dirichlet to Neumann map is defined also in other applications of graph theory, see, e.g., [52].

The general setup displayed above can be easily converted to the case of d-regular graphs, and the information accumulated for their spectral and eigenfunctions properties could be used in the scattering context. This program is now under study.

5. Summary and prospects

The three topics covered in the preceding sections are at the center of the research in quantum chaos, and the close conceptual and technical links with quantum chaos substantiate the claim expressed by the title of the present article – indeed, d-regular graphs are a paradigm model for quantum chaos. This model is as rich as the "quantum graphs" which where introduced a while ago, and turned out to be a very fertile domain in its own right, and provided several new insights to problems which are commonly addressed in "quantum chaos" [53, 54, 55]. As much of the research reported here is the result of work done in the past three years only, much remains to be done, and a few routes should be opened to explore further the potential stored in the d-regular graphs model.

i. The connection between graph combinatorics and spectral statistics should go much more deeply than what is known today. As was shown above, this research at the interface of the two fields stores potential advantages in both directions.

ii. Isospectral d-regular graphs are known to exist [56], and methods to build them systematically were proposed by several authors. It was suggested some time ago that the nodal domain counts can be used to distinguish between isospectral domains. Some analytical results on small and simple graphs, and numerical simulations pertaining for large graphs support the validity of this conjecture. Further research is needed to substantiate these claims in a rigorous way.

iii. The range of the graph model can be extended in a major way by assigning weights to the edges, and/or potentials to the vertices. The graph Laplacian would then read:

$$(\Delta \mathbf{f})_j = -\sum_i T_{j,i} A_{j,i} f_i + t_j f_j + V_j f_j, \qquad (77)$$

where the positive weights are denoted by $T_{j,i} = T_{i,j}$ and $t_j = \sum_i T_{i,j}$. The vertex potentials are V_i. Choosing the weights and potentials at random one could investigate the rôles of diagonal and off diagonal disorder on the spectrum and eigenfunctions, in the limit $V \to \infty$. The question whether a localization transition occurs is open.

iv. The spectra of quantum graphs with equal (or rationally related) edge lengths, are intimately related to the spectra of the underlying discrete graphs. In particular, the spectral properties of discrete d-regular graphs induce interesting features in their quantum analogues, which call for further research [4].

Acknowledgment

The author wishes to express his deep gratitude to Yehonatan Elon, Amit Godel and Idan Oren whose results, thesis and papers were cut-and-pasted to prepare this survey. The work was supported by the Minerva Center for non-linear Physics, the Einstein (Minerva) Center at the Weizmann Institute and the Wales Institute of Mathematical and Computational Sciences) (WIMCS). Grants from EPSRC (grant EP/G021287), ISF (grant 166/09) and BSF (grant 2006065) are acknowledged.

References

[1] Yehonatan Elon, *Eigenvectors of the discrete Laplacian on regular graphs-a statistical approach*, J. Phys. A: Math. Theor. **41**, 435203 (2008).

[2] Yehonatan Elon, *Gaussian Waves on the Regular Tree*, arXiv:0907.5065v2 [math-ph] (2009).

[3] PhD thesis submitted to the Feinberg School, The Weizmann Institute of Science (2010).

[4] MSc thesis submitted to the Feinberg School, The Weizmann Institute of Science (2009).

[5] Idan Oren, Amit Godel and Uzy Smilansky, *Trace formulae and spectral statistics for discrete Laplacians on regular graphs* (I), J. Phys. A: Math. Theor. **42**, 415101 (2009).

[6] Idan Oren and Uzy Smilansky, *Trace formulae and spectral statistics for discrete Laplacians on regular graphs* (II), J. Phys. A: Math. Theor. **43**, 225205 (2010).

[7] Uzy Smilansky *Exterior-Interior Duality for Discrete Graphs* J. Phys. A: Math. Theor. **42**, 035101 (2009).

[8] Uzy Smilansky, *Quantum Chaos on Discrete Graphs*, J. Phys. A: Math. Theor. **40**, F621–F630 (2007).

[9] D. Jakobson, S. Miller, I. Rivin and Z. Rudnick, *Level spacings for regular graphs*, IMA Volumes in Mathematics and its Applications **109**, 317–329 (1999).

[10] A.A. Terras, *Arithmetic Quantum Chaos*, IAS/Park City Mathematical Series **12**, 333–375 (2002).

[11] S. Hoory, N. Linial, and A. Wigderson *Expander Graphs and Their Applications*, Bulletin (New Series) of the American Mathematical Society **43**, Number 4, 439–561 (2006), S 0273-0979(06)01126-8.

[12] J. Dodziuk and W.S. Kandel *Combinatorial Laplacians and isoperimetric inequality*, in "From local times to global geometry, control and physics", K.D. Ellworthy ed., Pitman Research Notes in Mathematics Series, **150**, 68–74 (1986).

[13] B. Bollobas, Random Graphs, Academic Press, London (1985).

[14] Fan R.K. Chung, *Spectral Graph Theory*, Regional Conference Series in Mathematics **92**, American Mathematical Society (1997).

[15] J. Friedman, *Some geometric aspects of graphs and their eigenfunctions*, Duke Math. J. **69**, 487–525 (1993).

[16] N. Alon, *Eigenvalues and Expanders*, Combinatorica, **6**, 83–96 (1986).

[17] H. Kesten *Symmetric random walks on groups*, Trans. Am. Math. Soc. **92**, 336–354 (1959).

[18] McKay, B.D., *The expected eigenvalue distribution of a random labelled regular graph*, Linear Algebr. Appl. **40**, 203–216 (1981).

[19] M. Ram Murty, Ramanjuan Graphs, J. Ramanujan Math. Soc. **18**, 1–20 (2003).

[20] J.E. Avron, A. Raveh and B. Zur, *Adiabatic quantum transport in multiply connected systems*, Reviews of Modern Physics **60**, No. 4 (1988).

[21] O. Bohigas, M.-J. Giannoni, and C. Schmit *Characterization of chaotic quantum spectra and universality of level fluctuation laws*, Phys. Rev. Lett. **52**, 1–4 (1984).

[22] M.V. Berry and M. Tabor, *Level clustering in the regular spectrum*, Proc. Roy. Soc. A **356**, 375–394 (1977).

[23] O. Bohigas, *Random Matrix Theories and Chaotic Dynamics*, in *Chaos and Quantum Physics*, M.J. Giannoni, A. Voros and J. Zinn-Justin, editors, pp. 87–199, North-Holland (1989).

[24] Fritz Haake, *Quantum Signatures of Chaos*, Springer-Verlag Berlin and Heidelberg, (2001).

[25] S. Sodin, *The Tracy–Widom law for some sparse random matrices*, J. Stat. Phys. **136**, 834–841 (2009), arXiv:0903.4295v2 [math-ph].

[26] M.C. Gutzwiller, J. Math. Phys. **12**, 343 (1984).

[27] M. Gutzwiller, *Chaos in Classical and Quantum Mechanics*, Springer Verlag, New York (1991).

[28] M.V. Berry, *Semiclassical Theory of Spectral Rigidity*, Proceedings of the Royal Society of London. Series A, Mathematical and Physical Sciences, Vol. 400, No. 1819 (Aug. 8, 1985), pp. 229–251.

[29] N. Argaman, F. Dittes, E. Doron, J. Keating, A. Kitaev, M. Sieber and U. Smilansky. *Correlations in the Acions of Periodic Orbits Derived from Quantum Chaos*. Phys. Rev. Lett. **71**, 4326–4329 (1993).

[30] M. Sieber, K. Richter, *Correlations between Periodic Orbits and their Rôle in Spectral Statistics*, Physica Scripta, Volume T90, Issue 1, pp. 128–133.

[31] S. Heusler, S. Müller, P. Braun, and F. Haake, *Universal spectral form factor for chaotic dynamics*, J. Phys. A **37**, L31 (2004).

[32] H. Bass, *The Ihara–Selberg zeta function of a tree lattice*, Internat. J. Math. **3**, 717–797 (1992).

[33] L. Bartholdi, *Counting paths in graphs*. Enseign. Math **45**, 83–131 (1999).

[34] N. Alon, I. Benjamini, E. Lubetzky, S. Sodin, *Non-backtracking random walks mix faster*, arXiv:math/0610550v1.

[35] P. Mněv, *Discrete Path Integral Approach to the Selberg Trace Formula for Regular Graphs*, Commun. Math. Phys. **274**, 233–241 (2007).

[36] S. Janson, T. Łuczak and A. Ruciński,*Random Graphs*, John Wiley & Sons, Inc. (2000).

[37] N.C. Wormald, *The asymptotic distribution of short cycles in random regular graphs*, J. Combin. Theory, Ser. B **31**, 168–182 (1981).

[38] B. Bollobás, *A probabilistic proof of an asymptotic formula for the number of labelled regular graphs*, European J. Combin. **1**, 311–316 (1980).

[39] B.D. McKay, N.C. Wormald and B. Wysocka, *Short cycles in random regular graphs*, The electronic journal of combinatorics, **11**, R66.

[40] J.P. Keating and N.C. Snaith, *Random Matrix Theory and $\zeta(1/2 + it)$*, Commun. Math. Phys. **214**, 57–89 (2000).

[41] A.I. Shnirelman, *Ergodic properties of eigenfunctions*, Uspehi Mat. Nauk **29** (6(180)), 181–182 (1974).

[42] E.J. Heller, Phys. Rev. Lett. **53** 1515 (1984).

[43] M.V. Berry, J. Phys. A **10**, 2083–2091 (1977).

[44] R. Courant, Nachr. Ges. Wiss. Göttingen, Math. Phys. Kl. (1923).

[45] Galya Blum, Sven Gnutzmann, and Uzy Smilansky, *Nodal Domains Statistics: A Criterion for Quantum Chaos*, Phys. Rev. Lett. **88**,114101 (2002).

[46] E. Bogomolny and C. Schmit, Phys. Rev. Lett. **88**, 114102 (2002).

[47] Geoffrey Grimmett, Percolation, 2nd Edition, Grundlehren der mathematischen Wissenschaften, vol. 321, Springer, (1999).

[48] R. Blümel and U. Smilansky, *Random matrix description of chaotic scattering: Semi Classical Approach*. Phys. Rev. Lett. **64**, 241–244 (1990).

[49] E. Doron, U. Smilansky and A. Frenkel. *Chaotic scattering and transmission fluctuations*. Physica D **50**, 367–390 (1991).

[50] T. Kottos and U. Smilansky, Quantum Graphs: A simple model for Chaotic Scattering. J. Phys. A. **36**, 3501–3524 (2003).

[51] S. Fedorov and B. Pavlov, *Discrete wave scattering on star-graph*. J. Phys. A: Math. Gen. **39**, 2657–2671 (2006).

[52] E.B. Curtis and J.A. Morrow, *The Dirichlet to Neumann map for a resistor network* SIAM Journal on Applied Mathematics **51**, 1011–1029 (1991).

[53] T. Kottos and U. Smilansky, *Quantum Chaos on Graphs*, Phys. Rev. Lett. **79**, 4794–4797 (1997).

[54] T. Kottos and U. Smilansky, *Periodic orbit theory and spectral statistics for quantum graphs*, Annals of Physics **274**, 76–124 (1999).

[55] Sven Gnutzmann and Uzy Smilansky, *Quantum Graphs: Applications to Quantum Chaos and Universal Spectral Statistics*, Advances in Physics bf 55, 527–625 (2006).

[56] R. Brooks, Ann. Inst. Fourier **49**, 707–725 (1999).

Uzy Smilansky
Department of Physics of Complex Systems
The Weizmann Institute of Science
76100 Rehovot, Israel

and

School of Mathematics
Cardiff University
Cardiff, Wales, UK
e-mail: `Uzy.Smilansky@weizmann.ac.il`

Quantum Chaos, Random Matrix Theory, and the Riemann ζ-function

Paul Bourgade and Jonathan P. Keating

> **Abstract.** We review some connections between quantum chaos and the theory of the Riemann zeta function and the primes. Specifically, we give an overview of the similarities between the semiclassical trace formula that connects quantum energy levels and classical periodic orbits in chaotic systems and an analogous formula that connects the Riemann zeros and the primes. We also review the role played by Random Matrix Theory in both quantum chaos and the theory of the zeta function. The parallels we review are conjectural and still far from being understood, but the ideas have led to substantial progress in both areas.

Hilbert and Pólya put forward the idea that the zeros of the Riemann zeta function may have a spectral origin: the values of t_n such that $\frac{1}{2}+\mathrm{i} t_n$ is a non trivial zero of ζ might be the eigenvalues of a self-adjoint operator. This would imply the Riemann Hypothesis. From the perspective of Physics one might go further and consider the possibility that the operator in question corresponds to the quantization of a classical dynamical system.

The first significant evidence in support of this spectral interpretation of the Riemann zeros emerged in the 1950's in the form of the resemblance between the Selberg trace formula, which relates the eigenvalues of the Laplacian and the closed geodesics of a Riemann surface, and the Weil explicit formula in number theory, which relates the Riemann zeros to the primes. More generally, the Weil explicit formula resembles very closely a general class of Trace Formulae, written down by Gutzwiller, that relate quantum energy levels to classical periodic orbits in chaotic Hamiltonian systems.

The second significant evidence followed from Montgomery's calculation of the pair correlation of the t_n's (1972): the zeros exhibit the same repulsion as the eigenvalues of typical large unitary matrices, as noted by Dyson. Montgomery conjectured more general analogies with these random matrices, which were confirmed by Odlyzko's numerical experiments in the 1980's.

Later conjectures relating the statistical distribution of random matrix eigenvalues to that of the quantum energy levels of classically chaotic systems connect these two themes.

We here review these ideas and recent related developments: at the rigorous level strikingly similar results can be independently derived concerning number-theoretic L-functions and random operators, and heuristics allow further steps in the analogy. For example, the t_n's display Random Matrix Theory statistics in the limit as $n \to \infty$, while lower-order terms describing the approach to the limit are described by non-universal (arithmetic) formulae similar to ones that relate to semiclassical quantum eigenvalues. In another direction and scale, macroscopic quantities, such as the moments of the Riemann zeta function along the critical line on which the Riemann Hypothesis places the non-trivial zeros, are also connected with random matrix theory.

1. First steps in the analogy

This section describes the fundamental mathematical concepts (i.e., the Riemann zeta function and random operators) the connections between which are the focus of this survey: linear statistics (trace formulas) and microscopic interactions (fermionic repulsion). These statistical connections have since been extended to many other L-functions (in the Selberg class [47], over function fields [34]): for the sake of brevity we only consider the Riemann zeta function.

1.1. Basic theory of the Riemann zeta function

The Riemann zeta function can be defined, for $\sigma = \Re(s) > 1$, as a Dirichlet series or an Euler product:

$$\zeta(s) = \sum_{n=1}^{\infty} \frac{1}{n^s} = \prod_{p \in \mathcal{P}} \frac{1}{1 - \frac{1}{p^s}},$$

where $\mathcal{P}$ is the set of all prime numbers. The second equality is a consequence of the unique factorization of integers into prime numbers. A remarkable fact about this function, proved in Riemann's original paper, is that it can be meromorphically extended to the complex plane, and that this extension satisfies a functional equation.

Theorem 1.1. *The function ζ admits an analytic extension to $\mathbb{C} - \{1\}$ which satisfies the equation (writing $\xi(s) = \pi^{-s/2}\Gamma(s/2)\zeta(s)$)*

$$\xi(s) = \xi(1-s).$$

Proof. The gamma function is defined for $\Re(s) > 0$ by $\Gamma(s) = \int_0^\infty e^{-t}t^{s-1}\mathrm{d}t$, hence, substituting $t = \pi n^2 x$,

$$\pi^{-\frac{s}{2}}\Gamma\left(\frac{s}{2}\right)\frac{1}{n^s} = \int_0^\infty x^{\frac{s}{2}-1}e^{-n^2\pi x}\mathrm{d}x.$$

If we sum over n, the sum and integral can be exchanged for $\Re(s) > 1$ because of absolute convergence, hence

$$\pi^{-\frac{s}{2}}\Gamma\left(\frac{s}{2}\right)\zeta(s) = \int_0^\infty x^{\frac{s}{2}-1}w(x)\mathrm{d}x \tag{1}$$

for $w(x) = \sum_{n=1}^\infty e^{-n^2\pi x}$. The Poisson summation formula implies that the Jacobi theta function $\theta(x) = \sum_{n=-\infty}^\infty e^{-n^2\pi x}$ satisfies the functional equation

$$\theta\left(\frac{1}{x}\right) = \sqrt{x}\,\theta(x),$$

hence $2w(x)+1 = (2w(1/x)+1)/\sqrt{x}$. Equation (1) therefore yields, by first splitting the integral at $x = 1$ and substituting $1/x$ for x between 0 and 1,

$$\xi(s) = \int_1^\infty x^{\frac{s}{2}-1}w(x)\mathrm{d}x + \int_1^\infty x^{-\frac{s}{2}-1}w\left(\frac{1}{x}\right)\mathrm{d}x$$

$$= \int_1^\infty x^{\frac{s}{2}-1}w(x)\mathrm{d}x + \int_1^\infty x^{-\frac{s}{2}-1}\left(-\frac{1}{2} + \frac{\sqrt{x}}{2} + \sqrt{x}w(x)\right)\mathrm{d}x$$

$$\xi(s) = \frac{1}{s(s-1)} + \int_1^\infty \left(x^{-\frac{s}{2}-\frac{1}{2}} + x^{\frac{s}{2}-1}\right)w(x)\mathrm{d}x,$$

still for $\Re(s) > 1$. The right-hand side is properly defined on $\mathbb{C} - \{0,1\}$ (because $w(x) = O(e^{-\pi x})$ as $x \to \infty$) and invariant under the substitution $s \to 1 - s$: the expected result follows. □

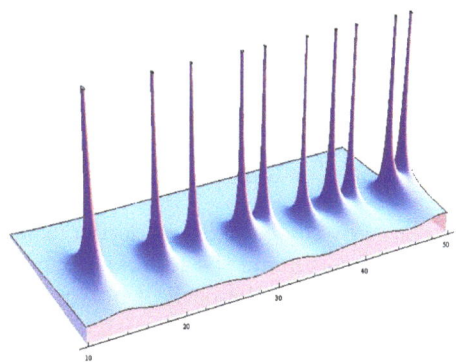

FIGURE 1. The first ζ zeros: $1/|\zeta|$ in the domain $-2 < \sigma < 2$, $10 < t < 50$.

From the above theorem, the zeta function admits trivial zeros at $s = -2$, $-4, -6, \ldots$ corresponding to the poles of $\Gamma(s/2)$. All non-trivial zeros are confined in the critical strip $0 \leq \sigma \leq 1$, and they are symmetrically positioned about the real axis and the critical line $\sigma = 1/2$. The Riemann hypothesis asserts that they all lie on this line.

One can define the argument of $\zeta(s)$ continuously along the line segments from 2 to $2+it$ to $1/2+it$. Then the number of such zeros ρ counted with multiplicities in $0 < \Im(\rho) < t$ is asymptotically (as shown by a calculus of residues)

$$\mathcal{N}(t) = \frac{t}{2\pi} \log \frac{t}{2\pi e} + \frac{1}{\pi} \arg \zeta \left(\frac{1}{2} + it \right) + \frac{7}{8} + \mathrm{O}\left(\frac{1}{t}\right). \tag{2}$$

In particular, the mean spacing between ζ zeros at height t is $2\pi/\log|t|$.

The fact that there are no zeros on $\sigma = 1$ led to the proof of the prime number theorem, which states that

$$\pi(x) \underset{x \to \infty}{\sim} \frac{x}{\log x}, \tag{3}$$

where $\pi(x) = |\mathcal{P} \cap [\![1, x]\!]|$. The proof makes use of the Van Mangoldt function, $\Lambda(n) = \log p$ if n is a power of a prime p, 0 otherwise: writing $\psi(x) = \sum_{n \leq x} \Lambda(n)$, (3) is equivalent to $\lim_{x \to \infty} \psi(x)/x = 1$, because obviously $\psi(x) \leq \pi(x) \log x$ and, for any $\varepsilon > 0$, $\psi(x) \geq \sum_{x^{1-\varepsilon} \leq p \leq x} \log p \geq (1-\varepsilon)(\log x)(\pi(x) + \mathrm{O}(x^{1-\varepsilon}))$. Differentiating the Euler product for ζ, if $\Re(s) > 1$,

$$-\frac{\zeta'}{\zeta}(s) = \sum_{n \geq 1} \frac{\Lambda(n)}{n^s},$$

which allows one to transfer the problem of the asymptotics of ψ to analytic properties of ζ: for $c > 0$, by a residues argument

$$\frac{1}{2\pi i} \int_{c-i\infty}^{c+i\infty} \frac{y^s}{s} \, ds = 0 \text{ if } 0 < y < 1, \; 1 \text{ if } y > 1,$$

hence

$$\begin{aligned}\psi(x) &= \sum_{n=2}^{\infty} \Lambda(n) \frac{1}{2\pi i} \int_{c-i\infty}^{c+i\infty} \frac{(x/n)^s}{s} \, ds \\ &= \frac{1}{2\pi i} \int_{c-it}^{c+it} \left(-\frac{\zeta'}{\zeta}(s)\right) ds \frac{x^s}{s} ds + \mathrm{O}\left(\frac{x \log^2 x}{t}\right),\end{aligned} \tag{4}$$

where the error term, created by the bounds restriction, is made explicit and small by the choice $c = 1 + 1/\log x$. Assuming the Riemann hypothesis for the moment, for any $1/2 < \sigma < 1$, one can change the path of integration from $c-it, c+it$ to $\sigma + it, \sigma - it$ by just crossing the pole at $s = 1$, with residue x:

$$\psi(x) = x + \frac{1}{2\pi i} \int_{\sigma-it}^{\sigma+it} \left(-\frac{\zeta'}{\zeta}(s)\right) ds \frac{x^s}{s} ds + \mathrm{O}\left(\frac{x \log^2 x}{t}\right).$$

Independently, still under the Riemann hypothesis, one can show the bound $\zeta'/\zeta(\sigma + it) = \mathrm{O}(\log t)$, implying $\psi(x) = x + \mathrm{O}(x^\theta)$ for any $\theta > 1/2$, by choosing $t = x$. What if we do not assume the Riemann hypothesis? The above reasoning can be reproduced, giving a worse error bound, provided that the integration path on the right-hand side of (4) can be changed crossing only one pole and making ζ'/ζ small by approaching sufficiently the critical axis. This is essentially what was proved independently in 1896 by Hadamard and La Vallée Poussin, who showed that $\zeta(\sigma+it)$ cannot be zero for $\sigma > 1-c/\log t$, for some $c > 0$. This finally yields

$$\pi(x) = \mathrm{Li}(x) + \mathrm{O}\left(xe^{-c\sqrt{\log x}}\right), \quad \text{where } \mathrm{Li}(x) = \int_0^x \frac{ds}{\log s},$$

while the Riemann hypothesis would imply $\pi(x) = \mathrm{Li}(x) + \mathrm{O}(\sqrt{x}\log x)$. It would also have consequences for the extreme size of the zeta function (the Lindelöf hypothesis): for any $\varepsilon > 0$

$$\zeta(1/2 + it) = \mathrm{O}(t^{\varepsilon}),$$

which is equivalent to bounds on moments of ζ discussed in Section 3.

1.2. The explicit formula

We now consider the first analogy between the zeta zeros and spectral properties of operators, by looking at linear statistics. Namely, we state and give key ideas underlying the proofs of Weil's explicit formula concerning the ζ zeros and Selberg's trace formula for the Laplacian on surfaces with constant negative curvature.

First consider the Riemann zeta function. For a function $f\colon (0,\infty) \to \mathbb{C}$ consider its Mellin transform $F(s) = \int_0^\infty f(s)x^{s-1}dx$. Then the inversion formula (where σ is chosen in the fundamental strip, i.e., where the image function F converges)

$$f(x) = \frac{1}{2\pi \mathrm{i}} \int_{\sigma-\mathrm{i}\infty}^{\sigma+\mathrm{i}\infty} F(s)x^{-s}ds$$

holds under suitable smoothness assumptions, in a similar way as the inverse Fourier transform. Hence, for example,

$$\sum_{n=2}^\infty \Lambda(n)f(n) = \sum_{n=2}^\infty \Lambda(n)\frac{1}{2\pi \mathrm{i}} \int_{2-\mathrm{i}\infty}^{2+\mathrm{i}\infty} F(s)n^{-s}ds = \frac{1}{2\pi \mathrm{i}} \int_{2-\mathrm{i}\infty}^{2+\mathrm{i}\infty} \left(-\frac{\zeta'}{\zeta}\right)(s)F(s)ds.$$

Changing the line of integration from $\Re(s) = 2$ to $\Re(s) = -\infty$, all trivial and non-trivial poles (as well as $s = 1$) are crossed, leading to the following explicit formula by Weil.

Theorem 1.2. *Suppose that f is $\mathscr{C}^2$ on $(0,\infty)$ and compactly supported. Then*

$$\sum_\rho F(\rho) + \sum_{n\geq 0} F(-2n) = F(1) + \sum_{p\in\mathcal{P}, m\in\mathbb{N}} (\log p)\, f(p^m),$$

where the first sum the first sum is over non-trivial zeros counted with multiplicities.

When replacing the Mellin transform by the Fourier transform, the Weil explicit formula takes the following form, for an even function h, analytic on $|\Im(z)| < 1/2 + \delta$, bounded, and decreasing as $h(z) = \mathrm{O}(|z|^{-2-\delta})$ for some $\delta > 0$.

Here, the sum is over all γ_n's such that $1/2 + i\gamma_n$ is a non-trivial zero, and $\hat{h}(x) = \frac{1}{2\pi}\int_{-\infty}^{\infty} h(y) e^{-ixy} dy$:

$$\sum_{\gamma_n} h(\gamma_n) - 2h\left(\frac{i}{2}\right) = \frac{1}{2\pi}\int_{\mathbb{R}} h(r) \left(\frac{\Gamma'}{\Gamma}\left(\frac{1}{4} + \frac{i}{2}r\right) - \log \pi\right) dr \qquad (5)$$
$$- 2 \sum_{p \in \mathcal{P}, m \in \mathbb{N}} \frac{\log p}{p^{m/2}} \hat{h}(m \log p).$$

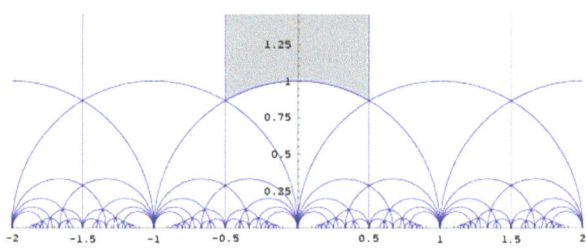

FIGURE 2. Geodesics and a fundamental domain (for the modular group) in the hyperbolic plane.

A formally similar relation holds in a different context, through Selberg's trace formula. In one of its simplest manifestations, this can be stated as follows. Let $\Gamma \backslash \mathbb{H}$ be an hyperbolic surface, where Γ is a subgroup of $\mathrm{PSL}_2(\mathbb{R})$, orientation-preserving isometries of the hyperbolic plane $\mathbb{H} = \{x+iy, y > 0\}$, the Poincaré half-plane with metric

$$d\mu = \frac{dxdy}{y^2}. \qquad (6)$$

The Laplace–Beltrami operator $\Delta = y^2(\partial_{xx} + \partial_{yy})$ is self-adjoint with respect to the invariant measure (6), i.e., $\int v(\Delta u)d\mu = \int (\Delta v)u d\mu$, so all eigenvalues of $-\Delta$ are real and positive. If $\Gamma \backslash \mathbb{H}$ is compact, the spectrum of $-\Delta$ restricted to a fundamental domain $\mathcal{D}$ of representatives of the conjugation classes is discrete, $0 = \lambda_0 < \lambda_1 < \cdots$ with associated eigenfunctions $u_1, u_2, \ldots$:

$$\begin{cases} (\Delta + \lambda_n)u_n = 0, \\ u_n(\gamma z) = u_n(z) \text{ for all } \gamma \in \Gamma, z \in \mathbb{H}. \end{cases}$$

To state Selberg's trace formula, we need, as previously, a function h analytic on $|\Im(z)| < 1/2 + \delta$, even, bounded, and decreasing as $h(z) = O(|z|^{-2-\delta})$, for some $\delta > 0$.

Theorem 1.3. *Under the above hypotheses, setting $\lambda_k = s_k(1-s_k)$, $s_k = 1/2 + ir_k$, then*

$$\sum_{k=0}^{\infty} h(r_k) = \frac{\mu(\mathcal{D})}{2\pi} \int_{-\infty}^{\infty} rh(r)\tanh(\pi r)dr + \sum_{p \in \mathcal{P}, m \in \mathbb{N}^*} \frac{\ell(p)}{2\sinh\left(\frac{m\ell(p)}{2}\right)} \hat{h}(m\ell(p)), \quad (7)$$

where $\hat{h}$ is the Fourier transform of h ($\hat{h}(x) = \frac{1}{2\pi}\int_{-\infty}^{\infty} h(y)e^{-ixy}dy$), $\mathcal{P}$ is now the set of all primitive[1] periodic orbits[2] and ℓ is the geodesic distance for the metric (6).

Sketch of proof. It is a general fact that the eigenvalue density function $d(\lambda)$ is linked to the Green function associated to λ (($\Delta + \lambda)G^{(\lambda)}(z, z') = \delta_{z-z'}$, where δ is the Dirac distribution at 0) through

$$d(\lambda) = -\frac{1}{\pi} \int_{\mathcal{D}} \Im\left(G^{(\lambda)}(z, z)\right) d\mu.$$

To calculate $G^{(\lambda)}$, we need to sum the Green function associated to the whole Poincaré half-plane over the images of z by elements of Γ (in the same way as the transition probability from z' to z is the sum of all transition probabilities to images of z):

$$G^{(\lambda)}(z, z') = \sum_{\gamma \in \Gamma} G_{\mathbb{H}}^{(\lambda)}(\gamma(z), z').$$

Thanks to the numerous isometries of $\mathbb{H}$, the geodesic distance for the Poincaré plane is well known. This yields an explicit form of the Green function, leading to

$$d(\lambda) = \frac{1}{2\sqrt{2}\pi^2} \sum_{\gamma \in \Gamma} \int d\mu(z) \int_{\ell(z,\gamma(z))}^{\infty} \frac{\sin(rs)}{\sqrt{\cosh s - \cosh \ell(z, \gamma(z))}} ds,$$

with $\lambda = 1/4 + r^2$. The mean density of states corresponds to $\gamma = \text{Id}$ and an explicit calculation yields

$$\langle d(\lambda) \rangle = \frac{\mu(\mathcal{D})}{4\pi} \tanh(\pi r).$$

It is not clear at this point how the primitive periodic orbits appear from the elements in Γ. The sum over group elements γ can be written as a sum over conjugacy classes $\bar{\gamma}$. This gives

$$d(\lambda) = \langle d(\lambda) \rangle + \sum_{\bar{\gamma}} d_{\bar{\gamma}}(\lambda)$$

where

$$d_{\bar{\gamma}}(\lambda) = \frac{1}{2\sqrt{2}\pi^2} \int_{\text{FD}(\bar{\gamma})} d\mu(z) \int_{\ell(z,\gamma(z))}^{\infty} \frac{\sin(rs)}{\sqrt{\cosh s - \cosh \ell(z, \gamma(z))}} ds,$$

where $\text{FD}(\bar{\gamma})$ is the fundamental domain associated to the subgroup S_γ of elements commuting with γ (independent of the representant of the conjugacy class). The subgroup S_γ is generated by an element γ_0:

$$S_\gamma = \{\gamma_0^m, m \in \mathbb{Z}\}.$$

---

[1] i.e., not the repetition of shorter periodic orbits.
[2] of the geodesic flow on $\Gamma\backslash\mathbb{H}$.

Then an explicit (but somewhat tedious) calculation gives

$$d_{\overline{\gamma}}(\lambda) = \frac{\ell^{(0)}(p)}{4\pi \sinh\left(\frac{s\ell(p)}{2}\right)} \cos(s\ell(p)),$$

where $\ell^{(0)}(p)$ (resp. $\ell(p)$) is defined by $2\cosh \ell^{(0)}(p) = \operatorname{Tr}(\gamma_0)$ (resp. $2\cosh \ell(p) = \operatorname{Tr}(\gamma)$). Independent calculation shows that $\ell^{(0)}(p)$ (resp. $\ell(p)$) is also the length between z and $\gamma_0(z)$ (resp. z and $\gamma(z)$). Hence they are the lengths of the unique (up to conjugation) periodic orbits associated to γ_0 (resp. γ). The above proof sketch can be made rigorous by integrating d with respect to a suitable test function h. □

The similarity between the explicit formulas (5) and (7) suggests that prime numbers may correspond to primitive orbits, with lengths $\log p, p \in \mathcal{P}$. This analogy remains when counting primes and primitive orbits. Indeed, as a consequence of Selberg's trace formula, the number of primitive orbits with length less than x is

$$|\{\ell(p) < x\}| \underset{x \to \infty}{\sim} \frac{e^x}{x},$$

and following the prime number theorem (3),

$$|\{\log(p) < x\}| \underset{x \to \infty}{\sim} \frac{e^x}{x}.$$

These connections are reviewed at much greater length in [5].

Finally, note that the signs of the oscillating parts are different between equation (5) and (7). One explanation suggested by Connes [14] is that the ζ zeros may not be in the spectrum of an operator but in its absorption: for an Hermitian operator with continuous spectrum along the whole real axis, they would be exactly the missing points where the eigenfunctions vanish.

1.3. Basic theory of random matrices eigenvalues

As we will see in the next section, the correlations between ζ zeros show striking similarities with those known to exist between the eigenvalues of random matrices. This adds further weight to the idea that there may be a spectral interpretation of the zeros and provides another link with the theory of quantum chaotic systems. We need first to introduce the matrices we will consider. These have the property that their spectrum has an explicit joint distribution, exhibiting a two-point repulsive interaction, like fermions. Importantly, these correlations have a *determinantal* structure.

If $\chi = \sum_i \delta_{X_i}$ is a simple point process on a complete separate metric space Λ, consider the point process

$$\Xi^{(k)} = \sum_{X_{i_1},\ldots,X_{i_k} \text{ all distinct}} \delta_{(X_{i_1},\ldots,X_{i_k})} \qquad (8)$$

on Λ^k. One can define in this way a measure $M^{(k)}$ on Λ^k by $M^{(k)}(\mathcal{A}) = \mathbb{E}\left(\Xi^{(k)}(\mathcal{A})\right)$ for any Borel set $\mathcal{A}$ in Λ^k. Most of the time, there is a natural measure λ on Λ, in our cases $\Lambda = \mathbb{R}$ or $(0, 2\pi)$ and λ is the Lebesgue measure. If $M^{(k)}$ is absolutely continuous with respect to λ^k, there exists a function ρ_k on Λ^k such that for any Borel sets $B_1, \ldots, B_k$ in Λ

$$M^{(k)}(B_1 \times \cdots \times B_k) = \int_{B_1 \times \cdots \times B_k} \rho_k(x_1, \ldots, x_k) \mathrm{d}\lambda(x_1) \ldots \mathrm{d}\lambda(x_k).$$

Hence one can think about $\rho_k(x_1, \ldots, x_k)$ as the asymptotic (normalized) probability of having exactly one particle in neighborhoods of the x_k's. More precisely under suitable smoothness assumptions, and for distinct points $x_1, \ldots, x_k$ in $\Lambda = \mathbb{R}$,

$$\rho_k(x_1, \ldots, x_k) = \lim_{\varepsilon \to 0} \frac{\mathbb{P}(\chi(x_i, x_i + \varepsilon) = 1, 1 \leq i \leq k)}{\prod_{j=1}^k \lambda(x_j, x_j + \varepsilon)}.$$

This is called the kth-order correlation function of the point process. Note that ρ_k is not a probability density. If χ consists almost surely of n points, it satisfies the integration property

$$(n-k)\rho_k(x_1, \ldots, x_k) = \int_\Lambda \rho_{k+1}(x_1, \ldots, x_{k+1}) \mathrm{d}\lambda(x_{k+1}). \qquad (9)$$

A particularly interesting class of point processes is the following.

Definition 1.4. If there exists a function $K : \mathbb{R} \times \mathbb{R} \to \mathbb{R}$ such that for all $k \geq 1$ and $(z_1, \ldots, z_k) \in \mathbb{R}^k$

$$\rho_k(z_1, \ldots, z_k) = \det \left(K(z_i, z_j)_{i,j=1}^k \right)$$

then χ is said to be a determinantal point process with respect to the underlying mesure λ and correlation kernel K.

The determinantal condition for *all* correlation functions looks very restrictive, but it is not: for example, if the joint density of all n particles can be written as a Vandermonde-type determinant, then so can the lower-order correlation functions, as shown by the following argument, standard in Random Matrix Theory. It shows that for a Coulomb gas at a specific temperature $(1/2)$ in dimension 1 or 2, all correlations functions are explicit, a very noteworthy feature.

Proposition 1.5. *Let* $\mathrm{d}\lambda$ *be a probability measure on* $\mathbb{C}$ *(eventually concentrated on a line) such that for the Hermitian product*

$$(f, g) \mapsto \langle f, g \rangle = \int f \bar{g} \mathrm{d}\lambda$$

polynomial moments are defined till order at least $n - 1$. *Consider the probability distribution with density*

$$F(x_1, \ldots, x_n) = c(n) \prod_{k<l} |x_l - x_k|^2$$

with respect to $\prod_{j=1}^n \mathrm{d}\lambda(x_j)$, *where* $c(n)$ *is the normalization constant. For this joint distribution,* $\{x_1, \ldots, x_n\}$ *is a determinantal point process with explicit kernel.*

Proof. Let P_k ($0 \leq k \leq n-1$) be monic polynomials with degree k. Thanks to Vandermonde's formula and the multilinearity of the determinant

$$\prod_{k<l}(x_l - x_k) = \sqrt{\prod_{k=0}^{n-1} \|P_k\|_{L^2(\lambda)}} \det\left(\frac{P_k(x_j)}{\|P_k\|_{L^2(\lambda)}}\right)_{k,j=1}^n.$$

Multiplying this identity with itself and using $\det(AB) = \det A \det B$ gives

$$F(x_1, \ldots, x_n) = \det\left(K_n(x_j, x_k)_{j,k=1}^n\right)$$

with $K_n(x,y) = c \sum_{k=0}^{n-1} \frac{P_k(x)\overline{P_k(y)}}{\|P_k\|_{L^2(\lambda)}^2}$, the constant c depending on λ, n and the P_i's. This shows that the correlation ρ_n has the desired determinantal form. The following lemma by Gaudin (see [38]) together with the integration property (9) shows that if the polynomials P_k's are orthogonal in $L^2(\lambda)$, then

$$\rho_l(x_1, \ldots, x_l) = \det\left(K_n(x_j, x_k)_{j,k=1}^l\right)$$

for all $1 \leq l \leq n$. The probability density condition $\int_\mathbb{R} \rho_1(x) d\lambda(x) = n$ implies $c = 1$, and finally the Christoffel–Darboux formula for orthogonal polynomials gives

$$K_n(x,y) = \sum_{k=0}^{n-1} \frac{P_k(x)P_k(y)}{\|P_k\|_{L^2(f)}^2} = \frac{1}{\|P_n\|_{L^2(f)}^2} \frac{P_n(x)\overline{P_{n-1}(y)} - P_{n-1}(x)P_n(y)}{x-y}, \quad (10)$$

which concludes the proof. □

Lemma 1.6. *Suppose that the function K satisfies, for some measurable set I, the semigroup relation $\int_I K(x,y)K(y,z)d\lambda(y) = K(x,z)$ for all x and z in I, and denote $n = \int_I K(x,x)d\lambda(x)$. Then for all k,*

$$\int_{(k+1)\times(k+1)} \det K(x_i, x_j) d\lambda(x_{k+1}) = (n-k) \det_{k\times k} K(x_i, x_j).$$

We apply the above discussion to the following examples, which are among the most studied random matrices. The first involves Hermitian matrices with Gaussian entries; the second relates to a compact group: uniformly distributed unitary matrices. Their spectrum is a determinantal point process, which implies a *repulsion* between the eigenvalues, similar to that between fermions. We illustrate this in the previous figure with one example of determinantal statistics (eigenvalues of a Haar-distributed unitary matrix, outer circle) together with, for comparison, Poisson distributed points (uniform independent points, inner circle), in dimension $n = 30$.

First, consider the so-called Gaussian unitary ensemble (GUE). This is the ensemble of random $n \times n$ Hermitian matrices with independent (up to symmetry) Gaussian entries: $M_{ij}^{(n)} = \overline{M_{ji}^{(n)}} = \frac{1}{\sqrt{n}}(X_{ij} + iY_{ij}), 1 \leq i < j \leq n$, where the X_{ij}'s and Y_{ij}'s are independent centered real Gaussians entries with mean 0 and variance $1/2$ and $M_{ii}^{(n)} = X_{ii}/\sqrt{n}$ with X_{ii} real centered Gaussians with variance 1, still independent. For this ensemble, the distribution of the eigenvalues has an explicit density

$$\frac{1}{Z_n} e^{-n \sum_{i=1}^{n} \lambda_i^2 / 2} \prod_{1 \leq i < j \leq n} |\lambda_i - \lambda_j|^2 \tag{11}$$

with respect to the Lebesgue measure. We denote by (h_n) the Hermite polynomials, more precisely the successive monic polynomials orthogonal with respect to the Gaussian weight $e^{-x^2/2}dx$, and the associated normalized functions

$$\psi_k(x) = \frac{e^{-x^2/4}}{\sqrt{\sqrt{2\pi}k!}} h_k(x).$$

Then from (10), the set of point $\{\lambda_1, \ldots, \lambda_n\}$ with law (11) is a determinantal point process with kernel (with respect to the Lebesgue measure on $\mathbb{R}$) given by

$$K^{\text{GUE}(n)}(x, y) = n \frac{\psi_n(x\sqrt{n})\psi_{n-1}(y\sqrt{n}) - \psi_{n-1}(x\sqrt{n})\psi_n(y\sqrt{n})}{x - y},$$

defined by continuity when $x = y$. The Plancherel–Rotach asymptotics for the Hermite polynomials implies that, as $n \to \infty$, $K^{\text{GUE}(n)}(x, x)/n$ has a non-trivial limit. More precisely, the empirical spectral distribution $\frac{1}{n} \sum \delta_{\lambda_i}$ converges in probability to the semicircle law (see, e.g., [1]) with density

$$\rho_{sc}(x) = \frac{1}{2\pi} \sqrt{(4 - x^2)_+}$$

with respect to the Lebesgue measure. This is the asymptotic behavior of the spectrum in the macroscopic regime. The microscopic interactions between eigenvalues also can be evaluated thanks to asymptotics of the Hermite orthogonal polynomials: for any $x \in (-2, 2)$, $u \in \mathbb{R}$,

$$\frac{1}{n\rho_{sc}(x)} K^{\text{GUE}(n)} \left(x, x + \frac{u}{n\rho_{sc}(x)} \right) \xrightarrow[n \to \infty]{} K(u) = \frac{\sin(\pi u)}{\pi u}, \tag{12}$$

leading to a repulsive correlation structure for the eigenvalues at the scale of the average gap: for example the two-point correlation function asymptotics are

$$\left(\frac{1}{n\rho_{sc}(x)} \right)^2 \rho_2^{\text{GUE}(n)} \left(x, x + \frac{u}{n\rho_{sc}(x)} \right) \xrightarrow[n \to \infty]{} r_2(u) = 1 - \left(\frac{\sin(\pi u)}{\pi u} \right)^2, \tag{13}$$

which vanishes at $u = 0$, while for independent points the asymptotic two points correlation function would be identically 1. A remarkable fact about the above sine kernel is that it appears universally in the limiting correlation functions of random Hermitian matrices with independent (up to symmetry) entries (the so-called Wigner ensemble). This was proved under very weak conditions on the entries in independent and complementary works by Tao-Vu and Erdös et al. (see [23]). In their result, a Wigner matrix is like a matrix from the GUE from the point of view of the variance normalization but with no Gaussianity condition. We just assume that the entries X_{ij}'s and Y_{ij}'s have a subexponential decay: for some constants c and c',

$$\mathbb{P}(|X_{ij}| \leq t^c) \leq e^{-t}, \ \mathbb{P}(|Y_{ij}| \leq t^c) \leq e^{-t}, \ t > c'.$$

Theorem 1.7. *Under the above hypothesis, denoting by $\rho_k^{\mathrm{Wig}(n)}$ the correlation functions associated with the eigenvalues of the Wigner matrix, for any u, ε such that $[u - \varepsilon, u + \varepsilon] \subset (-2, 2)$, and any continuous compactly supported $f : \mathbb{R}^k \to \mathbb{R}$*

$$\frac{1}{2\varepsilon} \int_{x-\varepsilon}^{x+\varepsilon} \int_{\mathbb{R}^k} \frac{f(u_1, \ldots, u_k)}{\rho_{sc}(x')^k} \rho_k^{\mathrm{Wig}(n)}$$
$$\times \left(x' + \frac{u_1}{n\rho_{sc}(x')}, \ldots, x' + \frac{u_k}{n\rho_{sc}(x')} \right) \mathrm{d}u_1 \ldots \mathrm{d}u_k \mathrm{d}x'$$

converges as $n \to \infty$ to (K is the above-mentioned sine kernel)

$$\int_{\mathbb{R}^k} f(u_1, \ldots, u_k) \det_{k \times k} (K(u_i - u_j)) \, \mathrm{d}u_1 \ldots \mathrm{d}u_k.$$

Note that both works leading to the above result, through very different, proceed by comparison with the explicit GUE asymptotics.

The second classical example of a matrix-related determinantal point process is that of the eigenvalues of uniformly distributed unitary matrices. For $u_n \sim \mu_{\mathrm{U}(n)}$ ($\mu_{\mathrm{U}(n)}$ is the Haar measure[3] on the unitary group) the density of the eigenangles $0 \leq \theta_1 < \cdots < \theta_n < 2\pi$, with respect to the Lebesgue measure on the corresponding simplex is

$$\frac{1}{(2\pi)^n} \prod_{j<k} |e^{i\theta_j} - e^{i\theta_k}|^2.$$

In this case, the polynomials in the recipe from Proposition 1.5 are those orthogonal with respect to the Hermitian product on the unit circle $\int_{\mathscr{C}} \bar{p} q \mathrm{d}z$. These are the monomials (X^k). Consequently, the correlation functions $\rho_k^{\mathrm{U}(n)}$, $1 \leq k \leq n$, are

---

[3]i.e., the unique left (or right) translation invariant measure on the compact group $\mathrm{U}(n)$: if x has distribution $\mu_{\mathrm{U}(n)}$ then for any fixed $a \in \mathrm{U}(n)$

$$ax \stackrel{\mathrm{law}}{=} x.$$

determinants based on the same kernel:
$$\rho_k^{U(n)}(\theta_1,\ldots,\theta_n) = \det_{k\times k}\left(K^{U(n)}(\theta_i - \theta_j)\right), \quad K^{U(n)}(\theta) = \frac{1}{2\pi}\frac{\sin(n\theta/2)}{\sin(\theta/2)}.$$

In an easier way than for the GUE, the limiting sine kernel again appears
$$\frac{2\pi}{n}K^{U(n)}\left(\frac{2\pi\theta}{n}\right) \xrightarrow[n\to\infty]{} K(\theta).$$

This microscopic description of the fermionic aspect of the eigenvalues also appears in a number theoretic context, as considered in the next section.

1.4. Repulsion of the zeta zeros

Being simply the Fourier transform of an interval, the sine kernel (12) appears in many different contexts in mathematics and physics. However it was a striking result when Montgomery discovered in the early 70s that it describes the pair correlation of the zeta zeros. During tea time in Princeton he mentioned his result to Dyson, who immediately recognized the limiting pair correlation r_2 for eigenvalues of the GUE, (13). This unexpected result gave new insight into the Hilbert–Pólya suggestion that the zeta zeros might linked to eigenvalues of a self-adjoint operator acting on a Hilbert space.

The Random-Matrix connection was tested numerically by Odlyzko [41] and found to provide a remarkably accurate model of the data. For example, supposing that all orders correlations of the ζ zeros coincide with determinants of the sine kernel, one expects that the histogram of the normalized spacings between

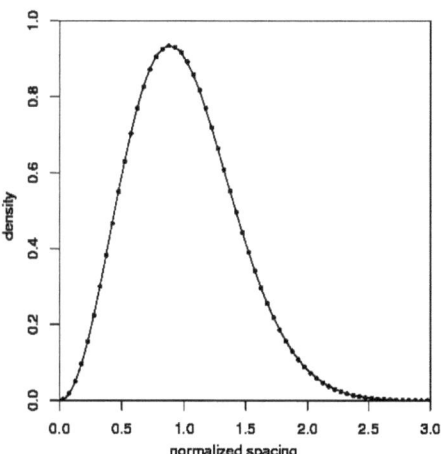

FIGURE 3. The distribution function of asymptotic gaps between eigenvalues ($\partial_s \det(\mathrm{Id} - K_{(0,s)})$) compared with the histogram of gaps between normalized ζ zeros, based on a billion (i.e., more precisely, 1,041,417,090) zeros of the zeta function, starting with zero number $10^{23} + 17{,}368{,}588{,}794$. Courtesy of © 2013 A. Odlyzko.

zeros converge to the distribution function of the same asymptotics related to the eigenvalues of the GUE or unitary group.

More precisely, we write as previously $1/2 \pm i\gamma_n$ for the zeta zeros counted with multiplicity, assume the Riemann hypothesis and the order $\gamma_1 \leq \gamma_2 \leq \cdots$ Let $\omega_n = \frac{\gamma_n}{2\pi} \log \frac{\gamma_n}{2\pi}$. From (2) we know that $\delta_n = \omega_{n+1} - \omega_n$ has a mean value 1 as $n \to \infty$, and its repartition function is expected to converge to

$$\frac{1}{n}|\{k \leq n : \delta_k < s\}| \xrightarrow[n \to \infty]{} -\partial_s \det(\mathrm{Id} - K_{(0,s)}),$$

where $K_{(0,s)}$ is the convolution operator acting on $\mathrm{L}^2(0,s)$ with kernel K. This comes from the inclusion-exclusion principle linking free intervals and correlation functions, in which the determinantal structure leads to a Fredholm determinant (see, e.g., [1]).

What exactly did Montgomery prove? Rather than mean spacings, a more precise understanding of the zeta zeros interactions relies on the study, as $t \to \infty$, of the spacings distribution function

$$\frac{1}{\mathcal{N}(t)}|\{(n,m) \in [\![1, \mathcal{N}(t)]\!]^2 : \alpha < \omega_n - \omega_m < \beta, n \neq m\}|,$$

where $\mathcal{N}(t)$ is the number of zeros till height t, and more generally the operator

$$r_2(f,t) = \frac{1}{\mathcal{N}(t)} \sum_{1 \leq j,k \leq \mathcal{N}(t), j \neq k} f(\omega_j - \omega_k).$$

If the ω_k's were asymptotically independently distributed (up to ordering), $r_2(f,x)$ would converge to $\int_\mathbb{R} f(y) dy$ as $x \to \infty$. That this is not the case follows from an important theorem due to Montgomery [39]:

Theorem 1.8. *Assume the Riemann hypothesis. Suppose f is a test function with the following property: its Fourier transform[4] is $\mathscr{C}^\infty$ and supported in $(-1,1)$. Then*

$$r_2(f,t) \xrightarrow[t \to \infty]{} \int_\mathbb{R} f(y) r_2(y) dy,$$

where $r_2(y) = 1 - \left(\frac{\sin(\pi y)}{\pi y}\right)^2$, as for Wigner or unitary random matrices.

An important conjecture due to Montgomery asserts that the above result holds with no condition on the support of the Fourier transform, but weakening the restriction even to $\mathrm{supp}\, \hat{f} \subset (-1-\varepsilon, 1+\varepsilon)$ for some $\varepsilon > 0$ seems out of reach with known techniques. The Montgomery conjecture would have important consequences for example in terms of second moments of primes in short intervals [25].

Sketch of proof of Montgomery's Theorem. Consider the function

$$F(\alpha, t) = \frac{1}{\frac{t}{2\pi} \log t} \sum_{0 \leq \gamma, \gamma' \leq t} t^{i\alpha(\gamma - \gamma')} \frac{4}{4 + (\gamma - \gamma')^2}.$$

---

[4]Contrary to the Weil and Selberg formulas (5) and (7), the chosen normalization here is $\hat{f}(x) = \int_{-\infty}^\infty f(y) e^{-i2\pi xy} dy$.

This is the Fourier transform of the normalized spacings, up to the factor $4/(4+(\gamma-\gamma')^2)$. This function naturally appears when counting the second-order moments

$$\int_0^t |G(s,t^\alpha)|^2 ds = F(\alpha,t)t\log t + O(\log^3 t), \quad G(s,x) = 2\sum_\gamma \frac{x^{i\gamma}}{1+(s-\gamma)^2}. \quad (14)$$

As G is a linear functional of the zeros, it can be written as a sum over primes by an appropriate explicit formula[5] like (5):

$$G(s,x) = -\sqrt{x}\left(\sum_{n\leq x} \Lambda(n)\left(\frac{x}{n}\right)^{-\frac{1}{2}+is} + \sum_{n>x}\Lambda(n)\left(\frac{x}{n}\right)^{\frac{3}{2}+is}\right)$$

$$+ x^{-1+is}(\log(|s|+2) + O(1)) + O\left(\frac{\sqrt{x}}{|s|+2}\right),$$

a fundamental formula due to Montgomery, which requires the Riemann hypothesis to yield the error term quoted. The moment (14) can therefore be expanded as a sum over primes, and the Montgomery–Vaughan inequality (Theorem 1.9) leads to

$$\int_0^t |G(s,t^\alpha)|^2 ds = (t^{-2\alpha}\log t + \alpha + o(1))t\log t.$$

These asymptotics can be proved by the Montgomery–Vaughan inequality, but only in the range $\alpha \in (0,1)$, which explains the support restriction in the hypotheses. Gathering both asymptotic expressions for the second moment of G yields $F(\alpha,t) = t^{-2\alpha}\log t + \alpha + o(1)$. Finally, by the Fourier inverse formula,

$$\frac{1}{\frac{t}{2\pi}\log t}\sum_{0\leq \gamma,\gamma'\leq t} f\left((\gamma-\gamma')\frac{\log t}{2\pi}\right)\frac{4}{4+(\gamma-\gamma')^2} = \int_\mathbb{R} F(\alpha,t)\hat{f}(\alpha)d\alpha.$$

If $\operatorname{supp}\hat{f} \subset (-1,1)$, this is approximately

$$\int_\mathbb{R} \hat{f}(\alpha)(t^{-2|\alpha|}+|\alpha|)d\alpha = \int_\mathbb{R} e^{-2|\alpha|}\hat{f}(\alpha/\log t)d\alpha + \int_\mathbb{R}|\alpha|\hat{f}(\alpha)d\alpha$$

$$= \hat{f}(0) + f(0) - \int_\mathbb{R}(1-|\alpha|)\hat{f}(\alpha)d\alpha + o(1) = f(0) + \int_\mathbb{R} f(x)\left(1-\left(\frac{\sin\pi x}{\pi x}\right)^2\right)dx + o(1),$$

by the Plancherel formula. □

Theorem 1.9. *Let (a_r) be complex numbers, (λ_r) distinct real numbers and $\delta_r = \min_{s\neq r}|\lambda_r - \lambda_s|$. Then*

$$\frac{1}{t}\int_0^t \left|\sum_r a_r e^{i\lambda_r s}\right|^2 ds = \sum_r |a_r|^2\left(1 + \frac{3\pi\theta}{t\delta_r}\right)$$

---

[5]The factor $4/(4+(\gamma-\gamma')^2)$ gives convergence properties necessary to the explicit formula.

for some $|\theta| < 1$. In particular,

$$\int_0^t \left|\sum_{n=1}^\infty \frac{a_n}{n^{is}}\right|^2 ds = t \sum_{n=1}^\infty |a_n|^2 + \mathrm{O}\left(\sum n|a_n|^2\right).$$

Montgomery's result has been extended in the following directions. Hejhal [31] proved that the triple correlations of the zeta zeros coincide with those of large Haar-distributed unitary matrices, and Rudnick and Sarnak [44] then showed that all correlations agree. These results are all restricted by the condition that the Fourier transform of f is supported on some compact set. To state the Rudnick–Sarnak result, we note as in [44]:

- $\mathcal{E}_t = \{\omega_i : i \le \mathcal{N}(t)\}$;
- f is a translation invariant function from $\mathbb{R}^n$ to $\mathbb{R}$ ($f(x+t(1,\ldots,1)) = f(x)$), symmetric and rapidly decreasing[6] on $\sum_1^k x_i = 0$;
- $r_n(f,t) = \frac{n!}{\mathcal{N}(t)} \sum_{S \subset \mathcal{E}_t, |S|=n} f(S)$, generalizing the previous definition of $r_2(f,t)$.

Theorem 1.10. *Assume the Riemann hypothesis[7] and that the Fourier transform of f is supported in $\sum_1^n |\xi_j| < 2$. Then*

$$r_n(f,t) \xrightarrow[t\to\infty]{} \int_{\mathbb{R}^n} f(x) \det_{n\times n}\left(\frac{\sin \pi(x_i - x_j)}{\pi(x_i - x_j)}\right) \delta_{x_1+\cdots+x_n} dx_1 \ldots dx_n.$$

Sketch of proof. The method employed by Rudnik and Sarnak makes use of smoothed statistics, namely

$$c_n(f,t,h) = \sum_{j_1,\ldots,j_n} h\left(\frac{\gamma_{j_1}}{t}\right) \ldots h\left(\frac{\gamma_{j_n}}{t}\right) f\left(\frac{\log t}{2\pi}\gamma_{j_1},\ldots,\frac{\log t}{2\pi}\gamma_{j_n}\right),$$

not assuming here that the indexes are necessarily distinct. This allows the use of two important ingredients:

- a Fourier transform to convert the nth-order statistics to linear ones:

$$c_n(f,t,h) = \int_{\mathbb{R}^n} \prod_{k=1}^n \sum_{j_k} h\left(\frac{\gamma_{j_k}}{t}\right) t^{-i\gamma_{j_k}\xi_k} d\mu(\xi), \qquad (15)$$

where $d\mu(\xi) = \Phi(\xi) \delta_{\xi_1+\cdots+\xi_n} d\xi_1 \ldots d\xi_n$ is the Fourier transform of f;
- Weil's explicit formula (5), or a variant, to transfer linear statistics over zeros to linear statistics over primes:

$$\sum_\gamma h(\gamma) = h\left(\frac{i}{2}\right) + h\left(-\frac{i}{2}\right) + \frac{1}{2\pi}\int_\mathbb{R} h(r)\left(\frac{\Gamma'}{\Gamma}\left(\frac{1}{2}+ir\right) + \frac{\Gamma'}{\Gamma}\left(\frac{1}{2}-ir\right)\right) dr$$

$$- \sum_n \frac{\Lambda(n)}{\sqrt{n}} \hat{h}(\log n) + \frac{\Lambda(n)}{\sqrt{n}} \hat{h}(-\log n). \qquad (16)$$

---

[6] i.e., faster than $|x|^{-\lambda}$ for any $\lambda > 0$.
[7] An unconditional result holds with smoothed test functions.

Substituting (16) into (15) and expanding the product, we end up with a sum of terms like

$$c_{r,s}(t) = \sum_{\mathbf{n}} \frac{\Lambda(n_1)\ldots\Lambda(n_{r+s})}{\sqrt{n_1\ldots n_{r+s}}} t^n$$

$$\int_{\mathbb{R}^n} \prod_{j=1}^{r} \hat{h}(t((\log t)\xi_j + \log n_j)) \prod_{j=r+1}^{r+s} \hat{h}(t((\log t)\xi_j - \log n_j)) \prod_{j>r+s} \hat{h}(t(\log t)\xi_j) d\mu(\xi).$$

As $\sum |\xi_j| < 2$, one can use the Montgomery–Vaughan inequality Theorem 1.9 to get the correct asymptotics: in the above sum the main contribution comes from choices of $\mathbf{n}$ such that

$$n_1\ldots n_r = n_{r+1}\ldots n_{r+s}, \tag{17}$$

i.e., the diagonal elements. The von Mongoldt function being supported on prime powers, the main contribution comes from the choice of prime n_j's, which implies $r = s$ by (17). We are therefore led to the asymptotics of

$$c_{r,r}(t) = \frac{t}{2\pi \log^{2r-1} t} \int_{\mathbb{R}} h(r)^n dr \sum_{p_1,\ldots,p_r \ll t} \sum_{\sigma \in \mathscr{S}_r} \frac{\log^2 p_1 \ldots \log^2 p_r}{p_1 \ldots p_r}$$

$$\Phi\left(-\frac{\log p_1}{\log t},\ldots,\frac{\log p_r}{\log t},\frac{\log p_{\sigma(1)}}{\log t},\ldots,\frac{\log p_{\sigma(r)}}{\log t},0,\ldots,0\right),$$

where $\mathscr{S}_r$ is the symmetric group with r elements. The equivalent of the above sum can be calculated thanks to the prime number theorem and integration by parts, leading to the estimate

$$c_n(f,t,h) \underset{t\to\infty}{\sim} \frac{t \log t}{2\pi} \int_{\mathbb{R}} h(r)^n dr \tag{18}$$

$$\left(\Phi(0) + \sum_{r=1}^{\lfloor n/2 \rfloor} \sum \int |v_1|\ldots|v_r| \Phi(v_1 e_{i_1,j_1},\ldots,v_r e_{i_r,j_r},0,\ldots,0) dv_1\ldots dv_r\right),$$

where the sum is over all choices of pairs of disjoint indices in $[\![1,n]\!]$ and $e_{i,j} = e_i - e_j$, (e_i) being an orthonormal basis of $\mathbb{C}^n$.

At this point, it is not clear how this is related to determinants of the sine kernel. This is a purely combinatorial problem: by inclusion-exclusion the asymptotics of $r_n(f,t,h)$ can be deduced from those of $c_m(f,t,h)$, for all m. Then it turns out that when writing

$$r_n(f,t,h) = \frac{t \log t}{2\pi} \int_{\mathbb{R}} h(r)^n \int_{\mathbb{R}^n} r(v)\Phi(v) dv (1 + o(1)),$$

the function r is exactly the Fourier transform of the determinant of the sine kernel, $\det_{n\times n} K(x_i - x_j)$.

For this last step, another way to proceed consists in making the same reasoning by replacing the zeta zeros by eigenvalues of a unitary matrix u, and computing expectations with respect to the Haar measure. The Fourier transform and explicit formula (rewriting linear statistics of eigenvalues as linear sums of $(\mathrm{Tr}(u^k))_k$) still hold. Diaconis and Shahshahani [22] proved that these traces converge in law to independent normal complex Gaussians as $n \to \infty$. This independence is equivalent to performing the above diagonal approximation (17) and allows one to get formula (18) in the context of random matrices. We independently know that the eigenvalues correlations are described by the sine kernel, which completes the proof. □

The scope of this analogy needs to be moderated: following [4, 8, 5], we will see in the next section that beyond leading order the two-point correlation function depends on the positions of the low ζ zeros, something that clearly contrasts with random matrix theory.

Moreover, Rudnick and Sarnak proved that the same fermionic asymptotics hold for any primitive L-function. However, we cannot expect that it holds for any L-function, because for example, for distinct primitive characters, the zeros of L_χ and $L_{\chi'}$ have no known link, so for the product of these L-functions, the zeros look like the superposition of two independent determinantal point processes. Systems with independent versus repelling eigenvalues are discussed in the next section.

2. Quantum chaology

Quantum chaos is concerned with the study of the quantum mechanics of classically chaotic systems. In reality, a quantum system is much less dependent on the initial conditions than a classical chaotic one, where orbits are generally divergent. This is the reason why M. Berry proposed the name *quantum chaology* instead of *quantum chaos*.

The statistics found for the ζ zeros can, in this context, be seen in a more general framework. Indeed, eigenvalue repulsion is conjectured to appear in the statistics of generic chaotic systems. In the same way as the appearance of the sine kernel in the description of the statistics of the ζ zeros is proved using the explicit formula, the Bohigas–Giannoni–Schmidt conjecture is intimately linked to a semiclassical asymptotic generalization of Selberg's trace formula due to Gutzwiller. When going from a trace formula to correlations, one deals with diagonal and non-diagonal terms (i.e., repeated or distinct orbits), and their relative magnitude is crucial. We will discuss below to which extent the diagonal terms dominate, and how to estimate the contribution of non-diagonal ones.

2.1. The Berry–Tabor and Bohigas–Giannoni–Schmit conjectures

One of the goals of quantum chaology is to exhibit characteristic properties of quantum systems which, in the semiclassical limit[8], reflect the regular or chaotic aspects of the underlying classical dynamics. For example, how does classical mechanics contribute to the distribution of the eigenvalues and the amplitudes of the eigenfunctions when the de Broglie wavelength tends to 0?

---

[8]The semiclassical limit corresponds to $\hbar \to 0$ in the Schrödinger equation; in many cases, including the examples considered here, this corresponds to the high-energy limit.

The examples we consider are two-dimensional quantum billiards[9]. For some billiards, the classical trajectories are integrable (regular) and for others they are chaotic. On the quantum side, the standing waves are described by the Helmholtz equation

$$-\frac{\hbar^2}{2m}\Delta\psi_n = \lambda_n\psi_n,$$

where the spectrum is discrete as the domain is compact, with ordered eigenvalues $0 \leq \lambda_1 \leq \lambda_2 \ldots$, and appropriate Dirichlet or Neumann boundary conditions. The questions about quantum billiards one is interested in include: how does $|\psi_n|^2$ get distributed in the domain and what is the asymptotic distribution of the λ_n's as $n \to \infty$?

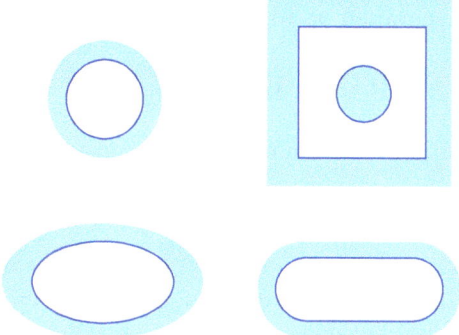

FIGURE 4. Regular (left) and chaotic (right) billiards. Upper right: Sinai's billiard. Modified after O. Bohigas [11].

One fundamental result due to Schnirelman [45] states that the quantum eigenfunctions become equidistributed with respect to the Liouville measure[10] ν, as $n \to \infty$, along a subsequence $(n_k)_{k\geq 0}$ of density one: for any measurable set I in the domain $\mathcal{D}$

$$\lim_{k\to\infty} \frac{\int_I |\psi_{n_k}|^2 dxdy}{\int_\mathcal{D} |\psi_{n_k}|^2 dxdy} = \frac{\nu(I)}{\nu(\mathcal{D})}.$$

This is referred to as *quantum ergodicity*. A stronger equipartition notion, *quantum unique ergodicity* [44], states that the above limit holds over $\mathbb{N}$, with no exceptional eigenfunctions. This is proved in very few cases. The systems where it has been proved include holomorphic cusp forms (related to billiards on $\mathbb{H}$), thanks to the work of Holowinsky and Soundararajan [32]. To satisfy quantum unique ergodicity, a system needs to avoid the problem of *scars*: for some chaotic systems, some

---
[9] A billiard is a compact connected set with nonempty interior, with a generally piecewise regular boundary, so that the classical trajectories are straight lines reflecting with equal angles of incidence and reflection.
[10] i.e., the Lebesgue mesure in our Euclidean case.

eigenfuntions (a negligible fraction of them) present an enhanced modulus near the short classical periodic orbits.

In great generality, according to the *semiclassical eigenfunction hypothesis*, the eigenstates should concentrate on those regions explored by a generic orbit as $t \to \infty$: for integrable systems the motion concentrates onto invariant tori while for the ergodic ones the whole energy surface is filled in a uniform way.

Concerning eigenvalue statistics, the situation is still complicated and somehow mysterious: there is a conjectural dichotomy between the chaotic and integrable cases.

First, in 1977, Berry and Tabor [3] put forward the conjecture that for a generic integrable system[11] the eigenvalues have the statistics of a Poisson point process, in the semiclassical limit. More precisely, by Weyl's law, we know that the number of such eigenvalues up to λ is

$$|\{i : \lambda_i \leq \lambda\}| \underset{\lambda \to \infty}{\sim} \frac{\text{area}(\mathcal{D})}{4\pi} \lambda. \tag{19}$$

To analyze the correlations between eigenvalues, consider the point process

$$\chi^{(n)} = \frac{1}{n} \sum_{i<n} \delta_{\frac{4\pi}{\text{area}(\mathcal{D})}(\lambda_{i+1}-\lambda_i)},$$

which has an expectation equal to 1 from (19). By the expected limiting Poissonian behavior, the spacing distribution converges to an exponential law: for any $I \subset \mathbb{R}^+$

$$\chi^{(n)}(I) \underset{n \to \infty}{\longrightarrow} \int_I e^{-x} \mathrm{d}x. \tag{20}$$

The limiting independence of the λ_j's also implies a variance of order n, like for any central limit theorem, in the above convergence. Note that the Berry–Tabor conjecture was rigorously proved for many integrable systems in the sense of *almost all* systems in certain families. One unconditional result concerns some fixed shifts on the torus: Marklof [37] proved that for a free particle on $\mathbb{T}^k$ with flux lines of strength $\alpha = (\alpha_1, \ldots, \alpha_k)$, if α is diophantine of type $\kappa < (k-1)/(k-2)$ and the components of $(\alpha, 1)$ are linearly independent over $\mathbb{Q}$, then the pair correlation of eigenvalues is asymptotically Poissonian.

In the chaotic case, the situation is radically different, the variance when counting the energy levels is believed to be of order $\log n$, so much less than in (20): the eigenvalues are supposed to repel each other and their statistics are conjectured to be similar to those of a random matrix, from an ensemble depending on the symmetries properties of the system (e.g., time-reversibility). This is known as the Bohigas–Giannoni–Schmidt Conjecture [10] (but see also [3]).

---

[11]There are exceptions, obvious or less obvious, many of them already known by Berry and Tabor[3], which is the reason why one expects the Poissonian behavior for *generic* systems.

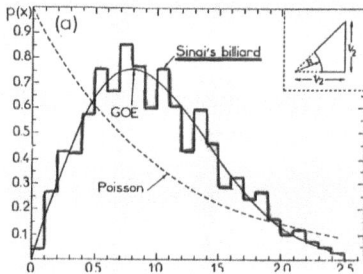

FIGURE 5. Energy levels for Sinai's billiard compared to those of the Gaussian Orthogonal Ensemble and Poissonian statistics. From [10] with permission from © 1984 The American Physical Society.

Numerical experiments were performed in [10] giving a correspondence between the eigenvalue spacings statistics for Sinai's billiard and those of the Gaussian Orthogonal Ensemble[12]. Dyson's reaction to these conjecture and experiments was the following[13].

This is a beautiful piece of work. It is extraordinary that such a simple model shows the GOE behavior so perfectly. I agree completely with your conclusions. I would say that the result is not quite surprising but certainly unexpected ... I once suggested to a student at Haverford that he build a microwave cavity and observe the resonances to see whether they follow the GOE distribution. So far as I know, the experiment was never done ... I always thought the cavity would have to be a complicated shape with many angles. I did not imagine that something as simple as the Sinai region would work.

A theoretical understanding of this conjecture proposed in [2] is related to correlations between classical periodic orbits, via the Gutzwiller trace formula explained in the next section. Our purpose consists in understanding the role of orbits of the classical motion to give insight into the derivation of the correlations of the ζ zeros [6, 7].

2.2. Periodic orbit theory

Consider a set of positive eigenvalues (λ_n) and the counting function

$$N(\lambda) = \sum_{n=1}^{\infty} \mathbb{1}_{\lambda_n < \lambda}.$$

In typical situations, this can be decomposed into a mean term and fluctuations,

$$N(\lambda) = \langle N(\lambda) \rangle + N^{\text{fl}}(\lambda).$$

For example, in the case of a quantum billiard on a domain $\mathcal{D}$, as previously discussed, the mean term is independent of whether the classical dynamics is regular

---

[12] i.e., the same eigenvalues spacings as for a random symmetric matrix with Gaussian entries.
[13] In a letter to O. Bohigas, 1983.

or chaotic and is given by
$$\langle N(\lambda) \rangle = \frac{\text{area}(\mathcal{D})}{4\pi} \lambda,$$
as shown by (19) and the fluctuating part, with mean zero, encodes independence (integrable) or repulsion (chaotic) for the energy levels. In another context, when counting the imaginary parts of the non-trivial ζ zeros, formula (2) implies
$$\langle N(t) \rangle = \frac{t}{2\pi} \log \frac{t}{2\pi e} + \frac{7}{8} + O\left(\frac{1}{t}\right),$$
$$N^{\text{fl}}(t) = \frac{1}{\pi} \arg \zeta\left(\frac{1}{2} + it\right) = \frac{1}{\pi} \Im\left(\log \zeta\left(\frac{1}{2} + it\right)\right).$$
The fact that this fluctuating part has mean zero can be seen as a byproduct of the central limit theorem (38).

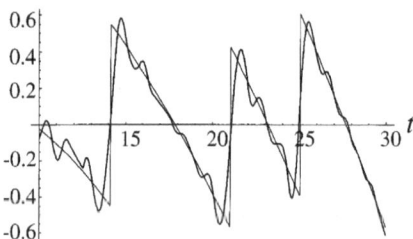

FIGURE 6. $N^{\text{fl}}(t)$ (thin line) compared with the truncated expansion (thick line) from (21) with the first 50 primes and all m. From[5] with permission from © 1999 SIAM.

The Euler product expression for ζ is not known to hold for $\sigma \in (1/2, 1)$, but we write formally
$$N^{\text{fl}}(t) = -\frac{1}{\pi} \sum_p \Im\left(\log\left(1 - \frac{e^{-it \log p}}{\sqrt{p}}\right)\right)$$
$$= -\frac{1}{\pi} \sum_{\mathcal{P}, \mathbb{N}^*} \frac{e^{-\frac{1}{2}m \log p}}{m} \sin(tm \log p). \tag{21}$$

As shown in Figure 6, truncating this expansion provides meaningful results.

We want to place the above fluctuation formulae in a more general context. Consider a dynamical system with coordinates $\mathbf{q} = (q_1, \ldots, q_d)$ and momenta $\mathbf{p} = (p_1, \ldots, p_d)$. The trajectories are generated by a Hamiltonian $H(\mathbf{q}, \mathbf{p})$. On the quantum side, $\mathbf{q}$ and $\mathbf{p}$ are operators with commutator $[\mathbf{q}, \mathbf{p}] = i\hbar$, so H is an operator whose eigenvalues are the quantum energy levels. For quantum billiards, H is independent of $\mathbf{q}$ in $\mathcal{D}$. We are interested in the situation where the energy is the only conserved quantity and neighboring trajectories diverge exponentially: the system is chaotic.

As seen in Section 1, the explicit formula (5) states that the ζ zeros have a distribution formally similar to the eigenvalues of the hyperbolic Laplacian, through the Selberg trace formula (7). This admits a semiclassical (i.e., asymptotic) generalization, originally derived in [27]. For a periodic orbit p, we denote the action by $S_p(\lambda) = \oint \mathbf{p} \cdot d\mathbf{q}$ and the period by $T_p = \partial_\lambda S_p$. The monodromy matrix M_p describes the exponential divergence of deviations from p of nearby geodesics. The Maslov index μ_p is related to the winding number of the invariant Lagrangian (stable and unstable) manifolds around the orbit: it describes the topological stability. The Maslov index of the m-repetition of the orbit p is equal to $m\mu_p$.

Gutwiller's trace formula. *With the preceding notation,*

$$N^{\mathrm{fl}}(\lambda) \underset{\lambda \to \infty}{\sim} \frac{1}{\pi} \sum_{\mathcal{P}, \mathbb{N}^*} \frac{\sin\left(m\, S_p(\lambda) - \frac{m\pi\mu_p}{2}\right)}{m\sqrt{|\det(M_p^m - \mathrm{Id})|}}, \qquad (22)$$

where $\mathcal{P}$ is the set of primitive orbits and m is the index of their repetitions.

To some extent, this formula should be considered as natural.

- The energy levels counted by N are associated with stationary states, i.e., time-independent objects. Their asymptotics correspond to the phase space structures invariant under translations along geodesics, by the correspondence principle. In our chaotic situation, there are two types of invariant manifolds, the whole surface (by ergodicity), leading to the term $\langle N(\lambda)\rangle$ and the periodic orbits which correspond to the fluctuations $N^{\mathrm{fl}}(\lambda)$.
- The exactness of the trace formula for manifolds of constant negative curvature (Selberg's trace formula) is analogous to the exact formula for the heat kernel in the Euclidean space. In the more general context of Riemannian manifolds, the heat kernel estimates are known only for short times and in terms of the geodesic distance: $p(x, y, t) \underset{t \to 0}{\sim} c\exp(-\ell(x,y)^2/2t)/t^{d/2}$, where the constant c involves the deviations from the geodesic via the Van Vleck-Morette determinant, analogously to $\det(M_p^m - \mathrm{Id})$ in the Gutzwiller trace formula.

Sketch of proof. Writing $d(\lambda) = d N(\lambda)/d\lambda$, we begin in the same way as for the Selberg trace formula, writing

$$d(\lambda) = -\frac{1}{\pi} \int \Im\left(G^{(\lambda)}(\mathbf{x}, \mathbf{x})\right) d\mathbf{x},$$

where $G^{(\lambda)}$ is the Green function associated with the energy λ. The mean eigenvalue density $\langle d(\lambda)\rangle$ corresponds to the small (minimal distance) trajectories between $\mathbf{x}$ and $\mathbf{y}$ as $\mathbf{y} \to \mathbf{x}$, and the fluctuating part $d^{\mathrm{fl}}(\lambda)$ is related to all other geodesics between $\mathbf{x}$ and itself, for example all repeated maximal circles in the spherical situation. A key assumption about the Green function is that it admits the expansion

$$G^{(\lambda)}(\mathbf{x}, \mathbf{y}) = \sum_{\text{geodesics}} A(\mathbf{x}, \mathbf{y}) e^{i S(\mathbf{x}, \mathbf{y})/\hbar},$$

where A can be developed as a series in $\hbar$, the sum is over all geodesics from $\mathbf{x}$ to $\mathbf{y}$, and $S(\mathbf{x},\mathbf{y}) = \int \mathbf{p}\cdot d\mathbf{q}$ depends on the trajectory and λ. This formula is justified by inserting $A(\mathbf{x},\mathbf{y})e^{iS(\mathbf{x},\mathbf{y})/\hbar}$ into the Schrödinger equation. Consequently,

$$d^{\mathrm{fl}}(\lambda) = \frac{1}{\pi}\int \Im\left(\sum_{\text{non-trivial geodesics}} A(\mathbf{x},\mathbf{x})e^{iS(\mathbf{x},\mathbf{x})/\hbar}\right) d\mathbf{x}.$$

A saddle point approximation can be performed as $\hbar \to 0$. On any critical point, $(\partial_{\mathbf{x}} S + \partial_{\mathbf{y}} S)_{\mathbf{x}=\mathbf{y}} = 0$, but $\partial_{\mathbf{x}} S = \mathbf{p}_f$ and $\partial_{\mathbf{y}} S = \mathbf{p}_i$, the momenta at the final and initial points respectively. Consequently, on the saddle, the momenta must be identical at the beginning and the end of the geodesic: the trajectory is periodic. The second derivatives, leading to the constant coefficients in the saddle point approximation, are related to the monodromy matrix, corresponding to the linear approximation between initial and final perturbations along the periodic orbit p:

$$d\begin{pmatrix}\mathbf{q}_f \\ \mathbf{p}_f\end{pmatrix} = \mathrm{M}_p\, d\begin{pmatrix}\mathbf{q}_i \\ \mathbf{p}_i\end{pmatrix}.$$

Moreover, when performing the saddle-point method, the Maslov index appears because it counts, roughly speaking, the number of caustics along the trajectory. All results together, with periodic orbits seen as repetitions of primitive periodic orbits, explain the origin of the main terms in (22). □

An approximation of the determinant can be performed for long orbits, in terms of the Liapunov (instability) exponent of the orbit, noted λ_p, and the (large) period $T_p = \partial_\lambda S_p$: $\det(\mathrm{M}_p^m - \mathrm{Id}) \approx e^{m\lambda_p T_p}$, so the Gutzwiller trace formula takes the form

$$\mathrm{N}^{\mathrm{fl}}(\lambda) \underset{\lambda\to\infty}{\sim} \frac{1}{\pi}\sum_{\mathcal{P},\mathbb{N}^*}\frac{e^{-\frac{1}{2}m\lambda_p T_p}}{m}\sin\left(m\,S_p(\lambda) - \frac{m\pi\mu_p}{2}\right). \qquad (23)$$

A comparison between formulas (21) and (23) yields the following formal definition of action, period and stability in the prime number context [5].

	Eigenvalues	Quantum energy levels	Zeta zeros
Asymptotics		$\hbar \to 0$	$t \to \infty$
Actions		$\frac{m S_p}{\hbar}$	$mt\log p$
Periods		$m\,T_p$	$m\log p$
Stabilities		λ_p	1

2.3. Diagonal approximation

The link between the eigenvalue counting functions discussed above and the correlation functions is formally given by

$$\begin{aligned}r_n^{(\lambda)}(x_1,\ldots,x_n) &= \langle d(\cdot+x_1)\ldots d(\cdot+x_n)\rangle \\ &= \langle d\rangle^n + r_n^{(\lambda,\mathrm{diag})}(x_1,\ldots,x_n) + r_n^{(\lambda,\mathrm{off})}(x_1,\ldots,x_n),\end{aligned} \qquad (24)$$

where $d(\lambda) = \frac{\partial N(\lambda)}{\partial \lambda}$ is the eigenvalues density, $r_n^{(\lambda)}$ is the correlation function of order n when considering eigenvalues up to height λ, and the terms $r_n^{(\lambda,\mathrm{diag})}, r_n^{(\lambda,\mathrm{off})}$ will be made explicit in the next few lines. The above formula makes sense once integrated with respect to a smooth enough test function, where for convenience no repetition between distinct eigenvalues is performed:

$$r_n^{(\lambda)}(f) := \frac{n!}{N(\lambda)} \sum_{S \subset \mathcal{E}_\lambda, |S|=n} f(S) = \int r_n^{(\lambda)}(x) f(x) \mathrm{d}x,$$

where $\mathcal{E}_\lambda$ is the set of eigenvalues up to height λ. (24) together with Gutzwiller's trace formula (22) allows one to calculate the correlation functions from the density, including for the ζ zeros, as in [6, 7]. We describe this approach, and show how it provides a heuristic justification for Montgomery's Conjecture, and also how it yields lower-order corrections to the random matrix limit for all orders of correlation functions. In order to be explicit, we focus on the two-point correlation function, $n = 2$.

The results from the previous paragraph can be written, with suitable coefficients $A_{p,m}$,

$$d(\lambda) = \langle d(\cdot) \rangle + \sum_{p,m} A_{p,m} e^{im S_p(\lambda)/\hbar},$$

which yields, once inserted in (24),

$$r_2^{(\lambda)}(x_1, x_2) \approx \langle d(\cdot) \rangle^2 + \sum_{p_i, m_i} A_{p_1, m_1} \overline{A_{p_2, m_2}} \langle e^{\frac{i}{\hbar}(m_1 S_{p_1}(\cdot + x_1) - m_2 S_{p_2}(\cdot + x_2))} \rangle.$$

The terms S_p can be expanded in terms of λ, with first derivative $\partial_\lambda S_p(\lambda) = T_p$, so the correlation function takes the form

$$r_2^{(\lambda)}(x_1, x_2) \qquad (25)$$
$$\approx \langle d(\cdot) \rangle^2 + \sum_{p_i, m_i} A_{p_1, m_1} \overline{A_{p_2, m_2}} \langle e^{\frac{i}{\hbar}(m_1 S_{p_1}(\cdot) - m_2 S_{p_2}(\cdot))} \rangle \cdot e^{\frac{i}{\hbar}(m_1 T_{p_1} x_1 - m_2 T_{p_2} x_2)}.$$

The main difficulty to evaluate this sum consists in an appropriate expectation for $\langle e^{\frac{i}{\hbar}(m_1 S_{p_1}(\cdot) - m_2 S_{p_2}(\cdot))} \rangle$. A first approximation consists in keeping only diagonal elements, i.e., orbits with exactly the same action: for distinct trajectories, averaging gives a 0 contribution in the large energy limit.

Let us consider this diagonal approximation in the ζ context (21). The height t dependence disappears when averaging and a direct calculation gives

$$r_2^{(\mathrm{diag})}(x_1, x_2) \approx \frac{\Re}{2\pi^2} \sum_{\mathcal{P}, \mathbb{N}^*} \frac{\log^2 p}{p^m} e^{i(x_1 - x_2) m \log p}.$$

The prime number theorem and a series expansion yield

$$r_2^{(\mathrm{diag})}(x_1, x_2) \underset{x_2 \to x_1}{\sim} -\frac{1}{2\pi^2 (x_1 - x_2)^2}.$$

This corresponds to the $r_2^{(\mathrm{diag})}$ part in the following decomposition of the two-point limiting correlation function associated to random matrices:

$$r_2(x) = 1 - \left(\frac{\sin(\pi x)}{\pi x}\right)^2 = \langle d \rangle^2 + r_2^{(\mathrm{diag})}(x) + r_2^{(\mathrm{off})}(x),$$

$$\langle d \rangle = 1, \ r_2^{(\mathrm{diag})}(x) = -\frac{1}{2(\pi x)^2}, \ r_2^{(\mathrm{off})}(x) = \frac{e^{i2\pi x} + e^{-i2\pi x}}{4(\pi x)^2}. \tag{26}$$

Alternatively, the right-hand side of the sum over primes can be written exactly in terms of the second derivative of $\log \zeta(1 + \mathrm{i}(x_1 - x_2))$. The pole of the zeta function gives the contribution calculated above from the prime number theorem when $x_1 - x_2 \to 0$. The structure around the pole, coming from the first few zeros, then gives corrections to the random matrix expression. Hence the statistical properties of the high-lying zeros show universal random-matrix behaviour, with non-universal corrections, which vanish in the appropriate limit, related to the low-lying zeros. This important *resurgence* was discovered in [8] (see also [5] for illustrations and an extensive discussion).

It follows from the fact that the diagonal terms do not give the full expression for the two-point correlation function that the non-diagonal terms are important. More precisely, such terms make no contribution if

$$\frac{1}{\hbar}(m_1 \mathsf{S}_{p_1} - m_2 \mathsf{S}_{p_2}) \gg 1 \tag{27}$$

in the quantum context. The average gap between lengths of periodic orbits around ℓ is about $\ell e^{-c\ell}$ (the number of orbits grows exponentially with the length). Hence, one expects that the diagonal approximation at the energy level λ can be performed, in the semiclassical limit, for orbits up to height $\ell \ll (\log \lambda)/c$. In particular, one sees with disappointment that this number grows only logarithmically with the energy. In the number theory context, (27) becomes

$$(m_1 \log p_1 - m_2 \log p_2) t \gg 1,$$

and the above discussion shows that we have to tackle the problem of close prime powers.

2.4. Beyond diagonal approximation

Keeping all off-diagonal contributions in (25) for the 2-point correlation of ζ yields

$$r_2^{(t,\mathrm{off})}(x_1, x_2) \approx \sum_{n_1 \neq n_2} \frac{\Lambda(n_1)\Lambda(n_2)}{4\pi^2 \sqrt{n_1 n_2}} \langle e^{\mathrm{i}t \log(n_1/n_2) + \mathrm{i}(x_1 \log n_1 - x_2 \log n_2)} \rangle,$$

where Λ is Van Mangoldt's function. Setting $x = x_1 - x_2$ and $n_1 = n_2 + r$, then expanding all functions in r yields

$$r_2^{(t,\mathrm{off})}(x) \approx \frac{1}{4\pi^2} \sum_{n,r} \frac{\Lambda(n)\Lambda(n+r)}{n} \langle e^{\mathrm{i}t\frac{r}{n} + \mathrm{i}x \log n} \rangle, \tag{28}$$

and the main difficulty therefore becomes the evaluation of the correlations between values of the Van Mangoldt function. No unconditional results are known about it. We make use of the following conjecture, from [29].

The Hardy–Littlewood conjecture. *For any even r, $\frac{1}{n}\sum_{k\leq n}\Lambda(k)\Lambda(k+r)$ has a limit as $n\to\infty$, equal to*

$$\alpha(r) = c\prod_{p|r,\,p>2}\frac{p-1}{p-2}, \quad \text{with} \quad c = 2\prod_{p>2}\left(1-\frac{1}{(p-1)^2}\right).$$

If r is odd, this limit is 0.

Note that the prime number theorem can be stated as

$$\lim_{n\to\infty}\frac{1}{n}\sum_{k\leq n}\Lambda(n) = 1.$$

Therefore, if the functions $\Lambda(k)$ and $\Lambda(k+r)$ were independent in the large k limit, $\alpha(r)$ would be 1. The Hardy–Littlewood conjecture states that this asymptotic independence holds up to arithmetical constraints. Assuming the accuracy of this conjecture yields, from (28),

$$r_2^{(t,\text{off})}(x) \approx \frac{1}{4\pi^2}\sum_{n,r}\alpha(r)e^{itr/n+ix\log n}.$$

This expression can be simplified to get

$$r_2^{(t,\text{off})}(x) \approx \frac{1}{4\pi^2}|\zeta(1+ix)|^2\left(\frac{t}{2\pi}\right)^{ix}\prod_{p}\left(1-\frac{1-p^{ix}}{(p-1)^2}\right).$$

In the limit of small x, $x = \frac{2\pi u}{\log(t/2\pi)}$ (consistent with the required normalization in Montgomery's Theorem 1.8), this expression can be shown to become

$$\frac{e^{i2\pi u}+e^{-i2\pi u}}{4(\pi u)^2},$$

the part that was missing in the decomposition (26). This provides strong heuristic support for Montgomery's conjecture. For more details see [6, 7, 8, 5]. Note that once again the approach to the random matrix limit is controlled by $\zeta(1+ix)$.

Note that the above method, inspired by the periodic orbit theory in quantum chaos, allows one to obtain error terms for the asymptotic pair correlation of the ζ zeros [9]. Taking into account the second-order correction to the sine kernel, one gets after scaling

$$r_2^{(t)}(x) = 1 - \left(\frac{\sin(\pi x)}{\pi x}\right)^2 - \frac{\beta}{\pi^2\langle d\rangle^2}\sin^2(\pi x) - \frac{\delta}{2\pi^2\langle d\rangle^3}\sin(2\pi x) + \mathrm{O}(\langle d\rangle^{-4}), \quad (29)$$

where

$$\langle d\rangle = \frac{1}{2\pi}\log\left(\frac{t}{2\pi}\right),$$

and β, δ are numerical constants given by

$$\beta = \gamma_0 + 2\gamma_1 + \sum_p \frac{\log^2 p}{(p-1)^2} \approx 1.573,$$

$$\delta = \sum_p \frac{\log^3 p}{(p-1)^2} \approx 2.315,$$

the γ_k's being the Stieljes constants $\gamma_k = \lim_{m \to \infty} \left(\sum_{j=1}^m \frac{\log^k j}{j} - \frac{\log^{k+1} m}{m+1} \right)$. This second-order formula gives remarkably accurate results, as shown in Figure 7. Finally, the above discussion can be applied to many other ζ statistics. For exam-

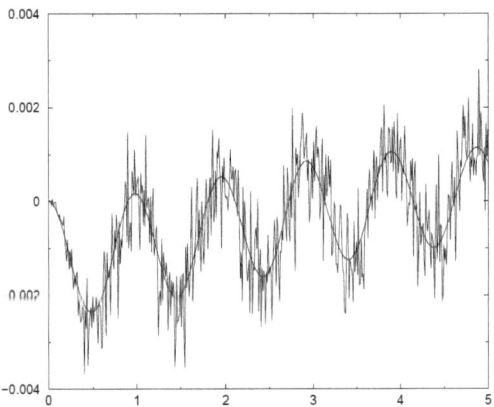

FIGURE 7. Difference $r_2^{(t)}(x) - r_2(x)$, x in $(0,5)$, for $2 \cdot 10^8$ Riemann zeros near the 10^{23}th zero, graph from [9]. Smooth line from formula (29), oscillating one from Olyzko's numerical data. From [9] with permission from © 2006 Progress of Theoretical Physics.

ple, the variance saturation of the counting function of the eigenvalues from the GUE admits a ζ counterpart, observed in Berry's original work [4]. This variance for $\mathcal{N}(t+\delta) - \mathcal{N}(t-\delta)$ has a universal behavior when δ is small enough and an arithmetic influence otherwise.

3. Macroscopic statistics

In the previous sections, the local fluctuations of zeros of L-functions were shown to be intimately linked to eigenvalues of quantum chaotic systems via random matrix theory. Another type of statistic was considered around 2000, providing even more striking evidence of a connection: at a macroscopic scale, i.e., statistics over all zeros, one also observes a close relationship with Random Matrix Theory. The main example concerns the moments of ζ.

3.1. Motivations for moments

As already mentioned, amongst the many consequences of the Riemann hypothesis, the Lindelöf conjecture asserts that ζ has a subpolynomial growth along the critical axis: $|\zeta(1/2+it)| = O(t^\varepsilon)$ for any $\varepsilon > 0$. One of the number theoretic consequences of these bounds would be $p_{n+1} - p_n = O(p_n^{1/2+\varepsilon})$ for any $\varepsilon > 0$, where p_n is the n th prime number. An apparently weaker conjecture (because it deals with mean values) concerns the moments of ζ: for any $k \in \mathbb{N}$ and $\varepsilon > 0$,

$$I_k(t) = \frac{1}{t}\int_0^t \left|\zeta\left(\frac{1}{2}+is\right)\right|^{2k} ds \ll t^\varepsilon.$$

This is actually equivalent to the Lindelöf hypothesis thanks to a good unconditional upper bound on the derivative: $\zeta'(1/2 + is) = O(s)$. More precise estimates were proved by Heath-Brown [30] for the following lower bound, and by Soundararajan [49], conditionally on the Riemann hypothesis, for the upper bound:

$$(\log t)^{k^2} \ll I_k(t) \ll (\log t)^{k^2+\varepsilon}$$

for any $\varepsilon > 0$. Unconditional equivalents are known only for $k=1$ and $k=2$: Hardy and Littlewood [28] obtained in 1918 that

$$I_1(t) \underset{t\to\infty}{\sim} \sum_{n\leq t} \frac{1}{n} \underset{t\to\infty}{\sim} \log t,$$

and Ingham [33] proved the $k=2$ case

$$I_2(t) \underset{t\to\infty}{\sim} 2 \sum_{n\leq t} \frac{d_2(n)^2}{n} \underset{t\to\infty}{\sim} \frac{1}{2\pi^2}(\log t)^4,$$

where the coefficients $d_k(n)$ are defined by $\zeta(s)^k = \sum_n d_k(n)/n^s$, $\Re(s) > 1$. Then, a precise analysis led Conrey and Ghosh [18] to conjecture

$$I_3(t) \underset{t\to\infty}{\sim} 43 \sum_{n\leq t} \frac{d_3(n)^2}{n}$$

and Conrey and Gonek [20] to

$$I_4(t) \underset{t\to\infty}{\sim} 24024 \sum_{n\leq t} \frac{d_4(n)^2}{n}.$$

Is it true that

$$I_k(t) \underset{t\to\infty}{\sim} c_k \sum_{n\leq t} \frac{d_k(n)^2}{n}$$

for some integer c_k, and what should c_k be? It is known thanks to the behavior of ζ at 1 and Tauberian theorems that

$$\sum_{n\leq t} \frac{d_k(n)^2}{n} \underset{t\to\infty}{\sim} \frac{H_\mathcal{P}(k)}{\Gamma(k^2+1)}(\log t)^{k^2}, \quad H_\mathcal{P}(k) = \prod_{p\in\mathcal{P}}\left(1-\frac{1}{p}\right)^{k^2} {}_2F_1\left(k,k,1,\frac{1}{p}\right).$$

What c_k should be for general k remained mysterious until the idea that such macroscopic statistics may be related to the corresponding ones for random matrices [35]. Searching the limits and deepness of this connection is maybe, more than the direct number-theoretic consequences, the main motivation for the study of these particular statistics.

3.2. The moments conjecture

The general equivalent for the ζ moments, proposed by Keating and Snaith, takes the following form. It coincides with all previous results and conjectures, corresponding to $k = 1, 2, 3, 4$. Note the difficulty to test it numerically, because of the $\log t$ dependence. However, as we will see later in this section, a complete expansion in terms of powers of $\log t$ was also proposed, which completely agrees with numerical experiments, giving strong support for the following asymptotics.

Conjecture 3.1. *For every* $k \in \mathbb{N}^*$

$$I_k(t) \underset{t\to\infty}{\sim} H_{\mathrm{Mat}}(k) H_{\mathcal{P}}(k) (\log t)^{k^2}$$

with the notation

$$H_{\mathcal{P}}(k) = \prod_{p \in \mathcal{P}} \left(1 - \frac{1}{p}\right)^{k^2} {}_2F_1\left(k, k, 1, \frac{1}{p}\right)$$

for the previously mentioned arithmetic factor, and the matrix factor

$$H_{\mathrm{Mat}}(k) = \prod_{j=0}^{k-1} \frac{j!}{(j+k)!}. \tag{30}$$

Note that the above term makes sense for imaginary k (it can be expressed by means of the Barnes G-function) so more generally this conjecture may be stated for any $\Re k \geq -1/2$.

Let us outline a few of key steps involved in understanding the origins of this conjecture. First suppose that $\sigma > 1$. Then the absolute convergence of the Euler product and the linear independence of the $\log p$'s ($p \in \mathcal{P}$) over $\mathbb{Q}$ make it possible to show that

$$\frac{1}{t} \int_0^t ds \, |\zeta(\sigma + is)|^{2k} \underset{t\to\infty}{\sim} \prod_{p \in \mathcal{P}} \frac{1}{t} \int_0^t \frac{ds}{\left|1 - \frac{1}{p^s}\right|^{2k}} \underset{t\to\infty}{\longrightarrow} \prod_{p \in \mathcal{P}} {}_2F_1\left(k, k, 1, \frac{1}{p^{2\sigma}}\right). \tag{31}$$

This asymptotic independence of the factors corresponding to distinct primes gives the intuition underpinning part of the arithmetic factor. Note that this equivalent of the kth moment is guessed to hold also for $1/2 < \sigma \leq 1$, which would imply the Lindelöf hypothesis (see Titchmarsh [50]). Moreover, the factor $(1 - 1/p)^{k^2}$ in $H_{\mathcal{P}}(k)$ can be interpreted as a compensator to allow the RHS in (31) to converge on $\sigma = 1/2$.

In another direction, when looking at the Dirichlet series instead of the Euler product, the Riemann zeta function on the critical axis ($\Re(s) = 1/2, \Im(s) > 0$) is the (not absolutely) convergent limit of the partial sums

$$\zeta_n(s) = \sum_{k=1}^{n} \frac{1}{k^s}.$$

Conrey and Gamburd [19] showed that

$$\lim_{n \to \infty} \lim_{t \to \infty} \frac{1}{t(\log n)^{k^2}} \int_0^t \left| \zeta_n \left(\frac{1}{2} + it \right) \right|^{2k} dt = H_{\text{Sq}}(k) H_{\mathcal{P}}(k),$$

where $H_{\text{Sq}}(k)$ is a factor distinct from $H_{\text{Mat}}(k)$ and linked to counting magic squares. So the arithmetic factor appears when considering the moments of the partial sums.

The matrix factor, which is consistent with numerical experiments, comes from the idea [35] (motivated by Montgomery's theorem) that a good approximation for the zeta function is the determinant of a unitary matrix. Thanks to Selberg's integral[14], the Mellin–Fourier for the determinant of a $n \times n$ random unitary matrix ($Z_n(u, \phi) = \det(\mathrm{Id} - e^{-i\phi} u)$) with respect to the Haar measure $\mu_{U(n)}$ is

$$P_n(s,t) = \mathbb{E}_{\mu_{U(n)}} \left(|Z_n(u,\phi)|^t e^{is \arg Z_n(u,\phi)} \right) = \prod_{j=1}^{n} \frac{\Gamma(j)\Gamma(t+j)}{\Gamma(j + \frac{t+s}{2})\Gamma(j + \frac{t-s}{2})}. \quad (32)$$

The closed form (32) implies in particular

$$\mathbb{E}_{\mu_{U(n)}} \left(|Z_n(u,\phi)|^{2k} \right) \underset{n \to \infty}{\sim} H_{\text{Mat}}(k) n^{k^2}.$$

This leads one to introduce $H_{\text{Mat}}(k)$ in the conjectured asymptotics of $I_k(T)$. This matrix factor is supposed to be universal: it should for example appear in the asymptotic moments of Dirichlet L-functions.

However, these explanations are not sufficient to understand clearly how these arithmetic and matrix factors must be combined to get the Keating–Snaith conjecture. A clarification of this point is the purpose of the three following paragraphs.

The hybrid model. Gonek, Hughes and Keating [26] gave an explanation for the moments conjecture based on a particular factorization of the zeta function.

Let $s = \sigma + it$ with $\sigma \geq 0$ and x a real parameter. Let $u(x)$ be a nonnegative $\mathscr{C}^\infty$ function of mass 1, supported on $[e^{1-1/x}, e]$, and set $U(z) = \int_0^\infty u(x) E_1(z \log x) dx$, where $E_1(z)$ is the exponential integral $\int_z^\infty (e^{-w}/w) \, dw$. Let also

$$P_x(s) = \exp \left(\sum_{n \leq x} \frac{\Lambda(n)}{n^s \log n} \right)$$

---

[14] More about the Selberg integrals and its numerous applications can be found in [24].

where Λ is Van Mangoldt's function ($\Lambda(n) = \log p$ if n is an integral power of a prime p, 0 otherwise), and

$$Z_x(s) = \exp\left(-\sum_{\rho_n} U\left((s-\rho_n)\log x\right)\right)$$

where $(\rho_n, n \geq 0)$ are the non-trivial ζ zeros. Then unconditionally, for any given integer m,

$$\zeta(s) = P_x(s)Z_x(s)\left(1 + O\left(\frac{x^{m+2}}{(|s|\log x)^m}\right) + O\left(x^{-\sigma}\log x\right)\right),$$

where the constants in front of the O only depend on the function u and m: this is a hybrid formula for ζ, with both a Euler and Hadamard product. The P_x term corresponds to the *arithmetic* factor of the moments conjecture, while the Z_x term corresponds to the *matrix* factor. More precisely, this decomposition suggests a proof for Conjecture 3.1 along the following steps.

First, for a value of the parameter x chosen such that $x = O(\log(t)^{2-\varepsilon})$, the following splitting conjecture states that the moments of zeta are well approximated by the product of the moments of P_x and Z_x (they are sufficiently *independent*):

$$\frac{1}{t}\int_0^t ds \left|\zeta\left(\frac{1}{2}+is\right)\right|^{2k} \underset{t\to\infty}{\sim} \left(\frac{1}{t}\int_0^t ds \left|P_x\left(\frac{1}{2}+is\right)\right|^{2k}\right)$$
$$\times \left(\frac{1}{t}\int_0^t ds \left|Z_x\left(\frac{1}{2}+is\right)\right|^{2k}\right).$$

Assuming that the above result is true, we then need to approximate the moments of P_x and Z_x. Concerning P_x [26] proves that

$$\frac{1}{t}\int_0^t ds \left|P_x\left(\frac{1}{2}+is\right)\right|^{2k} = H_\mathcal{P}(k)\left(e^\gamma \log x\right)^{k^2}\left(1 + O\left(\frac{1}{\log x}\right)\right).$$

Finally, an additional conjecture about the moments of Z_x would be the last step in the moments conjecture:

$$\frac{1}{t}\int_0^t ds \left|Z_x\left(\frac{1}{2}+is\right)\right|^{2k} \underset{t\to\infty}{\sim} H_{\mathrm{Mat}}(k)\left(\frac{\log t}{e^\gamma \log x}\right)^{k^2}. \tag{33}$$

The reasoning which leads to this supposed asymptotic is the following. First of all, the function Z_x is not as complicated as it seems, because as x tends to ∞, the function u tends to the Dirac measure at point e, so $Z_x\left(\frac{1}{2}+is\right) \approx \prod_{\gamma_n}(i(t-\gamma_n)e^\gamma \log x)$. The ordinates γ_n (where ζ vanishes) are supposed to have many statistical properties identical to those of the eigenangles of a random element of $U(n)$. In order to make an adequate choice for n, we recall that the γ_n have spacing $2\pi/\log t$ on average, whereas the eigenangles have mean gap $2\pi/n$: thus n should be chosen to be about $\log t$. Then the random matrix moments lead to conjecture (33).

Multiple Dirichlet series. In a completely different direction, Diaconu, Goldfeld and Hoffstein [21] proposed an explanation of the Conjecture 3.1 relying only on a supposed meromorphy property of the multiple Dirichlet series

$$\int_1^\infty \zeta(s_1 + \varepsilon_1 \mathrm{i}t) \ldots \zeta(s_{2m} + \varepsilon_{2m}\mathrm{i}t) \left(\frac{2\pi e}{t}\right)^{k\mathrm{i}t} t^{-w} \mathrm{d}t,$$

with $w, s_k \in \mathbb{C}$, $\varepsilon_k = \pm 1$ ($1 \leq k \leq 2m$). They make no use of any analogy with random matrices to predict the moments of ζ, and recover the Keating–Snaith conjecture. Important tools in their method are a group of approximate functional equations for such multiple Dirichlet series and a Tauberian theorem to connect the asymptotics as $w \to 1^+$ and the moments $\int_1^t$.

An intriguing question is whether their method applies or not to predict the joint moments of ζ,

$$\frac{1}{t}\int_1^t \mathrm{d}t \prod_{j=1}^\ell \left|\zeta\left(\frac{1}{2} + \mathrm{i}(t + s_j)\right)\right|^{2k_j}$$

with $k_j \in \mathbb{N}^*$, ($1 \leq j \leq \ell$), the s_j's being distinct elements in $\mathbb{R}$. If such a conjecture could be stated, independently of any considerations about random matrices, this would be an accurate test for the correspondence between random matrices and L-functions. For such a conjecture, one expects that it agrees with the analogous result on the unitary group, which is a special case of the Fisher–Hartwig asymptotics of Toeplitz determinants first proven by Widom [51]:

$$\mathbb{E}_{\mu_{U(n)}} \left(\prod_{j=1}^\ell \left|\det(\mathrm{Id} - e^{\mathrm{i}\phi_j} u)\right|^{2k_j}\right)$$

$$\underset{n \to \infty}{\sim} \prod_{1 \leq i < j \leq \ell} |e^{\mathrm{i}\phi_i} - e^{\mathrm{i}\phi_j}|^{-2k_i k_j} \prod_{j=1}^\ell \mathbb{E}_{\mu_{U(n)}} \left(\left|\det(\mathrm{Id} - e^{\mathrm{i}\phi_j} u)\right|^{2k_j}\right)$$

$$\underset{n \to \infty}{\sim} \prod_{1 \leq i < j \leq \ell} |e^{\mathrm{i}\phi_i} - e^{\mathrm{i}\phi_j}|^{-2k_i k_j} \prod_{j=1}^\ell H_{\mathrm{Mat}}(k_j) n^{k_j^2},$$

the ϕ_j's being distinct elements modulo 2π.

Joint moments. Strong evidence supporting Conjecture 3.1 was obtained in [17]: an extension is given to the joint moments of the Riemann zeta function with entries shifted by constants along the critical axis. In particular, this generalized moment conjecture gives a complete expansion of $I_k(t)$ in terms of powers of $\log t$ which agrees remarkably with numerical tests.

More precisely, we denote briefly $z = (z_1, \ldots, z_{2k})$ and introduce the Euler product

$$a_k(z) = \prod_{p \in \mathcal{P}, 1 \leq i,j \leq k} \left(1 - \frac{1}{1 + p^{z_i - z_{j+k}}}\right) \int_0^1 \mathrm{d}\theta \prod_{j=1}^k \frac{1}{\left(1 - \frac{e^{\mathrm{i}2\pi\theta}}{p^{1/2+z_j}}\right)\left(1 - \frac{e^{-\mathrm{i}2\pi\theta}}{p^{1/2-z_{j+k}}}\right)},$$

and $g(z) = a_k(z) \prod_{1 \le i,j \le k} \zeta(1 + z_i - z_{j+k})$. Then we define $P_k(x, (\alpha_i)_1^{2k})$ as the integral (the path of integration being defined as small circles surrounding the poles α_i)

$$e^{-\frac{x}{2} \sum_1^k (\alpha_i - \alpha_{i+k})} \frac{(-1)^k}{(k!)^2 (i 2\pi)^{2k}} \oint \cdots \oint e^{\frac{x}{2} \sum_1^k (z_i - z_{i+k})}$$
$$\times \frac{g(z) \Delta^2(z)}{\prod_{1 \le i,j \le 2k} (z_i - \alpha_j)} dz_1 \ldots dz_{2k},$$

where Δ is the Vandermonde determinant $\Delta(z) = \prod_{i<j}(z_j - z_i)$. Then the complete moments conjecture from [17] is that, for any $\varepsilon > 0$,

$$\int_0^t \prod_{i=1}^k \zeta\left(\frac{1}{2} + is + \alpha_i\right) \prod_{j=k+1}^{2k} \zeta\left(\frac{1}{2} + is - \alpha_j\right) ds$$
$$= \int_0^t P_k\left(\log \frac{s}{2\pi}, (\alpha_i)_1^{2k}\right) \left(1 + O\left(s^{-\frac{1}{2}+\varepsilon}\right)\right) ds. \tag{34}$$

One can prove that for $\alpha_1 = \cdots = \alpha_{2k} = 0$, P_k is polynomial in x with degree k^2 with leading coefficient as expected from Conjecture 3.1. Where does the general moments conjecture come from? A very similar formula was obtained by the same authors [16] concerning the characteristic polynomial of a random unitary matrix, noted here $Z_u(\alpha) = \det(\text{Id} - e^{-\alpha}u)$:

$$\mathbb{E}_{\mu_{U(n)}} \left(\prod_{i=1}^k Z_u(\alpha_i) \prod_{j=k+1}^{2k} Z_{u^\dagger}(-\alpha_j) \right) = e^{-\frac{n}{2} \sum_{i=1}^k (\alpha_i - \alpha_{k+i})} \frac{(-1)^k}{(k!)^2 (i 2\pi)^{2k}}$$
$$\oint \cdots \oint e^{\frac{n}{2} \sum_{i=1}^k (z_i - z_{k+i})} \prod_{1 \le i,j \le k} \frac{1}{1 - e^{z_{k+j} - z_i}} \frac{\Delta^2(z)}{\prod_{1 \le i,j \le 2k} (z_i - \alpha_j)} dz_1 \ldots dz_{2k}.$$

From this formal analogy between the joint moments of unitary matrices and that of the Riemann zeta function one gets strikingly accurate numerical results. For example, the following numerical data from [17] compare the conjectural moments asymptotics when $k = 3$ (writing $P_3(x)$ for $P_3(x, (0, \ldots, 0))$), on an interval I

$$\int_I P_3\left(\frac{\log s}{2\pi}\right) ds \tag{35}$$

with the numerical computation for

$$\int_I \left|\zeta\left(\frac{1}{2} + is\right)\right|^6 ds. \tag{36}$$

Integration domain I	Full moment conjecture (35)	Numerics (36)	Ratio
[1300000, 1350000]	80188090542.5	80320710380.9	1.001654
[1350000, 1400000]	81723770322.2	80767881132.6	.988303
[1400000, 1450000]	83228956776.3	83782957374.3	1.006656
[0, 2350000]	3317437762612.4	3317496016044.9	1.000017

3.3. Gaussianity for ζ and characteristic polynomials

The explicit computation (32) of the mixed Mellin–Fourier transform of the characteristic polynomial allows one to prove the following central limit theorem [35]: as $n \to \infty$

$$\frac{\log Z_n}{\sqrt{\log n}} \xrightarrow{\text{law}} \mathcal{N}_1 + i\mathcal{N}_2, \qquad (37)$$

where $\mathcal{N}_1$ and $\mathcal{N}_2$ are standard independent real Gaussian random variables. A similar result holds, unconditionally, for the logarithm of the Riemann zeta function. This was shown by Selberg[15] [47].

This indicates that the correspondence between random matrices and L-functions is not uniquely observable through the microscopic repulsion, but also at a macroscopic level, like in the moment conjecture.

Theorem 3.2. *Writing ω for a uniform random variable on $(0,1)$,*

$$\frac{\log \zeta\left(\frac{1}{2} + i\omega t\right)}{\sqrt{\frac{1}{2}\log\log t}} \xrightarrow[t \to \infty]{\text{law}} \mathcal{N}_1 + i\mathcal{N}_2. \qquad (38)$$

It may be of interest, to understand the mixing properties of primes, to give ideas of the proof of this central limit theorem.

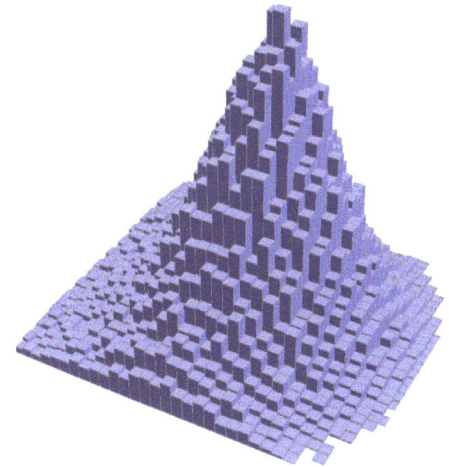

FIGURE 8. Histogram of 10^5 values of $\log \zeta$ around 10^6.

---

[15] Note that it predates the analogous result about random matrices (37), an unusual situation.

Sketch of proof. Suppose the Euler product of ζ holds for $1/2 \leq \Re(s) \leq 1$: then $\log \zeta(s) = -\sum_{p \in \mathcal{P}} \log(1 - p^{-s})$ can be approximated by $\sum_{p \in \mathcal{P}} p^{-s}$. Let $s = 1/2 + i\omega t$ with ω uniform on $(0, 1)$. As the $\log p$'s are linearly independent over $\mathbb{Q}$, the terms $\{p^{-i\omega t} \mid p \in \mathcal{P}\}$ can be viewed as independent uniform random variables on the unit circle as $t \to \infty$, hence it is a natural thought that a central limit theorem might hold for $\log \zeta(s)$.

The crucial point to get such arithmetical central limit theorems is the approximation by sufficiently short Dirichlet series. More precisely, the explicit formula for ζ'/ζ, by Landau, gives such an approximation ($x > 1$, s distinct from 1, the zeroes ρ and $-2n$, $n \in \mathbb{N}$):

$$\frac{\zeta'}{\zeta}(s) = -\sum_{n \leq x} \frac{\Lambda(n)}{n^s} + \frac{x^{1-s}}{1-s} - \sum_{\rho} \frac{x^{\rho-s}}{\rho - s} + \sum_{n=1}^{\infty} \frac{x^{-2n-s}}{2n + s},$$

from which we get an approximate formula for $\log \zeta(s)$ by integration. However, the sum over the zeros is not absolutely convergent, hence this formula is not sufficient. Selberg found a slight change in the above formula, that makes a great difference because all infinite sums are now absolutely convergent: under the above hypotheses, if

$$\Lambda_x(n) = \begin{cases} \Lambda(n) & \text{for } 1 \leq n \leq x, \\ \Lambda(n) \frac{\log \frac{x^2}{n}}{\log n} & \text{for } x \leq n \leq x^2, \end{cases}$$

then

$$\frac{\zeta'}{\zeta}(s) = -\sum_{n \leq x^2} \frac{\Lambda_x(n)}{n^s} + \frac{x^{2(1-s)} - x^{1-s}}{(1-s)^2 \log x} + \frac{1}{\log x} \sum_{\rho} \frac{x^{\rho - s} - x^{2(\rho - s)}}{(\rho - s)^2}$$
$$+ \frac{1}{\log x} \sum_{n=1}^{\infty} \frac{x^{-2n-s} - x^{-2(2n+s)}}{(2n+s)^2}.$$

Assuming the Riemann hypothesis, the above formulas give a simple expression for $(\zeta'/\zeta)(s)$ for $\Re(s) \geq 1/2$: for $x \to \infty$, all terms in the infinite sums converge to 0 because $\Re(\rho - s) < 0$. By subtle arguments, Selberg showed that, although RH is necessary for the *almost sure* coincidence between ζ'/ζ and its Dirichlet series, it is not required in order to get a good L^k approximation. In particular, for any $k \in \mathbb{N}^*$, $0 < a < 1$, there is a constant $c_{k,a}$ such that for any $1/2 \leq \sigma \leq 1$, $t^{a/k} \leq x \leq t^{1/k}$,

$$\frac{1}{t} \int_1^t \left| \log \zeta(\sigma + is) - \sum_{p \leq x} \frac{p^{-is}}{p^\sigma} \right|^{2k} ds \leq c_{k,a}.$$

In the following, we only need the case $k = 1$ in the above formula: with ω uniform on $(0, 1)$, $\log \zeta\left(\frac{1}{2} + i\omega t\right) - \sum_{p \leq t} \frac{p^{-i\omega t}}{\sqrt{p}}$ is bounded in L^2, and after normalization by $\frac{1}{\sqrt{\log \log t}}$, it converges in probability to 0. Hence, the central limit theorem for

$\log \zeta$ is equivalent to

$$\frac{1}{\sqrt{\log\log t}} \sum_{p\leq t} \frac{p^{-i\omega t}}{\sqrt{p}} \xrightarrow[t\to\infty]{\text{law}} \mathcal{N}_1 + i\mathcal{N}_2.$$

The proof of the above result proceeds in two steps.

Firstly, the length of the Dirichlet series can still be decreased, thanks to the Montgomery–Vaughan inequality, Theorem 1.9. From the properties of linear independence of primes, there is a constant $c > 0$ independent of p with $\min_{p'\neq p} |\log p - \log p'| > \frac{c}{p}$, so for any $m_t < t$

$$\frac{1}{t}\int_0^t \left| \sum_{m_t < p < t} \frac{p^{-is}}{\sqrt{p\log\log t}} \right|^2 ds \leq \sum_{m_t < p < t} \frac{1}{p\log\log t} \left(1 + 3\pi c\frac{p}{t}\right).$$

By the prime number theorem this goes to 0 provided that, as $t \to \infty$, $\frac{\log\log m_t}{\log\log t} \to 1$. If the above condition is satisfied, we therefore just need to prove

$$\sum_{p\leq m_t} \frac{p^{i\omega t}}{\sqrt{p\log\log t}} \xrightarrow{\text{law}} \mathcal{N}_1 + i\mathcal{N}_2. \tag{39}$$

Secondly, the classical central limit theorem states that

$$\sum_{p\leq m_t} \frac{e^{i\omega_p}}{\sqrt{p\log\log t}} \xrightarrow{\text{law}} \mathcal{N}_1 + i\mathcal{N}_2, \tag{40}$$

when the ω_p's are independent uniform random variables on $(0, 2\pi)$. The $\log p$'s being linearly independent over $\mathbb{Q}$, it is well known that as $t \to \infty$ any given finite number of the $p^{i\omega t}$'s are asymptotically independent and uniform on the unit circle. The problem here is that the number of these random variables increases as they become independent. If this number increases sufficiently slowly ($\log m_t/\log t \to 0$), one can show that (40) implies (39). This is a result about the mixing time of T^s, the translation on the torus $\mathbb{T}^n$ with vector $s(\log p_1, \ldots, \log p_n)$, where both s and n go to ∞: the mean in time (39) is very close to the mean in space (40). The method to prove it consists in computing the moments of the time average, and show that only the diagonal terms contribute (i.e., the terms corresponding to the space average). $\square$

A natural attitude in probability, to prove a central limit theorem such as (37), consists in identifying independent random variables. However, from the uniform (Haar) measure on $U(n)$, such an identification is not straightforward.

To tackle this problem we need to understand how one can generate the Haar measure as a product of independent transformations, to deduce identities in law concerning the characteristic polynomials of random matrices [12]. Let us take the example of a uniformly distributed element of $O(3)$. It seems natural to proceed as follows:

- $O(e_1)$ is uniform on the unit sphere;
- $O(e_2)$ is uniform on the unit circle orthogonal to $O(e_1)$;

- $O(e_3)$ is uniform on $\{O(e_1) \wedge O(e_2), -O(e_1) \wedge O(e_2)\}$.

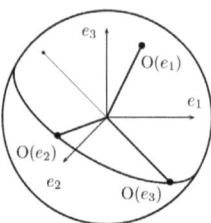

From [13] with permission from © 2009 P. Bourgade.

The lines hereafter are a formalization of the above simple idea, written here for the unitary group. For any $0 \leq k \leq n$, note $\mathcal{H}_k := \{u \in \mathrm{U}(n) \mid u(e_j) = e_j, 1 \leq j \leq k\}$, the subgroup of $\mathrm{U}(n)$ stabilizing of the first k basis vectors, and μ_k its Haar measure (in particular $\mu_0 = \mu_{\mathrm{U}(n)}$). Moreover, let p_k be the projection $u \mapsto u(e_k)$. A sequence $(\nu_1, \ldots, \nu_n)$ of probability measures on $\mathrm{U}(n)$ is said *coherent* with the Haar measure $\mu_{\mathrm{U}(n)}$ if for all $1 \leq k \leq n$, $\nu_k(\mathcal{H}_{k-1}) = 1$ and $p_k(\nu_k) = p_k(\mu_{k-1})$.

A result of [12] asserts that if $(\nu_1, \ldots, \nu_n)$ is coherent with $\mu_{\mathrm{U}(n)}$, then one has the equality of measures

$$\mu_{\mathrm{U}(n)} = \nu_1 \times \nu_2 \times \cdots \times \nu_n, \tag{41}$$

which means that to generate a uniform unitary matrix, one can proceed by multiplying n independent unitary matrices from the embedded subgroups $(\mathcal{H}_k)$, provided that their first non-trivial row is uniform on spheres of increasing size.

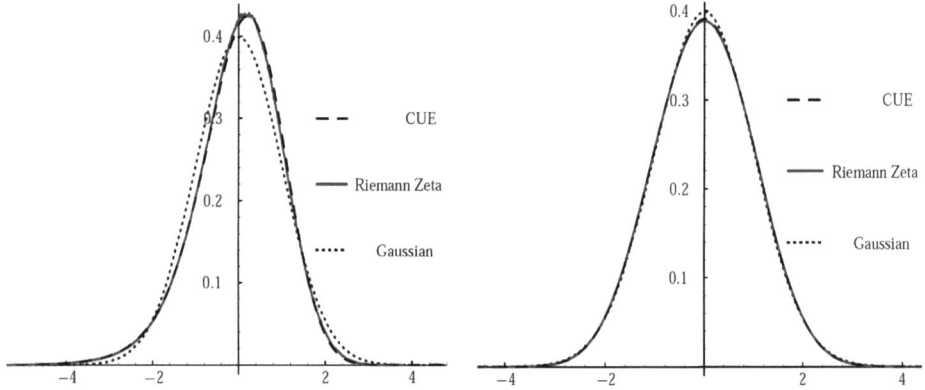

From [35] with permission from © 2000 Springer Science+Business Media.

If one chooses reflections[16] $r_1, \ldots, r_n$ for these independent transformations the projection of (41) by the determinant takes a remarkably easy form, because

---

[16] i.e., $\mathrm{rank}(r_k - \mathrm{Id}) = 1$ or 0.

of the algebraic identity

$$\det\left(\operatorname{Id}-r\begin{pmatrix}1 & 0 \\ 0 & u\end{pmatrix}\right) = (1 - \langle e_1, r(e_1)\rangle)\det(\operatorname{Id}-u)$$

when r is a reflection. Iterating this formula in (41) yields, for $u \sim \mu_{U(n)}$,

$$\det(\operatorname{Id}-u) \overset{\text{law}}{=} \prod_{k=1}^{n} \left(1 - e^{i\omega_k}\sqrt{\beta_{1,k-1}}\right), \tag{42}$$

where all random variables are independent, ω_k uniform on $(0, 2\pi)$, and β a beta random variable[17] with indicated parameters. Indeed, the first coordinate of a unit complex vector uniformly distributed on a k-dimensional complex sphere is known to be distributed like $e^{i\omega_k}\sqrt{\beta_{1,k-1}}$. The decomposition (42) also raises the (open) question of an analogue of these unexpected independent random variables in a number-theoretic context.

It also gives a direct proof of (32) as well as a derivation of the central limit theorem (37), by a simple central limit theorem for sums of independent random variables. Moreover, this decomposition yields a speed of convergence, by Berry–Esseen type theorems, and that this convergence rate is better for the imaginary part than for the real one (because $\Im\log\left(1 - e^{i\omega_k}\sqrt{\beta_{1,k-1}}\right)$ has a symmetric density, which is not the case for $\Re\log\left(1 - e^{i\omega_k}\sqrt{\beta_{1,k-1}}\right)$). This can be compared to the previous histograms of $\log \zeta$ values, around height 10^{20}, from [35] and based on numerical data in [41]. They illustrate the Selberg limit theorem (38), by comparing $\Re(\log \zeta)$ (up) and $\Im(\log \zeta)$ (down) with the Gaussian distribution and the density of $\log\det(\operatorname{Id}-u)$ where $u \sim \mu_{U(42)}$ (the Circular Unitary Ensemble, CUE). The dimension and height are chosen to satisfy $n \approx \log t$. This shows a better agreement between U(42) and ζ statistics than with the Gaussian, and a better convergence speed for $\arg \zeta$ than for $\log|\zeta|$, like in the unitary case.

Finally, the correspondence $n \leftrightarrow \log t$ holds not only for the matrix dimension compared to the height along the critical axis, but also for small shifts away from the unit circle and the critical axis. For example, the same techniques as above yield the following phase transition in the normalization, from a constant if sufficiently close to the critical line till a lower one when going further.

If $\varepsilon_n \to 0$, $\varepsilon_n \gg 1/n$, and $u_n \sim \mu_{U(n)}$, then

$$\frac{\log\det(\operatorname{Id}-e^{-\varepsilon_n}u_n)}{\sqrt{-\frac{1}{2}\log\varepsilon_n}} \overset{\text{law}}{\longrightarrow} \mathcal{N}_1 + i\mathcal{N}_2$$

The required normalisation becomes $\sqrt{\frac{1}{2}\log n}$ if $\varepsilon_n \ll 1/n$.

If $\varepsilon_t \to 0$, $\varepsilon_t \gg 1/\log t$, and ω is uniform on $(0, 1)$, then

$$\frac{\log\zeta\left(\frac{1}{2} + \varepsilon_t + i\omega t\right)}{\sqrt{-\frac{1}{2}\log\varepsilon_t}} \overset{\text{law}}{\longrightarrow} \mathcal{N}_1 + i\mathcal{N}_2$$

The required normalisation becomes $\sqrt{\frac{1}{2}\log\log t}$ if $\varepsilon_t \ll 1/\log t$.

---

[17] $\beta_{a,b}$ is a random variable supported on $(0,1)$ with measure $\frac{\Gamma(a+b)}{\Gamma(a)\Gamma(b)}x^{a-1}(1-x)^{b-1}dx$.

3.4. Families of L-functions

Another type of statistics concerns families of L-functions, i.e., averages over a set of functions at a specific point instead of a mean along the critical axis for a given L-function. In this context, the limiting statistics are linked to distinct compact groups, including:

- $U(n)$, the group of $n \times n$ unitary matrices, involved in many analogies as we have seen;
- $SO(2n)$, the special orthogonal group, orthogonal $2n \times 2n$ matrices u with $\det(u) = 1$;
- $USp(2n)$, the unitary symplectic group, $2n \times 2n$ unitary matrices satisfying $u^t J u = J$, where $J = \begin{pmatrix} 0 & \mathrm{Id}_n \\ -\mathrm{Id}_n & 0 \end{pmatrix}$.

A result by Katz–Sarnak [34] states that, irrespective of the choice of the three above groups, for any k the kth consecutive spacings measures

$$\mu_k^{(u)}[a,b] = \frac{1}{n} \left| \left\{ 1 \leq j \leq n : \frac{n}{2\pi}(\theta_{j+k} - \theta_j) \in [a,b] \right\} \right|$$

are the same in the limit $n \to \infty$, where $0 \leq \theta_1 \leq \ldots$ are the ordered eigenangles of u. These local statistics do not depend on the group in question for large dimensions. This corresponds to the universality of the GUE statistics for zeros at large height along the critical axis, for example for Dirichlet L-functions or L-functions attached to elliptic curves. Consequently, to make a statistical distinction amongst families, one needs to look at low-lying zeros, i.e., close to the symmetry point $1/2$. They correspond to eigenvalues close to 1 on the unit circle.

One example of a family of L-functions is the following, attached to real quadratic Dirichlet characters. For a prime p, let $\chi_d(p) = 0$ if $p \mid d$, 1 if $p \nmid d$ and d is a square modulo p, and -1 otherwise. Then define the Dirichlet L-function for $\Re(s) > 1$ by

$$L^{(\mathrm{Dir})}(s, \chi_d) = \prod_{\mathcal{P}} \frac{1}{1 - \frac{\chi_d(p)}{p^s}} = \sum_{n=1}^{\infty} \frac{\chi_d(n)}{n^s},$$

where the definition of χ_d is extended to $\mathbb{N}$ and satisfies the multiplicative property $\chi_d(mn) = \chi_d(m)\chi_d(n)$. Then $L^{(\mathrm{Dir})}$ can be meromorphically extended to $\mathbb{C}$, satisfies a functional equation, and its central statistics $L^{(\mathrm{Dir})}(1/2, \chi_d)$ are supposedly linked to the Haar measure of the unitary symplectic group. For this group, the eigenvalues are symmetrically positioned about the real axis (like for the zeros of $L^{(\mathrm{Dir})}$) and with eigenangles density proportional to

$$\prod_{1 \leq j < k \leq n} (\cos \theta_j - \cos \theta_k)^2 \prod_{j=1}^{n} \sin^2 \theta_j. \tag{43}$$

The conjecture of Conrey–Farmer [15] and Keating–Snaith [36] is that, for any integer k (they extend it to real k thanks to the Barnes function),

$$\frac{1}{D^*}\sum_{|d|\leq D}^{*} L(1/2,\chi_d)^k \underset{D\to\infty}{\sim} \mathcal{H}_{\text{Mat}}^{(\text{USp})}(k)\, \mathcal{H}_{\mathcal{P}}^{(\text{Dir})}(k) \left(\frac{1}{2}\log D\right)^{\frac{k(k+1)}{2}},$$

where $\sum^*$ means that the summation is restricted to fundamental discriminants[18], D^* is the number of terms in the sum, and

$$\mathcal{H}_{\text{Mat}}^{(\text{USp})}(k) = \lim_{n\to\infty}\frac{1}{n^{\frac{k(k+1)}{2}}}\mathbb{E}_{\mu_{\text{USp}(k)}}\left(|\det(\text{Id}-u)|^k\right) = 2^{\frac{k(k+1)}{2}}\prod_{j=1}^{k}\frac{j!}{(2j)!},$$

$$\mathcal{H}_{\mathcal{P}}^{(\text{Dir})}(k) = \prod_{p}\frac{\left(1-\frac{1}{p}\right)^{\frac{k(k+1)}{2}}}{1+\frac{1}{p}}\left(\frac{\left(1-\frac{1}{\sqrt{p}}\right)^{-k}+\left(1+\frac{1}{\sqrt{p}}\right)^{-k}}{2}+\frac{1}{p}\right).$$
(44)

The limit in (44) can be performed either by the Weyl integration formula (43) and Selberg integrals asymptotics, or by a decomposition as a product of independent random variables like (42).

Examples of L-functions families featuring orthogonal symmetry and $SO(2n)$ statistics can be found in [15]. This includes for example L-functions associated with elliptic curves and twisted by Dirichlet characters. For all families, with unitary, orthogonal or symplectic type, the conjectured asymptotics of low-lying zeros are relevant with numerical calculations [43].

Citing Katz and Sarnak, *we believe that the further understanding of the source of such symmetries holds the key to finding a natural spectral interpretation of the zeros.*

References

[1] G.W. Anderson, A. Guionnet, O. Zeitouni, *An Introduction to Random Matrices*, Cambridge University Press, 2009.

[2] N. Argaman, F.M. Dittes, E. Doron, J.P. Keating, A. Kitaev, M. Sieber, U. Smilansky, *Correlations in the actions of periodic orbits derived from quantum chaos*, Phys. Rev. Lett. **71**, 4326–4329 (1993).

[3] M.V. Berry, M. Tabor, *Level clustering in the regular spectrum*, Proc. Roy. Soc. Lond. A **356**, 375–394 (1977).

[4] M.V. Berry, *Semiclassical formula for the number variance of the Riemann zeros*, Nonlinearity **1**, 399–407 (1988).

[5] M.V. Berry, J.P. Keating, *The Riemann zeros and eigenvalue asymptotics*, SIAM Review **41**, 236–266 (1999).

---

[18] d is a fundamental discriminant if any decomposition $d = d_0 f^2$, with d_0 a discriminant and $f \in \mathbb{N}$, implies $f = 1$.

[6] E.B Bogomolny, J.P. Keating, *Random matrix theory and the Riemann zeros I: three and four-point correlations*, Nonlinearity **8**, 1115–1131 (1995).

[7] E.B. Bogomolny, J.P. Keating, *Random matrix theory and the Riemann zeros II: n-point correlations*, Nonlinearity **9**, 911–935 (1996).

[8] E.B. Bogomolny, J.P. Keating, *Gutzwiller's trace formula and spectral statistics: Beyond the diagonal approximation*, Phys. Rev. Lett. **77**, 1472–1475 (1996).

[9] E. Bogomolny, O. Bohigas, P. Leboeuf, A.G. Monastra, *On the spacing distribution of the Riemann zeros: corrections to the asymptotic result*, J. Phys. A: Math. Gen. **39**, 10743–10754 (2006).

[10] O. Bohigas, M.-J. Giannoni, C. Schmidt, *Characterization of chaoic quantum spectra and universality of level fluctuation laws*, Phys. Rev. Lett. **52**, 1–4 (1984).

[11] O. Bohigas (1991), *Random Matrix Theories and Chaotic Dynamics*, pp. 87–199 of Chaos et Physique Quantique/Chaos and Quantum Physics, eds. M.-J. Giannoni, A. Voros and J. Zinn-Justin (Elsevier Science Publishers).

[12] P. Bourgade, C.P. Hughes, A. Nikeghbali, M. Yor, *The characteristic polynomial of a random unitary matrix: a probabilistic approach*, Duke Mathematical Journal **145**, 1, 45–69 (2008).

[13] P. Bourgade, *On random matrices and L-functions*, Ph.D. thesis, 2009.

[14] A. Connes, *Trace formula in non-commutative Geometry and the zeros of the Riemann zeta function*, Selecta Math. (N.S.) **5**, no. 1, 29–106 (1999).

[15] J.B. Conrey, D.W. Farmer, *Mean values of L-functions and symmetry*, International Mathematics Research Notices, 883–908, 2000.

[16] J.B. Conrey, D.W. Farmer, J.P. Keating, M.O. Rubinstein, and N.C. Snaith, *Autocorrelation of random matrix polynomials*, Commun. Math. Phys. **237** 3, 365–395 (2003).

[17] J.B. Conrey, D.W. Farmer, J.P. Keating, M.O. Rubinstein, N.C. Snaith, *Integral moments of L-functions*, Proc. London Math. Soc. **91**, 33–104 (2005).

[18] J.B. Conrey, A. Ghosh, *On mean values of the zeta function, iii*, Proceedings of the Amalfi Conference in Analytic Number Theory, Università di Salerno, 1992.

[19] J.B. Conrey, A. Gamburd, *Pseudomoments of the Riemann zeta-function and pseudomagic squares*, Journal of Number Theory **117**, 263–278 (2006), Issue 2.

[20] J.B. Conrey, S.M. Gonek, *High moments of the Riemann zeta function*, Duke Math. J. **107**, 577–604 (2001).

[21] A. Diaconu, D. Goldfeld, J. Hoffstein, *Multiple Dirichlet Series and Moments of Zeta and L-Functions*, Compositio Mathematica **139**, N. 3, 297–360 (64), December (2003).

[22] P. Diaconis, M. Shahshahani, *On the eigenvalues of random matrices*, Studies in applied probability, J. Appl. Probab. **31A**, 49–62 (1994).

[23] L. Erdös, J. Ramirez, B. Schlein, T. Tao, V. Vu, H.T. Yau, *Bulk universality for Wigner hermitian matrices with subexponential decay*, Math. Res. Lett. **17** no. 4, 667–674 (2010).

[24] P.J. Forrester, S. Ole Warnaar, *The importance of the Selberg integral*, Bull. Amer. Math. Soc. **45**, 489–534 (2008).

[25] D.A. Goldston, H.L. Montgomery, *Pair correlation of zeroes and primes in short intervals*, Analytic Number Theory and Diophantine Problems, Birkhäuser, Boston, Mass. 1987, 183–203.

[26] S.M. Gonek, C.P. Hughes, J.P. Keating, *A hybrid Euler–Hadamard product formula for the Riemann zeta function*, Duke Math. J. **136**, 3, 507–549 (2007).

[27] M.C. Gutzwiller, *Periodic orbits and lassical quantization conditions*, J. Math. Phys. **12**, 343–358 (1971).

[28] G.H. Hardy, J.E. Littlewood, *Contributions to the theory of the Riemann zeta-function and the theory of the distributions of primes*, Acta Mathematica **41** (1918).

[29] G.H. Hardy, J.E. Littlewood, *Some Problems of Partitio Numerorum III, On the Expression of a Number as a Sum of Primes*, Acta Math. **44**, 1–70 (1923).

[30] D.R. Heath-Brown, *Fractional moments of the Riemann zeta-function*, J. London Math. Soc. **24**, 65–78 (1981).

[31] D. Hejhal, *On the triple correlation of the zeroes of the zeta function*, IMRN **7**, 293–302 (1994).

[32] R. Holowinsky, K. Soundararajan, *Mass equidistribution for Hecke eigenforms*, Ann. of Math. (2) **172** no. 2, 1517–1528 (2010).

[33] A.E. Ingham, *Mean-value theorems in the theory of the Riemann zeta-function*, Proceedings of the London Mathematical Society **27**, 278–290 (1926).

[34] N.M. Katz, P. Sarnak, *Random Matrices*, Frobenius Eigenvalues and monodromy, American Mathematical Society Colloquium Publications, 45. American Mathematical Society, Providence, Rhode island, 1999.

[35] J.P. Keating, N.C. Snaith, *Random Matrix Theory and $\zeta(1/2 + it)$*, Comm. Math. Phys. **214**, 57–89 (2000).

[36] J.P. Keating, N.C. Snaith, *Random matrix theory and L-functions at $s = 1/2$*, Comm. Math. Phys. **214**, 91–110 (2000).

[37] J. Marklof, *Pair correlation densities of inhomogeneous quadratic forms II*, Duke mathematical journal **115**, No. 3, 409–434 (2002).

[38] M.L. Mehta, *Random matrices*, Third edition, Pure and Applied Mathematics Series **142**, Elsevier, London, 2004.

[39] H.L. Montgomery, *The pair correlation of zeros of the zeta function*, Analytic number theory (Proceedings of Symposium in Pure Mathemathics **24** (St. Louis Univ., St. Louis, Mo., 1972), American Mathematical Society (Providence, R.I., 1973), pp. 181–193.

[40] H.L. Montgomery, R.C. Vaughan, *Hilbert's inequality*, J. London Math. Soc. (2), **8**, 73–82 (1974).

[41] A.M. Odlyzko, *On the distribution of spacings between the zeros of the zeta function*, Math. Comp. **48**, 273–308 (1987).

[42] B. Riemann, *Über die Anzahl der Primzahlen unter einer gegebenen Grösse*, Monatsberichte der Berliner Akademie, Gesammelte Werke, Teubner, Leipzig, 1892.

[43] M. Rubinstein, *Evidence for a spectral interpretation of the zeros of L-functions*, PhD thesis, Princeton University.

[44] Z. Rudnick, P. Sarnak, *Zeroes of principal L-functions and random matrix theory*, Duke Math. J. **81**, no. 2, 269–322 (1996). A celebration of John F. Nash.

[45] A.I. Schnirelman, *Ergodic properties of eigenfunctions*, Usp. Math. Nauk **29**, 181182 (1974).

[46] A. Selberg, *Über einen Satz von A. Gelfond*, Arch. Math. Naturvid. **44**, 159–171 (1941).

[47] A. Selberg, *Old and new conjectures and results about a class of Dirichlet series*, Proceedings of the Amalfi Conference on Analytic Number Theory (Maiori, 1989), 367–385, Univeristà di Salerno, Salerno, 1992.

[48] A. Soshnikov, *Determinantal random point fields*, Russ. Math. Surv. 55 (5), 923–975 (2000).

[49] K. Soundararajan, *Moments of the Riemann zeta-function*, Annals of Mathematics **170**, 981–993 (2009).

[50] E.C. Titchmarsh, *The Theory of the Riemann Zeta Function*, London, Oxford University Press, 1951.

[51] H. Widom, *Toeplitz determinants with singular generating functions*, Amer. J. Math. **95**, 333–383 (1973).

Paul Bourgade
Télécom ParisTech
23, Avenue d'Italie
F-75013 Paris, France
e-mail: `paul.bourgade@polytechnique.org`

Jonathan P. Keating
University of Bristol
University Walk, Clifton
Bristol BS8 1TW, UK
e-mail: `J.P.Keating@bristol.ac.uk`

Chaos in Microwave Resonators

Hans-Jürgen Stöckmann

Abstract. Chaotic billiards are a paradigm of quantum chaos studied theoretically in numerous papers. In flat microwave resonators with cross-sections mimicking the billiard shape there is a one-to-one correspondence between the stationary Schrödinger equation and the Helmholtz equation. This allows an experimental access to questions hitherto studied exclusively theoretically. In the article various aspects of quantum chaos are presented and illustrated by experimental results. It continues with a discussion of random matrices and the universal features of wave functions of chaotic billiards. Next, semiclassical quantum mechanics is introduced, establishing a link between the quantum-mechanical Green function and the classical trajectories. The article ends with a presentation of recent applications of wave-chaos research.

1. Introduction

Up to about 1990 the quantum mechanics of classically chaotic systems, shortly termed "quantum chaos", was essentially a domain of theory [1]. Only two classes of experimental result had been available at that time. First, there were the spectra of compound nuclei giving rise to the development of random matrix theory in the sixties of the last century, and second the experiments with highly excited hydrogen and alkali atoms in strong magnetic or strong radio frequency fields. The situation changed with the appearance of the various types of billiard experiments. After the first microwave study performed in the author's group [2] there were numerous experiments with classical waves on liquid surfaces, in plates, solids, and rods, with electrons in quantum dots, tunnelling barriers and quantum corrals, and with ultra-cold atoms confined in billiards formed by light walls. References and a more detailed account on the subject may be found in Ref. [3]. It will become clear in the following that the difference between classical waves and matter waves is not of relevance, since the universal features we shall discuss are common to all types of wave not necessarily quantum mechanically in origin. This is why "wave chaos" would be a better term to describe this field of research, and there are authors avoiding the term "quantum chaos" as a whole.

This article is organized as follows. After a short introduction into the subject in Section 2, concentrating on the inherent difficulties with the definition of chaos in quantum mechanics, in Section 3 the microwave technique is introduced. In the subsequent three sections various aspects of quantum chaos are presented and illustrated by experimental results. The selection was exclusively guided by the intension to provide experimental illustrations of essential theoretical results. There was not the intent to give an exhaustive overview on the subject. In Section 4 a short introduction into the concept of random matrices is given, illustrating the remarkable observation that a mayor part of the statistical properties of the spectra of chaotic systems can be obtained already if the underlying Hamiltonian is substituted by a matrix the elements of which are chosen at random, only obeying some constraints. In Section 5 universal features of the wave functions of chaotic billiards are discussed. Here a lot of exact results can be obtained from the simple assumption that at each point in the billiard the wave function may be looked upon as a superposition of waves, with the same modulus of the wave vector, entering randomly from all directions. In Section 6, semiclassical quantum mechanics is introduced, based on the disseminating papers by M. Gutzwiller, establishing a link between the quantum-mechanical Green function and the classical trajectories. The article ends with a presentation of recent applications of wave-chaos research.

2. From classical to quantum mechanics

To illustrate the difficulties one is facing with the concept of chaos in quantum mechanics, let us first consider an idealised billiard system, i.e., the classical dynamics of a single particle travelling frictionless through a box with infinitely high walls. For a circular billiard the trajectory is regular (see Figure 1(a)). There are two constants of motion, the total energy E, and the angular momentum L. Since there are two degrees of freedom as well, the system is integrable. Small uncertainties in the initial conditions, such as the distance between two neighbouring trajectories, will therefore increase only linearly in time. The situation is qualitatively different

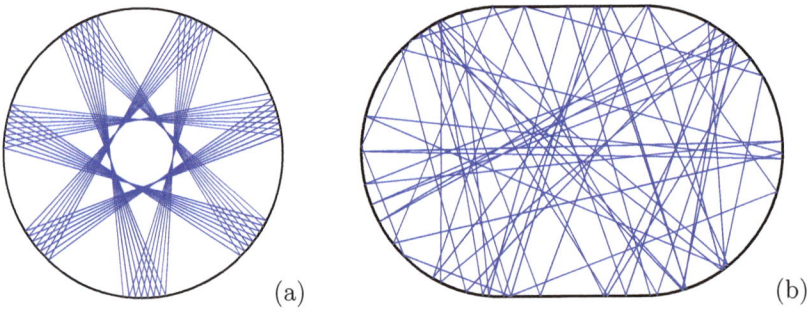

FIGURE 1. Classical trajectories in a circular (a) and a stadium (b) billiard.

for the stadium billiard, the guinea pig in billiard research (see Figure 1(b)). There is only one constant of motion left, the total energy E, and the distance between neighbouring trajectories increases exponentially with time. The stadium billiard thus is chaotic.

In quantum mechanics this distinction between integrable and chaotic systems does not work. The initial conditions are defined only within the limits of the uncertainty relation

$$\Delta x \, \Delta p \leq \frac{1}{2}\hbar, \qquad (1)$$

and the concept of trajectories looses its significance. One may even ask whether quantum chaos does exist at all. Since the Schrödinger equation is linear, a quantum mechanical wave packet can be constructed from the eigenfunctions by the superposition principle. But if the wave packet once has been generated, its evolution for arbitrarily long times is available without any problem. There is no room left for chaos.

On the other hand the correspondence principle demands that there *must* be a relation between *linear* quantum mechanics and *nonlinear* classical mechanics at least in the regime of large quantum numbers. This apparent contradiction has been resolved by semiclassical quantum mechanics, derived in essential parts by M. Gutzwiller in a series of papers (see Ref. [4] for a review). The theory relates classical trajectories to quantum mechanical spectra and wave functions, and this defines the program of quantum chaos research, namely to look for the fingerprints of classical chaos in the quantum mechanical properties of the system.

Billiards are ideally suited systems for this purpose. The numerical calculation of the classical trajectories is elementary, and the stationary Schrödinger equation reduces to a simple wave equation

$$-\frac{\hbar^2}{2m}\left(\frac{\partial^2}{\partial x^2} + \frac{\partial^2}{\partial y^2}\right)\psi_n = E_n \psi_n. \qquad (2)$$

The potential appears only in the boundary condition $\psi_n|_S = 0$, where S is the surface of the billiard. Though billiard systems are conceptually simple, they nevertheless show the full complexity of non-linear systems.

A further advantage is the equivalence of the stationary Schrödinger equation with the time-independent wave equation, the Helmholtz equation

$$-\left(\frac{\partial^2}{\partial x^2} + \frac{\partial^2}{\partial y^2}\right)\psi_n = k_n^2 \psi_n, \qquad (3)$$

where ψ_n now is the amplitude of the wave field.

The equivalence of the stationary Schrödinger equation and the Helmholtz equation opens the opportunity to study questions and to test theories, originally motivated by quantum mechanics, by means of classical waves. The boundary conditions for the classical and the corresponding quantum mechanical systems may differ, but this is not of relevance for the questions to be treated in this article.

FIGURE 2. Chladni giving a public demonstration of his sound figures. © Unknown artist. We tried to trace the rights holder. If someone has legitimate claims, please contact the publisher.

The first experiment of this type dates back already more then 200 years. At the end of the 18th century E. Chladni developed a technique "to make sound visible", by decorating the nodal lines of vibrating plates with grains of sand. Chladni never achieved to get a permanent position at an university and earned his livings by giving demonstrations of his experiments to the public [5]. Figure 2 gives an example. On occasion of a stay in Paris 1808, he got an invitation by Napoleon to give a private performance in the Tuileries [6]. There is a very vivid report by Chladni on this visit [7]:

> "When I entered, he welcomed me, standing in the centre of the room, with the expressions of his favour. Napoleon showed much interest in my

experiments and explanations and asked me, as an expert in mathematical questions, to explain all topics thoroughly, so that I could not take the matter too easy. He was well informed that one is not yet able to apply a calculation to irregularly shaped areas, and that, if one were successful in this respect, it could be useful for applications to other subjects as well."

The last remark had been really visionary! Who could have imagined at that time that Chladni's experiments in a sense mean the starting point of quantum chaos research?

3. Microwave billiards

Modern experimental billiard studies started with microwave resonators [2]. Figure 3 (left) shows a typical set-up. The cavity is formed by a bottom plate sup-

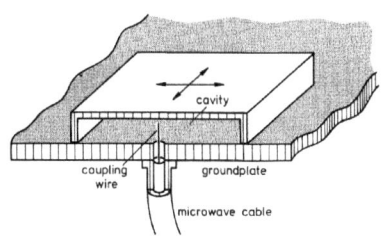 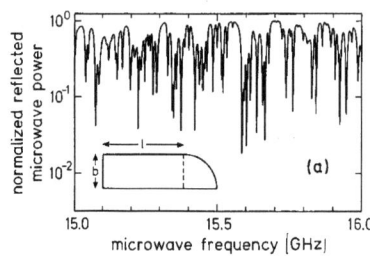

FIGURE 3. Microwave set-up to study spectra and wave functions (left), and a typical microwave reflection spectrum (right). From [2] & [9] with permission from © 1990, 1992 The American Physical Society.

porting the entrance antenna, and by an upper part whose position can be moved with respect to the lower one. As long as a maximum frequency $\nu_{\max} = c/2d$ is not exceeded, where d is the height of the resonator and c the velocity of light, the system can be considered as quasi-two-dimensional. In this situation the electromagnetic wave equations reduce to the scalar Helmholtz equation (3), where ψ_n corresponds to the electric field pointing perpendicularly from the bottom to the top plate. Since the electric field component parallel to the wall must vanish, we have the condition $\psi_n|_S = 0$ on the outer circumference S of the resonator. We have thus arrived at a complete equivalence between a two-dimensional quantum billiard and the corresponding quasi-two-dimensional microwave resonator, including the boundary conditions. As an example Figure 3 (right) shows the reflection spectrum of a microwave resonator of the shape of a quarter stadium [2]. Each minimum in the reflection corresponds to an eigenfrequency of the resonator.

A detailed consideration shows that the measurement directly yields the components of the scattering matrix S, where the diagonal element S_{nn} corresponds to the reflection amplitude at the nth antenna, and the off-diagonal matrix element S_{nm} to the transmission amplitude between antennas n and m. For isolated resonances the components of the scattering matrix are given by

$$S_{ij}(\vec{r}_i, \vec{r}_j, k) = \delta_{ij} - 2\imath\gamma \sum_n \frac{\bar{\psi}_n^*(\vec{r}_i)\bar{\psi}_n(\vec{r}_j)}{k^2 - \bar{k}_n^2 + \imath\Gamma_n}, \qquad (4)$$

where $\bar{k}_n$ and $\bar{\psi}_n(\vec{r}_i)$ are the nth k-eigenvalue and eigenfunction at the position of antenna i. The bar denotes that both quantities are slightly changed as compared to the closed system. γ is a factor describing the antenna coupling, assumed to be equal for all antennas for the sake of simplicity. In addition the resonances acquire a line width Γ_n. Apart from these modifications the microwave measurement yields directly the Green function of the system, and thus the complete quantum mechanical information. Equation (4) is the quantum-mechanical expression of the scattering matrix. The electromagnetical line widths γ_n are related to the Γ_n by $\Gamma_n = k\gamma_n$ [8]. This is a consequence of the different dispersion relations $\omega \sim k$ and $\omega \sim k^2$ for the electro-magnetic and the quantum-mechanical case, respectively. Equation (4) shows that the depth of a resonance contains the information

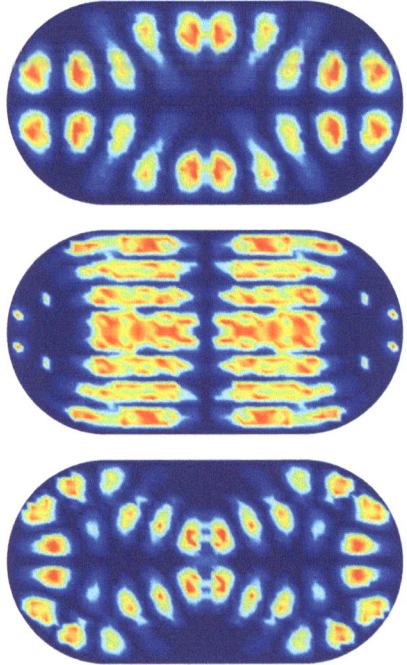

FIGURE 4. Wave functions in a stadium-shaped microwave resonator. The figure shows $|\psi_n(\vec{r})|^2$ in a colour plot. From [9] with permission from © 1992 The American Physical Society.

on the wave functions $\psi_n(\vec{r})$ at the antenna positions. By scanning with the antenna through the billiard $\psi_n(\vec{r})$ may thus be spatially resolved. Figure 4 shows a number stadium wave functions obtained in this way [9]. All wave functions show the phenomenon of scarring, meaning that the wave function amplitudes are not distributed more or less homogeneously over the area, but concentrate along classical periodic orbits. The phenomenon had been first described and termed by E. Heller [10]. The figure could give the impression that scarred wave functions are dominating, but this is only true for the lowest eigenvalues. With increasing energy the fraction of scarred wave functions tends to zero.

From the Fourier transform of the Green function the propagator is obtained, either the electro-magnetic or quantum-mechanical one, depending on the used dispersion relation [11]. Thus also the study of pulse propagation becomes possible.

Microwave billiards have a number of advantages as compared to nuclei: (a) typical wave lengths are of the order of mm to cm, resulting in very convenient sizes for the used resonators, (b) shapes of the resonators, coupling strengths to antennas etc. can be perfectly controlled, (c) parameter variations, e.g., of the coupling strength, the position of an impurity, or of one length can be easily achieved, and (d) last but not least, we have seen that in microwave systems the complete scattering matrix is obtainable, including the phases. This is extraordinary, in standard scattering experiments, such as in in nuclear physics, usually only reduced information, such as scattering cross-sections, is available, resulting in a complete loss of the phase information. This is why a number of predictions of scattering theory have been tested not in nuclei but microwave billiards.

4. Random matrices

In the spring times of nuclear physics in the midst of the last century there appeared a vast amount of experimental results on cross-sections, partial cross-section etc., obtained by bombarding target nuclei with light projectiles. Nearly nothing was known at that time on the origin of the nuclear forces. Could it be expected under such circumstances to obtain any relevant information at all from these irregularly looking spectra without any recognisable pattern? Here one idea showed up to be extremely useful, notwithstanding its obviously oversimplifying nature: If nothing is known on the nuclear Hamiltonian H, just let us take its matrix elements in some basis as random numbers, with only some global constraints, e.g., by taking the matrix H symmetric for systems with, or Hermitian for systems without time-reversal symmetry, and by fixing the variance of its matrix elements. Assuming basis invariance of the distribution of the matrix elements one immediately sees that the matrix elements are uncorrelated and Gaussian distributed [12]. There are three Gaussian ensembles, the orthogonal one (GOE) for time-reversal invariant systems with integer spin, the unitary one (GUE) for systems with broken time-reversal symmetry, and the symplectic one (GSE) for time-reversal invariant systems with half-integer spin. Here "orthogonal" etc. refers to the invariance properties of the respective ensembles. Recently it has been shown by M. Zirnhauer [13] that in fact there are seven additional ensembles, which become relevant, whenever pair-wise creation and annihilation of particles is involved, as it is the case in elementary particle physics and superconductivity. For all Gaussian ensembles exact expressions for many quantities of interest can be calculated explicitly, such as spectral correlation functions, eigenvalue spacings distributions etc.

The quantity most often studied in this context is the distribution of level spacings $p(s)$ normalised to a mean level spacing of one. For 2×2 matrices this quantity can be easily calculated, yielding for the GOE the famous Wigner surmise

$$p(s) = \frac{\pi}{2} s \exp\left(-\frac{\pi}{4} s^2\right). \tag{5}$$

For large matrices Equation (5) is still a good approximation with errors on the percent level. Figure 5 shows level spacings distributions for a variety of chaotic systems, all showing the same behaviour. Such observations had been the motivation for Bohigas, Giannoni and Schmitt [14] to formulate, what is today known as the BGS conjecture: *"Spectra of time-reversal-invariant systems whose classical analogs are K systems show the same fluctuation properties as predicted by GOE (alternative stronger conjectures that cannot be excluded would apply to less chaotic systems, provided that they are ergodic). If the conjecture happens to be true, it will then have been established the* universality of the laws of level fluctuations *in quantal spectra already found in nuclei and to a lesser extent in atoms. Then, they should also be found in other quantal systems, such as molecules, hadrons, etc."*. The paper had been extremely influential and initiated a complete program of quantum chaos research. To be fair one should mention that there had been another paper on the same subject [15] somewhat earlier, which, however, at that time did not find the attendance it would have deserved.

The replacement of H by a random matrix means to abandon any hope to learn more about nuclei from the spectra but some average quantities such as the mean level spacings. Of course this is not the end of the story: there *are* techniques to extract also individual system properties. We shall come back to this aspect in Section 6. But the loss of individual features in the spectra on the other hand means that it might be worthwhile to look for *universal* features being common to all chaotic systems. This approach showed up to be extremely fruitful. It allowed to apply results originally obtained for nuclei to many other systems as well, in particular quantum-dot systems [16] and microwave billiards [3].

In addition to the level spacings distribution in particular spectral correlations related to the spectral auto-correlation function $C(E) = \langle \rho(E_2) \rho(E_1) \rangle - \langle \rho(E_2) \rangle \langle \rho(E_1) \rangle$ are considered, where $E = E_2 - E_1$, and the brackets denote a spectral average. Quantities often studied in literature are number variance and spectral rigidity, see, e.g., Ref. [3] for details. Here another object shall be considered, the spectral form factor $K(t)$, since it will be of importance for the understanding of the relation between random matrix theory and semiclassical quantum

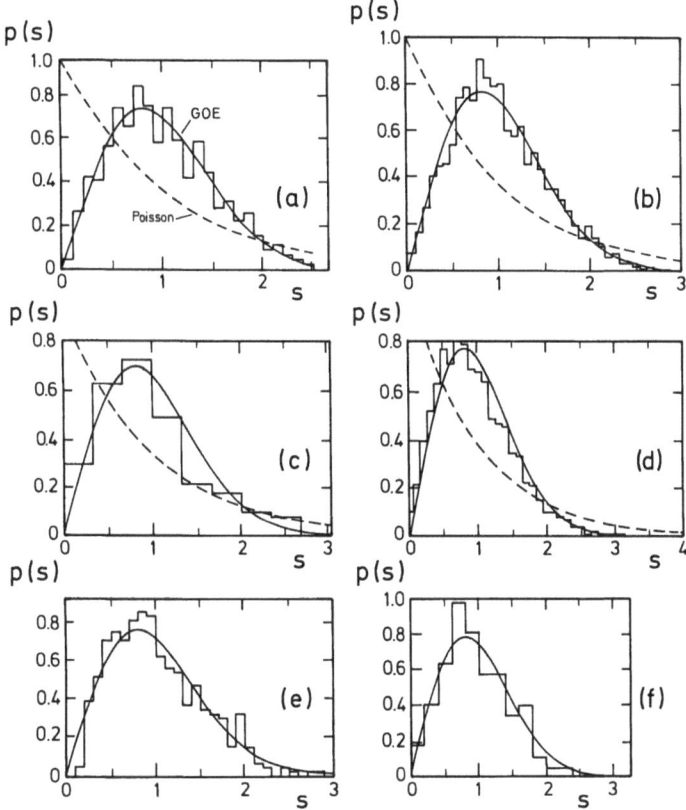

FIGURE 5. Level spacing distribution for a Sinai billiard (a), a hydrogen atom in a strong magnetic field (b), the excitation spectrum of a NO_2 molecule (c), the acoustic resonance spectrum of a Sinai-shaped quartz block (d), the microwave spectrum of a three-dimensional chaotic cavity (e), and the vibration spectrum of a quarter-stadium shaped plate (f). In all cases a Wigner distribution is found though only in the first three cases the spectra are quantum mechanically in origin. From [3] with permission from © 1999 Cambridge University Press.

mechanics to be discussed in Section 6. $K(t)$ is obtained from the Fourier transform of the spectral auto-correlation function. Random matrix theory yields explicit expressions. For later use the results for the GOE and the GUE shall be given (see, e.g., Ref. [1]):

$$K_{\text{GOE}}(t) = \begin{cases} 2t - t\ln(1+2t) \\ 2 - t\ln\frac{2t+1}{2t-1} \end{cases}, \quad K_{\text{GUE}}(t) = \begin{cases} t & t < 1 \\ 1 & t > 1 \end{cases}. \quad (6)$$

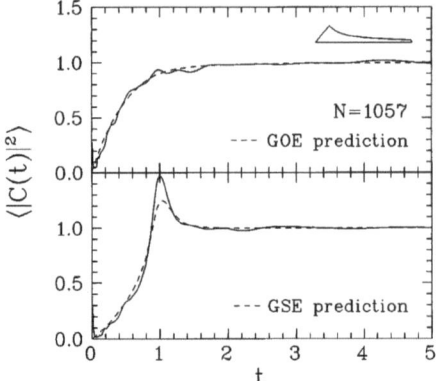

FIGURE 6. Spectral form factor for the spectrum of a microwave hyperbola billiard (top), and for the subspectrum obtained by considering only every second resonance (bottom). From [17] with permission from © 1997 The American Physical Society.

The time is given in units of the Heisenberg time $t_H = \hbar/\langle\Delta E\rangle$, where $\langle\Delta E\rangle$ is the mean level spacings.

Figure 6 shows as an experimental example the spectral form factor for a hyperbola billiard obtained in the group of A. Richter [17]. In the upper part of the figure $K(t)$ for the *complete* spectrum is shown. There is a good agreement with random matrix predictions from the GOE. This is consistent with the fact that microwave billiard systems are time-reversal invariant, and there is no spin. Spectra showing GSE statistics have not yet been studied experimentally, but there is the remarkable fact that GSE spectra can be generated by taking only every second level of a GOE spectrum [12]. Exactly this had been dome with the spectrum of the hyperbola billiard to obtain the spectral form factor in the lower part of the figure, being in perfect agreement with the expected GSE behaviour.

5. The random plane wave approximation

In a disseminating paper on wave functions in the stadium billiard McDonald and Kaufman noticed [18] that for most wave functions the amplitudes ψ are Gaussian distributed,

$$P_\psi(\psi) = \sqrt{\frac{A}{2\pi}} \exp\left(-\frac{A\psi^2}{2}\right), \qquad (7)$$

where A is the billiard area. Exceptions are scarred wave functions of the type shown in Figure 4. For the intensities $\rho = |\psi|^2$ of the wave function follows a Porter–Thomas distribution,

$$P_\rho(\rho) = \sqrt{\frac{A}{2\pi\rho}} \exp\left(-\frac{A\rho}{2}\right). \qquad (8)$$

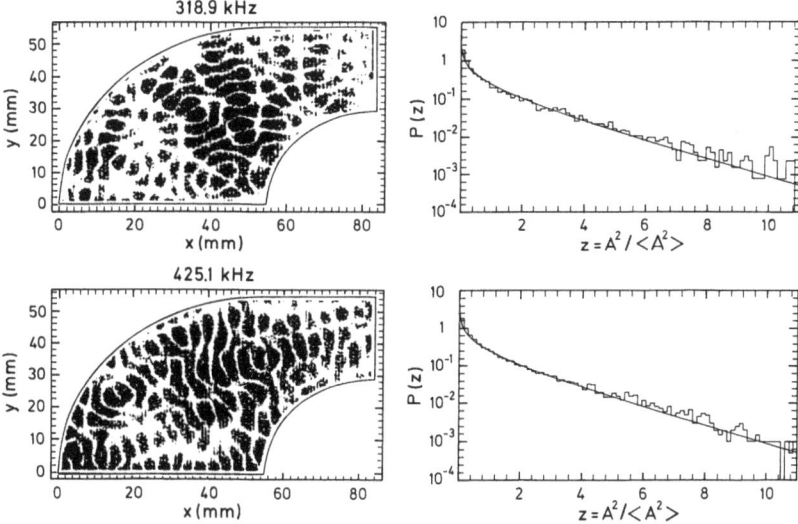

FIGURE 7. Vibration amplitude pattern for two eigenfrequencies of a plate of a quarter Sinai-stadium billiard (left column) and corresponding distribution function for the squared amplitudes. The solid line corresponds to a Porter–Thomas distribution, see Equation (8). From [19] with permission from © 2003 The American Physical Society.

This phenomenology is not restricted to quantum-mechanical systems. Figure 7 shows two vibration patterns for a plate of the shape of a quarter Sinai-stadium, together with the distribution functions for the squared amplitudes [19]. In both cases a perfect agreement with a Porter–Thomas distribution is found.

These findings can be explained under the assumption that at any point of the billiard the wave function may be considered as a superposition of plane waves [21]

$$\psi(k,\vec{r}) = \sum_n a_n e^{i\vec{k}_n \vec{r}}, \qquad (9)$$

entering from random directions $\vec{k}_n/k$, and random amplitudes a_n, but with the modulus $k = |\vec{k}_n|$ of the wave vector fixed. The Gauss distributions of the wave function amplitudes are then an immediate consequence of the central limit theorem. A more thorough foundation of the ansatz can be found in Ref. [22].

V. Doya et al. [20] succeeded in a nice demonstration of the model by sending laser light through a glass fiber with a D-shaped cross-section. The light output directly at the end of the fiber exhibits a chaotically looking speckle pattern, see Figure 8 (left). The right-hand part shows the corresponding far-field pattern, obtained by just putting an additional lens into the optical path. The far-field concentrates on a ring with $|k| = $ const., thus illustrating the validity of the model.

It had been mentioned already that most of the concepts developed in quantum chaos research may be adopted to all types of classical waves as well. Let

FIGURE 8. Near-field (left) and far-field (right) emission patterns after sending laser light through a D-shaped glass fiber (diameter $d = 107\,\mu m$). From [20] with permission from © 2002 The American Physical Society.

us take a three-dimensional microwave resonator as an example. In the general case the Maxwell equations cannot be reduced to the scalar wave equation (3) any longer, and there are altogether three boundary conditions which have to be

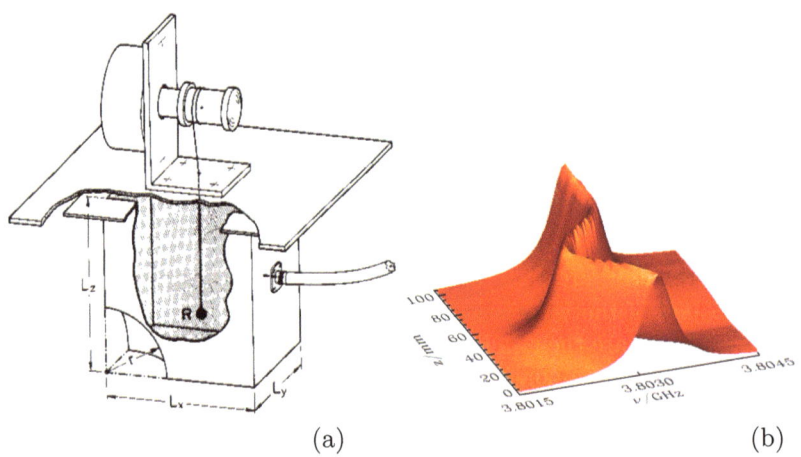

FIGURE 9. Experimental set-up to measure field distributions in a three-dimensional Sinai microwave resonator with $L_x = 96\,\text{mm}$, $L_y = 82\,\text{mm}$, $L_z = 106\,\text{mm}$, $r = 39\,\text{mm}$ (a) and frequency-shift spectrum for one resonance obtained by pulling the sphere perpendicularly through the resonator (b).

obeyed on metallic surfaces, namely

$$\nabla_\perp E|_S = 0, \qquad \nabla_\| B|_S = 0, \qquad (10)$$

where $\nabla_\perp$, $\nabla_\|$ are the components of the gradient, perpendicular and parallel to the surface.

The resonator used in the experiment consisted of a rectangular box with the octant of a sphere removed from one of the corners. It is thus a three-dimensional analog of a Sinai billiard. A metallic sphere suspended at a thread could be pulled through the resonator by means of a step motor. Figure 9 shows the experimental set-up, together with an example of a frequency shift measurement.

The fields have to adjust to the boundary conditions at the surface of the sphere, generating a frequency shift $\Delta\nu$ of the eigenfrequencies proportional to $-2\vec{E}^2 + \vec{B}^2$. Typical field distributions obtained in this way are shown in the left column of Figure 10 [23]. To visualise the distributions, a surface of constant frequency shift, i.e., of a constant value of $-2\vec{E}^2 + \vec{B}^2$, has been shaded. The examples shown allow an easy interpretation. The field distribution in the upper panel corresponds to a standing wave between the vertical planes, whereas the field distribution shown in the central row is scarred along a periodic orbit of the shape of a diamond. Finally, the field distribution in the bottom panel has a completely chaotic appearance.

To substantiate these observations, on the right column of Figure 10 the distributions of frequency shifts $\Delta\nu \sim -2\vec{E}^2 + \vec{B}^2$ are plotted. Under the assumption that all six field components $E_x, \ldots, B_z$ are uncorrelated and Gaussian distributed, the frequency shift distribution function can be calculated. The result is the solid line. A perfect agreement with the experiment is found for the chaotically looking field distribution of Figure 10(c), whereas for the two other examples the solid line is not able to describe the experimental findings.

What does this mean? At first sight the assumption of uncorrelated field components seems to be very strong. After all the components of $\vec{E}$ and $\vec{B}$ are intimately linked via the Maxwell equations. But if we assume again that the chaotic field distribution can be obtained by a superposition of plane waves, now of electromagnetic waves, the correlations disappear and one ends up exactly with the applied model of uncorrelated Gaussian distributed field components. The same model which had been applied with great success to quantum billiards, thus does work equally well for electromagnetic systems.

6. Semiclassical quantum mechanics

Before quantum mechanics had been established in its present form by Heisenberg, Schrödinger and others, it had been Bohr, and later Born and Sommerfeld, who developed a technique today known as semi-classical to calculate the spectrum of atomic hydrogen. At that time Einstein [24] argued that this approach must be a dead end, since semi-classical quantisation needs invariant tori in the phase space,

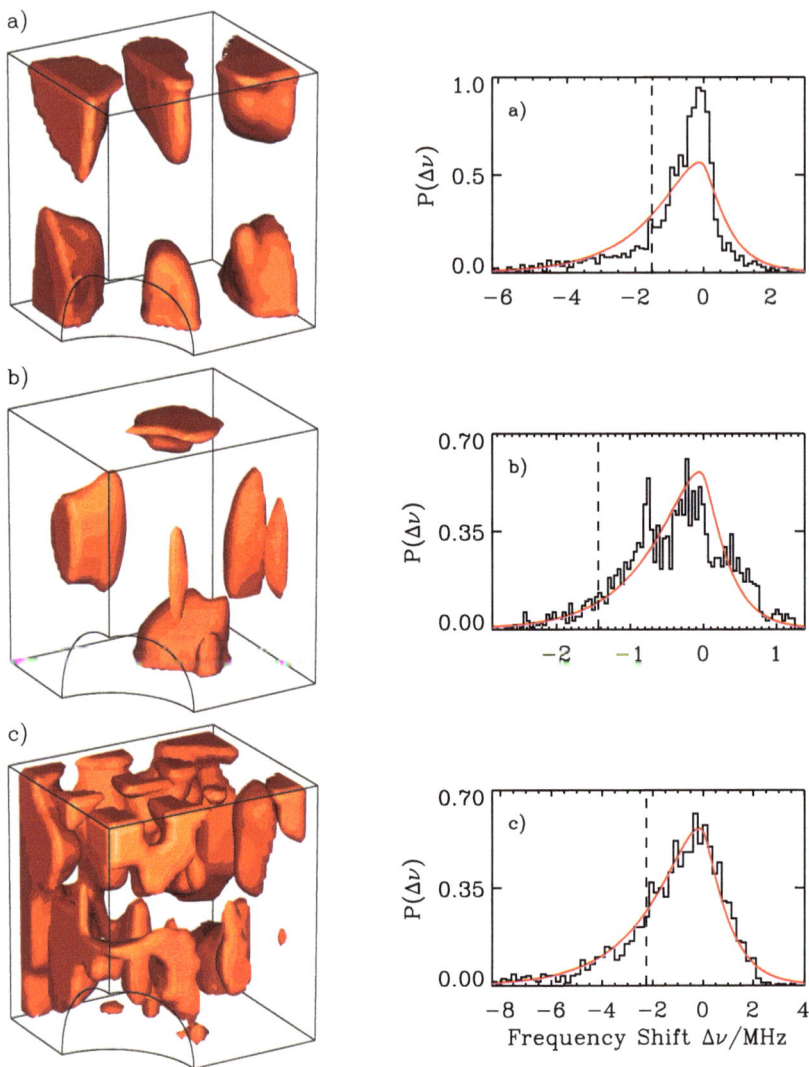

FIGURE 10. Field distributions for three typical eigenfrequencies at $\nu = 5.208\,\text{GHz}$ (a), $2.897\,\text{GHz}$ (b), $8.293\,\text{GHz}$ (c) for the three-dimensional Sinai resonator (left column), and corresponding frequency shift distribution (right column). From [23] with permission from © 1998 The American Physical Society.

preventing a semi-classical quantisation for non-integrable systems. This would mean that the technique applied hitherto would work for hydrogen-like systems only, since already the second element in the periodic system, Helium, is non-integrable. This had been one of the rare cases, where Einstein was wrong, though it needed half a century until M. Gutzwiller [4] showed in a series of papers that chaotic systems, too, allow for a semi-classical quantisation.

Starting point of his approach is Feynman's path integral for the quantum-mechanical propagator

$$K(q_A, q_B, t) = \int \mathcal{D}(q) W(q) \exp\left[\frac{i}{\hbar} \int_0^t L(q, \dot{q}) \, dt\right],\quad (11)$$

giving the probability amplitude for a particle to propagate in time t from q_A to q_B. The expression on the right-hand side means in a symbolical short-hand notation an integral over all pathes, classically allowed or forbidden, from q_A to q_B, where $W(q)$ takes into account the stability of the trajectory, and $L(q, \dot{q})$ is the Lagrange function. It is remarkable that both quantities are purely classical, quantum mechanics entering only via the $1/\hbar$ in the exponential. Expression (11) is not particularly useful for an explicit calculation of the propagator, since it involves a multi-dimensional integral, but it is very well suited as a starting point for a semi-classical approach.

In the semiclassical limit the phase factors are strongly fluctuating quantities, and only terms contribute significantly, where the phase is stationary, i.e., where the condition

$$\delta \int L(q, \dot{q}, t) dt = 0 \quad (12)$$

is met, where the variation is over all paths connecting q_A and q_B. But this is exactly Hamilton's principle of classical mechanics, from which the classical equations of motion can be derived. Expressed in other words: In the semi-classical limit only the classically allowed paths contribute to the path integral (11)!

This shall be demonstrated by two experimental examples. For the density of states Gutzwiller's approach yields his famous trace formula. It becomes particularly simple in billiard systems, if the wavenumber k is taken as the variable. In terms of k the density of states reads

$$\rho(k) = \rho_0(k) + \sum_n A_n e^{ikl_n}. \quad (13)$$

The first term varies smoothly with k and is given in its leading term by Weyl's formula

$$\rho_0(k) = \frac{A}{2\pi} k, \quad (14)$$

where A is the area of the billiard. The second term is heavily oscillating with k. The sum runs over all periodic orbits including repetitions. l_n is the length of the orbit, and A_n is a factor characterising the stability of the orbit.

The periodic orbit sum (13) divergent, and resummation techniques are needed to calculate the spectrum from the periodic orbits. But the inverse procedure, namely to extract the contributions of the different periodic orbits out of the spectra, is straightforward. For billiards the Fourier transform of the fluctuating part of the density of states,

$$\hat{\rho}_{osc}(l) = \int \rho_{osc}(k) e^{-\imath k l} \, dk$$
$$= \sum_n A_n \delta(l - l_n), \qquad (15)$$

directly yields the contributions of the orbits to the spectrum [2]. Each orbit gives rise to a delta peak at an l value corresponding to its length, and a weight corresponding to the stability factor of the orbit.

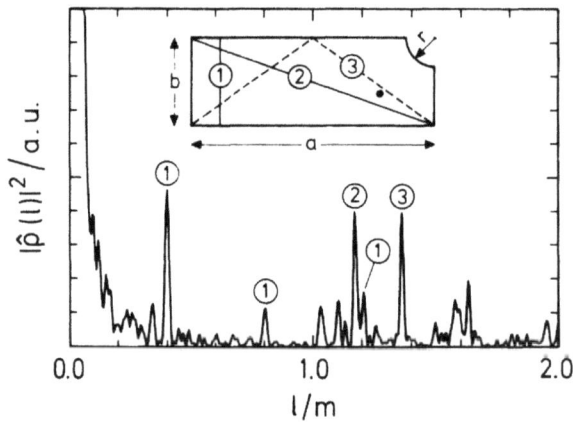

FIGURE 11. Squared modulus $|\hat{\rho}(t)|^2$ of the Fourier transform of the spectrum of a quarter Sinai billiard ($a = 56$ cm, $b = 20$ cm, $r = 7$ cm). Each resonance can be associated with a classical periodic orbit. From [2] with permission from © 1990 The American Physical Society.

Figure 11 shows for illustration the squared modulus of the Fourier transform of the spectrum of a microwave resonator shaped as a quarter Sinai billiard [2]. Each peak corresponds to a periodic orbit of the billiard. For the bouncing ball orbit, labelled by ①, three peaks associated with repeated orbits are clearly visible. The smooth part of the density of states is responsible for the increase of $|\hat{\rho}(k)|^2$ at small lengths.

In the second example results on the transport through a microwave "quantum dot", i.e., a resonator with two attached channels, shall be presented. For the transmission amplitude the semiclassical approach yields

$$t(k) = \sum_n A_n e^{\imath k l_n} . \qquad (16)$$

The sum is taken over all classical trajectories connecting the input and output channels. l_n now is the length of the nth trajectory with stability factor A_n, and k is the wavenumber. By taking the Fourier transform of the transmission, one

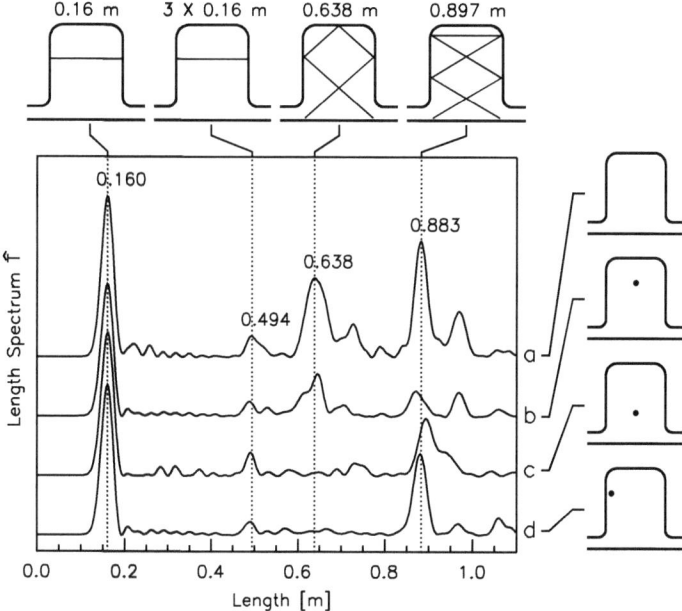

FIGURE 12. Length spectra $\hat{T} = |\hat{t}|^2$ obtained by a Fourier transformation of the transmission spectrum of a microwave "quantum dot", for the clean cavity (a), and with the impurity located at three different positions (indicated, b–d). Each maximum in the length spectrum corresponds to a trajectory connecting the entrance with the exit port. Whenever the absorber hits a classical trajectory, the corresponding resonance in the length spectrum is destroyed. From [25] with permission from © 2003 The American Physical Society.

obtains again the stability-weighted length spectrum, allowing for an immediate identification of the relevant transport pathes:

$$\hat{t}(l) = \frac{1}{2\pi} \int t(k) e^{-ikl} dk = \sum_n A_n \delta(l - l_n). \tag{17}$$

In Figure 12, length spectra $\hat{T} = |\hat{t}|^2$, such obtained are presented. For the empty the cavity (a), the length spectrum shows a number of peaks which can be associated with classical trajectories, shown in the upper part of the figure. With an absorber placed in the cavity, the magnitude of the length peaks depends sensitively on the absorber position position: whenever the absorber lies close to the semiclassical trajectory associated with a particular resonance in the length spectrum, the resonance is destroyed. If, on the other hand, the absorber misses the trajectory, the resonance remains intact. With this technique the length peaks can be associated unambiguously with the corresponding trajectories.

Semiclassical quantum mechanics relates the spectrum to the classical periodic orbits of the system, i.e., to *individual* system properties. In view of this fact one may wonder where the *universal* features discussed in Section 3 come in. To answer this question let us have a look onto the spectral form factor $K(t)$, introduced in Section 3 as the Fourier transform of the spectral autocorrelation function. Entering here with the trace formula, a semiclassical expression for $K(t)$ is obtained, which for billiards read

$$K(t) = \sum_{n,m} A_n^* A_m \delta\left[t - (t_n + t_m)/2\right] e^{ik(l_n - l_m)t}. \tag{18}$$

$K(t)$ shows peaks at times $(t_n + t_m)/2$, where $t_n = l_n/c$ is the period for the nth orbit. For short times all these peaks are well separated. It is essentially this what is seen in Figure 11. But with increasing time, because of the exponential proliferation of the long orbits in chaotic systems, the density of these peaks becomes so large that only an overall increase of $K(t)$ can be observed. This is the onset of the universal regime. In the so-called diagonal approximation it is argued that only terms with $n = m$ contribute, since the off-diagonal contributions are averaged out by the phase factor $e^{ik(l_n - l_m)t}$. For systems with time-reversal symmetry one has to consider in addition that all orbits come in pairs corresponding to clock- and counter-clock-wise propagation. Under this assumption Equation (18) yields $K_{\text{GOE}} = 2t$ and $K_{\text{GUE}} = t$ being just the leading terms of the expressions given in Equation (6) for $t < 1$. This had been shown already 30 years ago by M. Berry [26]. It needed 25 more years until it was recognised by M. Sieber and K. Richter [27] that there are off-diagonal terms giving non-negligible contributions. They are associated with pairs of orbits of the topology of the digit "8", one of them self-intersecting, the other one with an avoided crossing instead. The corresponding off-diagonal contributions give the next term in the series expansion of $K(t)$. For the GUE this contribution is zero since here the first-order contribution gives already the exact result (as long as $t < 1$). The program was completed by the Haake group [28] who showed that all orders in the expansion can be recovered by considering bundles of orbits with more and more crossings and avoided crossings. In a final step they achieved to extend the validity of the approach to times $t > 1$ [29]. This may be considered as the proof of the BGS conjecture, at least on the level of the spectral form factor.

7. Applications

Occasionally people doubt whether chaotic systems really need an extra quantum-mechanical treatment. The Schrödinger equation after all gives exact results both for regular and chaotic systems. Random matrix theory and the random-plane-wave approach, on the other hand mean caricatures of the true situation, and the semiclassical approach at best gives approximate results for high energies. Nowadays the numerical solution of the Schrödinger equation is no challenge any longer even for fairly complicated systems. Why should one resort to old-fashioned techniques which had been abandoned already 80 years ago, after the development of "correct" quantum mechanics had been been completed?

First, the semiclassical methods had not been developed as an alternative to the Schrödinger equation to calculate spectra of chaotic systems. This had been successful in exceptional cases only. And random matrix theory had been devised from the very beginning as a tool to understand the universal features of the spectra, but obviously not to get any information on the individual properties.

But the numerical solution of the Schrödinger equation means a black-box calculation, and the human brain is not adopted to perform fast Fourier transforms. This is why spectra as the one shown in Figure 3 seemingly do not contain any relevant information. But the brain is extremely good in identifying pathes and trajectories, and therefore the representation of the spectra in terms of classical trajectories, as shown in Figure 11 and Figure 12 allows an immediate suggestive interpretation.

All this is not just *l'art pour l'art*, as shall be demonstrated by two recent examples.

The relation between wave propagation and classical trajectories had become of practical importance in the development of microlasers. It had been found by Gmachl et al. [31] in quadrupolarly deformed dielectric disks that the strongest emission does not occurs at the points of largest curvature. J. Nöckel and D. Stone [32] proposed the classical phase space properties to be responsible for this at first sight counterintuitive behaviour. Again this shall be demonstrated by a microwave study. The left part of Figure 13 shows the snapshot of the pulse propagation in a dielectric quadrupole resonator made of teflon. The pulse had been

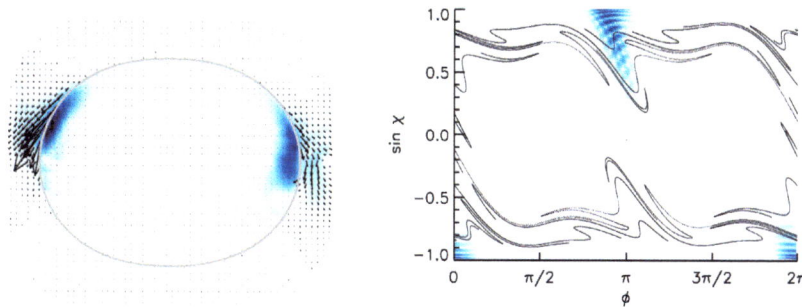

FIGURE 13. Snapshot of the pulse propagation in a dielectric quadrupole cavity made of teflon (length of the long axis $l = 113$ mm). The left figure shows the pulse intensity inside the teflon at the moment of strongest emission in a colour plot. In addition the Poynting vector is shown in the region outside of the teflon. The right figure shows the Husimi distribution of the pulse in a Poincaré plot. In addition the unstable manifold of the rectangular orbit is shown. From [30] with permission from © 2006 The American Physical Society.

generated by a Fourier transform from the experimentally determined scattering matrix $S_{12}(\vec{r}_1, \vec{r}_2, k)$, with $\vec{r}_1$ fixed, and $\vec{r}_2$ variable, as described in Section 3. The pulse starts as an outgoing circular wave from an antenna close to the boundary in the lower part of the cavity, but already after a short time only two pulses survive circulating clock- and counter-clockwise close to the border. For the figure a moment has been selected where there is a particularly strong emission to the outside. In the right part of the figure the same situation is shown in a Poincaré plot, with the polar angle as the abscissa, and the sine of the incidence angle as the ordinate. In a Poincaré plot each trajectory is mapped onto a sequence of points representing the reflections at the boundary. The intricate tongue-like structure represents the instable manifold of the rectangle. It had been obtained by starting a trajectory with a minute deviation from the instable rectangular periodic orbit. In addition the Husimi representation of the pulse in the right-hand part of the figure is shown. A Husimi distribution maybe looked upon as a decomposition od a quantum-mechanical wave function in terms of wave packets of minimum uncertainty, and is a convenient tool to establish a correspondence between wave propagation and classical trajectories (see, e.g., Ref. [33]). Now it becomes obvious why the strongest emission does *not* occur at the points of strongest emission. Teflon has an index of refraction of $n = 1.44$ meaning a $\sin \chi_{\text{crit}} = 0.69$ for the critical angle of total reflection. Thus the circulating pulses are trapped by total

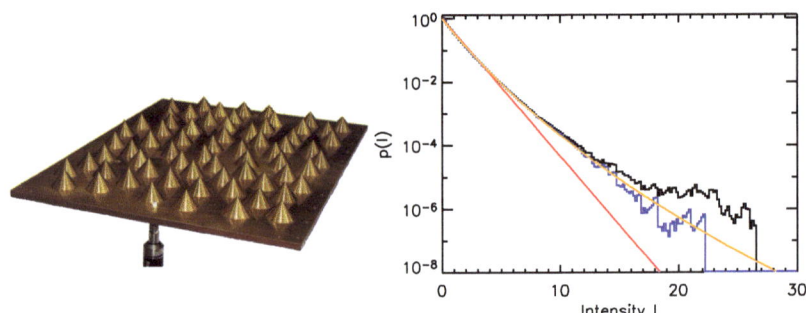

FIGURE 14. Experimental set-up for the freak wave study (left). The platform has a width of 260 mm, and a length of 360 mm. The source antenna is mounted in the bottom plate close to the border in the centre of the short side. The probe antenna is fixed in the top plate (not shown) and can be scanned in two horizontal dimensions to map the field distribution within the scattering arrangement. The right figure shows the probability distribution of intensities (black). If hot spot regions are excluded from the histogram, the found distribution is in accordance with multi-scattering theory, plotted in orange (light gray). The red (dark gray) line corresponds to the Rayleigh distribution. From [36] with permission from © 2010 The American Physical Society.

reflection, apart from a weak tunnelling escape due the curvature of the boundary. But whenever the critical line of total reflection is surpassed, there is a strong escape. This happens exactly in the region of the most pronounced tongues of the instable manifold of the rectangular orbit. It is thus this orbit, acting as a dynamical barrier, which is responsible for the observed emission behaviour [34]. In the following years "phase space engineering" has become an important tool towards the ultimate goal to construct a microcavity with unidirectional emission [35].

Another application was motivated by still poorly understood features in the propagation of ocean waves. It had been known for many years that the random plane waves model underestimates the probability for very high amplitudes, called freak or monster waves, by orders of magnitude. Usually non-linearities are blamed for these findings, but it had been found in a recent microwave study [36] that already in the linear regime there are strong deviations from the simple model. The left-hand part of Figure 14 shows the used set-up. Microwaves are propagating through an arrangement of randomly distributed cones acting as scatterers. The right-hand part of the figure shows the found intensity distribution. From the random plane wave approach one would expect a distribution according to $p(I) \sim \exp(-I/\langle I \rangle)$, known as Rayleigh's law. The actually found probability densities for high intensities are larger by orders of magnitudes. First, there is a multiple scattering correction, which had been known already for some time [37]. But in addition there is another contribution which could be attributed to the formation of hot spots due to caustics in the potential landscape generated by the

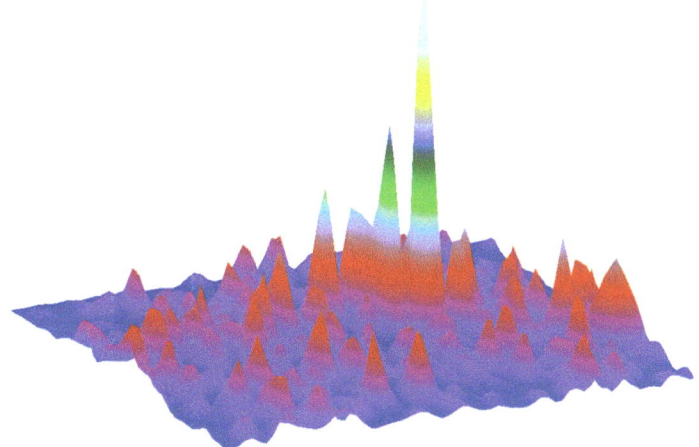

FIGURE 15. A "hot spot", observed at a frequency of 8.85 GHz. The experimental probability density for observing such a hot spot is one to two orders of magnitude larger than that expected from multiple scattering theory. From [36] with permission from © 2010 The American Physical Society.

cones. Figure 15 shows one of these hot spots. Without calling into question the importance of non-linearities for the generation of freak waves in principle, a better understanding of non-random features the linear regime is obviously needed.

The number of activities on the transport of different types of waves, light, seismic waves, water waves, sound waves etc. through disordered media is steadily increasing, many of them based on theories and techniques originally developed in wave and quantum chaos. After several decades of basic research the time for applications has come.

Most of the experimental examples presented in this article have been performed in the author's group at the university of Marburg. I want to thank all my coworkers for their excellent work. In particular I would like to mention my senior coworker U. Kuhl who shared the responsibility for the experiments with me over many years. The experiments had been funded by the Deutsche Forschungsgemeinschaft by numerous individual grants, and during the last three year via the research group 760 "Scattering systems with complex dynamics".

References

[1] F. Haake, *Quantum Signatures of Chaos. 2nd edition* (Springer, Berlin, 2001).
[2] H.-J. Stöckmann and J. Stein, Phys. Rev. Lett. **04**, 2215 (1990).
[3] H.-J. Stöckmann, *Quantum Chaos – An Introduction* (University Press, Cambridge, 1999).
[4] M.C. Gutzwiller, *Chaos in Classical and Quantum Mechanics, Interdisciplinary Applied Mathematics, Vol. 1* (Springer, New York, 1990).
[5] D. Ullmann, Eur. Phys. J. Special Topics **145**, 25 (2007).
[6] H.-J. Stöckmann, Eur. Phys. J. Special Topics **145**, 15 (2007).
[7] F. Melde, *Chladnis's Leben und Wirken* (N.G. Elwert'sche Verlagsbuchhandlung, Marburg, 1888).
[8] U. Kuhl, H.-J. Stöckmann, and R. Weaver, J. Phys. A **38**, 10433 (2005).
[9] J. Stein and H.-J. Stöckmann, Phys. Rev. Lett. **68**, 2867 (1992).
[10] E.J. Heller, Phys. Rev. Lett. **53**, 1515 (1984).
[11] J. Stein, H.-J. Stöckmann, and U. Stoffregen, Phys. Rev. Lett. **75**, 53 (1995).
[12] M.L. Mehta, *Random Matrices. 2nd edition* (Academic Press, San Diego, 1991).
[13] M.R. Zirnbauer, J. Math. Phys. **37**, 4986 (1996).
[14] O. Bohigas, M.J. Giannoni, and C. Schmit, Phys. Rev. Lett. **52**, 1 (1984).
[15] G. Casati, F. Valz-Gris, and I. Guarnieri, Lett. Nuov. Cim. **28**, 279 (1980).
[16] C.W.J. Beenakker, Rev. Mod. Phys. **69**, 731 (1997).
[17] H. Alt *et al.*, Phys. Rev. E **55**, 6674 (1997).
[18] S.W. McDonald and A.N. Kaufman, Phys. Rev. Lett. **42**, 1189 (1979).
[19] K. Schaadt, T. Guhr, C. Ellegaard, and M. Oxborrow, Phys. Rev. E **68**, 036205 (2003).

[20] V. Doya, O. Legrand, F. Mortessagne, and C. Miniatura, Phys. Rev. Lett. **88**, 014102 (2002).
[21] M.V. Berry, J. Phys. A **10**, 2083 (1977).
[22] S. Hortikar and M. Srednicki, Phys. Rev. Lett. **80**, 1646 (1998).
[23] U. Dörr, H.-J. Stöckmann, M. Barth, and U. Kuhl, Phys. Rev. Lett. **80**, 1030 (1998).
[24] A. Einstein, Verhandlungen der Deutschen Physikalischen Gesellschaft **19**, 82 (1917).
[25] Y.-H. Kim *et al.*, Phys. Rev. B **68**, 045315 (2003).
[26] M.V. Berry, Proc. R. Soc. Lond. A **400**, 229 (1985).
[27] M. Sieber and K. Richter, Phys. Scr. **T90**, 128 (2001).
[28] S. Müller *et al.*, Phys. Rev. Lett. **93**, 014103 (2004).
[29] S. Heusler *et al.*, Phys. Rev. Lett. **98**, 044103 (2007).
[30] R. Schäfer, U. Kuhl, and H.-J. Stöckmann, New J. of Physics **8**, 46 (2006).
[31] C. Gmachl *et al.*, Science **280**, 1556 (1998).
[32] J.U. Nöckel and A.D. Stone, Nature **385**, 45 (1997).
[33] A. Bäcker, S. Fürstberger, and R. Schubert, Phys. Rev. E **70**, 036204 (2004).
[34] H.G.L. Schwefel *et al.*, J. Opt. Soc. Am. B **21**, 923 (2004).
[35] J. Wiersig and M. Hentschel, Phys. Rev. Lett. **100**, 033901 (2008).
[36] R. Höhmann *et al.*, Phys. Rev. Lett. **104**, 093901 (2010).
[37] T.M. Nieuwenhuizen and M.C.W. van Rossum, Phys. Rev. Lett. **74**, 2674 (1995).

Hans-Jürgen Stöckmann
Fachbereich Physik der Philipps-Universität Marburg
Renthof 5
D-35032 Marburg, Germany
e-mail: `stoeckmann@physik.uni-marburg.de`

Anatomy of Quantum Chaotic Eigenstates

Stéphane Nonnenmacher

Abstract. We present one aspect of "Quantum Chaos", namely the description of high frequency eigenmodes of a quantum system, the classical limit of which is chaotic (we call such eigenmodes "chaotic eigenmodes"). A paradigmatic example is provided by the eigenmodes of the Laplace–Beltrami operator on a compact Riemannian manifold of negative curvature: the corresponding classical dynamics is the geodesic flow on the manifold, which is strongly chaotic (Anosov). Other well-studied classes of examples include certain Euclidean domains ("chaotic billiards"), or quantized chaotic symplectomorphisms of the two-dimensional torus.

We propose several levels of description, some of them allowing for mathematical rigor, others being more heuristic.

The macroscopic distribution of the eigenstates makes use of semiclassical measures, which are probability measures invariant w.r.t. the classical dynamics; these measures reflect the asymptotic "shape" of a sequence of high frequency eigenmodes. The quantum ergodicity theorem states that the vast majority of the eigenstates is associated with the "flat" (Liouville) measure. A major open problem addresses the existence of "exceptional" eigenmodes admitting different macroscopic properties.

The microscopic description deals with the structure of the eigenfunctions at the scale of their wavelengths. It is mainly of statistical nature: it addresses, for instance, the value distribution of the eigenfunctions, their short-distance correlation functions, the statistics of their nodal sets or domains. This microscopic description mostly relies on a random state (or random wave) Ansatz for the chaotic eigenmodes, which is far from being mathematically justified, but already offers interesting challenges for probabilists and harmonic analysts.

I am grateful to E. Bogomolny, who allowed me to reproduce several plots from [26]. The author has been partially supported by the Agence Nationale de la Recherche under the grant ANR-09-JCJC-0099-01. These notes were written while he was visiting the Institute of Advanced Study in Princeton, supported by the National Science Foundation under agreement No. DMS-0635607.

1. Introduction

These notes present a description of *quantum chaotic eigenstates*, that is bound states of quantum dynamical systems, whose classical limit is chaotic. The classical dynamical systems we will be dealing with are mostly of two types: geodesic flows on Euclidean domains ("billiards") or compact Riemannian manifolds, and canonical transformations on a compact phase space; the common feature is the "chaoticity" of the dynamics. The corresponding quantum systems will always be considered within the semiclassical (or high-frequency) régime, in order to establish a connection them with the classical dynamics. As a first illustration, we plot below two eigenstates of a paradigmatic system, the Laplacian on the *stadium billiard*, with Dirichlet boundary conditions[1].

The study of chaotic eigenstates makes up a large part of the field of *quantum chaos*. It is somewhat complementary with the contribution of J. Keating (who will focus on the statistical properties of quantum spectra, another major topic in quantum chaos). I do not include the study of eigenstates of quantum graphs (a recent interesting development in the field), since this question should be addressed in U. Smilansky's lecture. Although these notes are purely theoretical, H.-J. Stöckmann's lecture will show that the questions raised have direct experimental applications (his lecture should present experimentally measured eigenmodes of two- and three-dimensional "billiards").

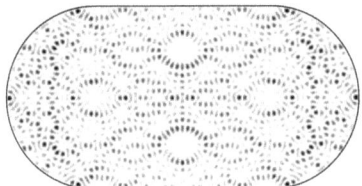 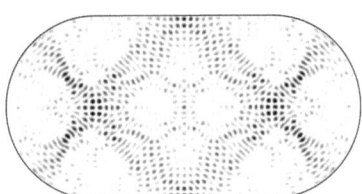

FIGURE 1. Two eigenfunctions of the Dirichlet Laplacian in the stadium billiard, with wavevectors $k = 60.196$ and $k = 60.220$ (see (9)). Large values of $|\psi(x)|^2$ correspond to dark regions, while nodal lines are white. While the left eigenfunction looks relatively "ergodic", the right one is *scarred* by two symmetric periodic orbits (see §4.2).

One common feature of the chaotic eigenfunctions (except in some very specific systems) is the absence of explicit, or even approximate, formulas. One then has to resort to indirect, rather unprecise approaches to describe these eigenstates. We will use various analytic tools or points of view: deterministic/statistical, macro/microscopic, pointwise/global properties, generic/specific systems. The level of rigour in the results varies from mathematical proofs to heuristics, generally supported by numerical experiments. The necessary selection of results reflects

---

[1]The eigenfunctions of the stadium plotted in this article were computed using a code nicely provided to me by E. Vergini, which uses the *scaling method* invented in [94].

my personal view or knowledge of the subject, it omits several important developments, and is more "historical" than sharply up-to-date. The list of references is thick, but in no way exhaustive.

These notes are organized as follows. We introduce in Section 2 the classical dynamical systems we will focus on (mostly geodesic flows and maps on the two-dimensional torus), mentioning their degree of "chaos". We also sketch the quantization procedures leading to quantum Hamiltonians or propagators, whose eigenstates we want to understand. We also mention some properties of the semiclassical/high-frequency limit. In Section 3 we describe the *macroscopic* properties of the eigenstates in the semiclassical limit, embodied by their *semiclassical measures*. These properties include the *quantum ergodicity* property, which for some systems (with arithmetic symmetries) can be improved to quantum *unique ergodicity*, namely the fact that all high-frequency eigenstates are "flat" at the macroscopic scale; on the opposite, some specific systems allow the presence of exceptionally localized eigenstates. In Section 4 we focus on more refined properties of the eigenstates, many of *statistical* nature (value distribution, correlation functions). Very little is known rigorously, so one has to resort to models of *random wavefunctions* to describe these statistical properties. The large values of the wavefunctions or Husimi densities are discussed, including the *scar phenomenon*. Section 5 discusses the most "quantum" or microscopic aspect of the eigenstates, namely their *nodal sets*, both in position and phase space (Husimi) representations. Here as well, the random state models are helpful, and lead to interesting questions in probability theory.

2. What is a quantum chaotic eigenstate?

In this section we first present a general definition of the notion of "chaotic eigenstates". We then focus our attention to geodesic flows on Euclidean domains or on compact Riemannian manifolds, which form the simplest systems proved to be chaotic. Finally we present some discrete time dynamics (chaotic canonical maps on the two-dimensional torus).

2.1. A short review of quantum mechanics

Let us start by recalling that classical mechanics on the phase space $T^*\mathbb{R}^d$ can be defined, in the Hamiltonian formalism, by a real-valued function $H(x,p)$ on that phase space, called the Hamiltonian. We will always assume the system to be autonomous, namely the function H to be independent of time. This function then generates the flow[2]

$$(x(t), p(t)) = \Phi_H^t(x(0), p(0)), \qquad t \in \mathbb{R},$$

---

[2] We always assume that the flow is complete, that is it does not blow up in finite time.

by solving Hamilton's equations:
$$\dot{x}_j(t) = \frac{\partial H}{\partial p_j}(x(t), p(t)), \quad \dot{p}_j = -\frac{\partial H}{\partial x_j}(x(t), p(t)). \qquad (1)$$

This flow preserves the symplectic form $\sum_j dp_j \wedge dx_j$, and the energy shells $\mathcal{E}_E = H^{-1}(E)$.

The corresponding quantum mechanical system is defined by an operator $\hat{H}_\hbar$ acting on the (quantum) Hilbert space $\mathcal{H} = L^2(\mathbb{R}^d, dx)$. This operator can be formally obtained by replacing coordinates x, p by operators:
$$\hat{H}_\hbar = H(\hat{x}_\hbar, \hat{p}_\hbar), \qquad (2)$$

where $\hat{x}_\hbar$ is the operator of multiplication by x, while the momentum operator $\hat{p}_\hbar = \frac{\hbar}{i}\nabla$, is conjugate to $\hat{x}$ through the $\hbar$-Fourier transform $\mathcal{F}_\hbar$. The notation (2) assumes that one has selected a certain ordering between the operators $\hat{x}_\hbar$ and $\hat{p}_\hbar$; in physics one usually chooses the fully symmetric ordering, also called the *Weyl quantization*: it has the advantage to make $\hat{H}_\hbar$ a self-adjoint operator on $L^2(\mathbb{R}^d)$. Quantization procedures can also be defined when the Euclidean space $\mathbb{R}^d$ is replaced by a compact manifold M. We will not describe it in any detail.

The quantum dynamics, which governs the evolution of the wavefunction $\psi(t) \in \mathcal{H}$ describing the system, is then given by the Schrödinger equation:
$$i\hbar \frac{\partial \psi(x,t)}{\partial t} = [\hat{H}_\hbar \psi](x,t). \qquad (3)$$

Solving this linear equation produces the propagator, that is the family of unitary operators on $L^2(\mathbb{R}^d)$,
$$U_\hbar^t = \exp(-i\hat{H}_\hbar/\hbar), \quad t \in \mathbb{R}.$$

Remark 1. In physical systems, Planck's constant $\hbar$ is a fixed number, which is of order 10^{-34} in SI units. However, if the system (atom, molecule, "quantum dot") is itself microscopic, the value of $\hbar$ may be comparable with the *typical action* of the system, in which case it is more natural to select units in which $\hbar = 1$. Our point of view throughout this work will be the opposite: we will assume that $\hbar$ is (very) small compared with the typical action of the system, and many results will be valid asymptotically, in the *semiclassical limit* $\hbar \to 0$.

2.2. Quantum-classical correspondence

At this point, let us introduce the crucial semiclassical property of the quantum evolution: it is called (in the physics literature) the quantum-classical correspondence, while in mathematics this result is known as Egorov's theorem. This property states that the evolution of observables approximately commutes with their quantization. For us, an observable is a smooth, compactly supported function on phase space $f \in C_c^\infty(T^*\mathbb{R}^d)$. The evolution of classical and quantum evolutions are defined by duality with that of particles/wavefunctions:
$$f(t) = f \circ \Phi_H^t, \qquad \hat{f}_\hbar(t) = U_\hbar^{-t} \hat{f}_\hbar U_\hbar^t.$$

The quantum-classical correspondence connects these two evolutions:
$$\forall t \in \mathbb{R}, \qquad f_\hbar(t) = \widehat{f(t)}_\hbar + \mathcal{O}(e^{\Gamma|t|}\hbar), \tag{4}$$
where the exponent $\Gamma > 0$ depends on the flow and on the observable f.

The most common form of dynamics is the motion of a scalar particle in an electric potential $V(x)$. It corresponds to the Hamiltonian
$$H(x,p) = \frac{|p|^2}{2m} + qV(x), \quad \text{quantized into} \quad \hat{H}_\hbar = -\frac{\hbar^2 \Delta}{2m} + qV(x). \tag{5}$$
We will usually scale the mass and electric charge to $m = q = 1$, keeping $\hbar$ small. Since the Hamilton flow (1) leaves each energy energy shell $\mathcal{E}_E$ invariant, we may restrict our attention to the flow on a single shell. We will be interested in cases where

1. the energy shell $\mathcal{E}_E$ is *bounded* in phase space (that is, both the positions and momenta of the particles remain finite at all times). This is the case if $V(x)$ is confining ($V(x) \to \infty$ as $|x| \to \infty$).
2. the flow on $\mathcal{E}_E$ is *chaotic* (and this is also the case on the neighbouring shells $\mathcal{E}_{E+\epsilon}$). "Chaos" is a vague word, which we will make more precise below.

The first condition implies that, provided $\hbar$ is small enough, the spectrum of $\hat{H}_\hbar$ is purely discrete near the energy E, with eigenstates $\psi_{\hbar,j} \in L^2(\mathbb{R}^d)$ (*bound states*). Besides, fixing some small $\epsilon > 0$ and letting $\hbar \to 0$, the number of eigenstates of $\hat{H}_\hbar$ with eigenvalues $E_{\hbar,j} \in [E - \epsilon, E + \epsilon]$ typically grows like $C\hbar^{-d}$. Under the second condition, the eigenstates with energies in this interval can be called *quantum chaotic eigenstates*.

Below we describe several degrees of "chaos", which regard the *long time properties* of the classical flow. These properties are relevant when describing the eigenstates of the quantum system, which form the "backbone" of the long time quantum dynamics. The main objective of quantum chaos consists in connecting, in a precise way, the classical and quantum long time (or time independent) properties.

2.3. Various levels of chaos

For most Hamiltonians of the form (5) (e.g., the physically relevant case of a hydrogen atom in a constant magnetic field), the classical dynamics on bounded energy shells $\mathcal{E}_E$ involves both regular and chaotic regions of phase space; one then speaks of a *mixed dynamics* on $\mathcal{E}_E$. The regular region is composed of a number of "islands of stability", made of quasiperiodic motion structured around stable periodic orbits; these islands are embedded in a "chaotic sea" where trajectories are unstable (they have a positive Lyapunov exponent). These notions of "island of stability" versus "chaotic sea" are rather poorly understood mathematically, but have received compelling numerical evidence [68]. The main conjecture concerning the corresponding quantum system, is that most eigenstates are either localized in the regular region, or in the chaotic sea [80]. To my knowledge this conjecture remains fully open at present, in part due to our lack of understanding of the classical dynamics.

For this reason, I will restrict myself (as most researches in quantum chaos do) to the case of systems admitting a *purely chaotic* dynamics on $\mathcal{E}_E$. I will allow various degrees of chaos, the minimal assumption being the *ergodicity* of the flow Φ_H^t on $\mathcal{E}_E$, with respect to the natural (Liouville) measure on $\mathcal{E}_E$. This assumption means that, for almost any initial position $\mathbf{x}_0 \in \mathcal{E}_E$, the time averages of any observable f converge to its phase space average:

$$\lim_{T \to \infty} \frac{1}{2T} \int_{-T}^{T} f(\Phi^t(\mathbf{x}_0)) \, dt = \int_{\mathcal{E}_E} f(\mathbf{x}) \, d\mu_L(\mathbf{x}) \stackrel{\text{def}}{=} \int f(\mathbf{x}) \, \delta(H(\mathbf{x}) - E) \, d\mathbf{x} \, . \quad (6)$$

A stronger chaotic property is the *mixing* property, or decay of time correlations between two observables f, g:

$$C_{f,g}(t) \stackrel{\text{def}}{=} \int_{\mathcal{E}_E} g \times (f \circ \Phi^t) \, d\mu_L - \int f \, d\mu_L \int g \, d\mu_L \xrightarrow{t \to \infty} 0 \, . \quad (7)$$

The rate of mixing depends on both the flow Φ^t and the regularity of the observables f, g. For very chaotic flows (Anosov flows, see §2.4.2) and smooth observables, the decay is exponential.

2.4. Geometric quantum chaos

In this section we give explicit examples of chaotic flows, namely geodesic flows in a Euclidean billiard, or on a compact manifold. The dynamics is then induced by the geometry, rather than a potential. Both the classical and quantum properties of these systems have been investigated a lot in the past 30 years.

2.4.1. Billiards. The simplest form of ergodic system occurs when the potential $V(x)$ is an infinite barrier delimiting a bounded domain $\Omega \subset \mathbb{R}^d$ (say, with piecewise smooth boundary), so that the particle moves freely inside Ω and bounces specularly at the boundaries. For obvious reasons, such a system is called a *Euclidean billiard*. All positive energy shells are equivalent to one another, up to a rescaling of the velocity, so we may restrict our attention to the shell $\mathcal{E} = \{(x,p), x \in \Omega, |p| = 1\}$. The long time dynamical properties only depend on the shape of the domain. For instance, in 2 dimensions, a rectangular, circular or elliptic billiards lead to an *integrable* dynamics: the flow admits two independent integrals of motion – in the case of the circle, the energy and the angular momentum. A convex billiard with a smooth boundary will always admit some stable "whispering gallery" stable orbits. On the opposite, the famous *stadium billiard* (see Figure 2) was proved to be ergodic by Bunimovich [32]. Historically, the first Euclidean billiard proved to be ergodic was the Sinai billiard, composed of one or several circular obstacles inside a square (or torus) [91]. These billiards also have positive Lyapunov exponents (meaning that almost all trajectories are exponentially unstable, see the left part of Figure 2). It has been shown more recently that these billiards are *mixing*, but with correlations decaying at polynomial or subexponential rates [34, 14, 73].

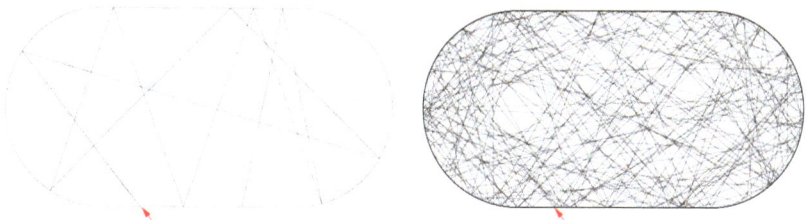

FIGURE 2. Left: Two trajectories in the stadium billiard, initially very close to one another, and then diverging fast (the red arrow shows the initial point). Right: Long evolution of one of the trajectories.

The quantization of the broken geodesic flow inside Ω is the (semiclassical) Laplacian with Dirichlet boundary conditions:

$$\hat{H}_\hbar = -\frac{\hbar^2 \Delta_\Omega}{2}. \tag{8}$$

Obviously, the parameter $\hbar$ only amounts to a rescaling of the spectrum: an eigenstate $\psi_\hbar$ of (8) with energy $E_\hbar \approx 1/2$ is also an eigenstate of $-\Delta_\Omega$ with eigenvalue $k^2 \approx \hbar^{-2}$. Hence, $\hbar$ represents the *wavelength* of $\psi_\hbar$, the inverse of its *wavevector* k. Fixing $E = 1/2$ and taking the semiclassical limit $\hbar \to 0$ is equivalent with studying the high-frequency or high-wavevector spectrum of $-\Delta_\Omega$.

The system (8) is often called a *quantum billiard*, although this operator is not only relevant in quantum mechanics, but in all sorts of wave mechanics (see H.-J. Stöckmann's lecture). Indeed, the scalar Helmholtz equation

$$\Delta \psi_j + k_j^2 \psi_j = 0, \tag{9}$$

may describe stationary acoustic waves in a cavity. This equation is also relevant to describe electromagnetic waves in a quasi-2D cavity, provided one is allowed to separate the different polarization components of the electric field.

Euclidean billiards thus form the simplest *realistic* quantized chaotic systems, for which the classical dynamics is well understood at the mathematical level. Besides, the spectrum of the Dirichlet Laplacian can be numerically computed up to large values of k using methods specific to the Euclidean geometry, like the scaling method [94]. For these reasons, these billiards have become a paradigm of quantum chaos studies.

2.4.2. Anosov geodesic flows. The strongest form of chaos occurs in systems (maps or flows) with the *Anosov property*, also called *uniformly hyperbolic systems* [5]. The first (and main) example of an Anosov flow is given by the *geodesic flow on a compact Riemannian manifold* (M, g) *of negative curvature*, generated by the free particle Hamiltonian $H(x, p) = |p|_g^2 /2$. Uniform hyperbolicity – which is induced by the negative curvature of the manifold – means that at each point $\mathbf{x} \in \mathcal{E}$ the tangent space $T_\mathbf{x}\mathcal{E}$ splits into the vector $X_\mathbf{x}$ generating the flow, the unstable subspace $E_\mathbf{x}^+$ and the stable subspace $E_\mathbf{x}^-$. The stable (resp. unstable) subspace is

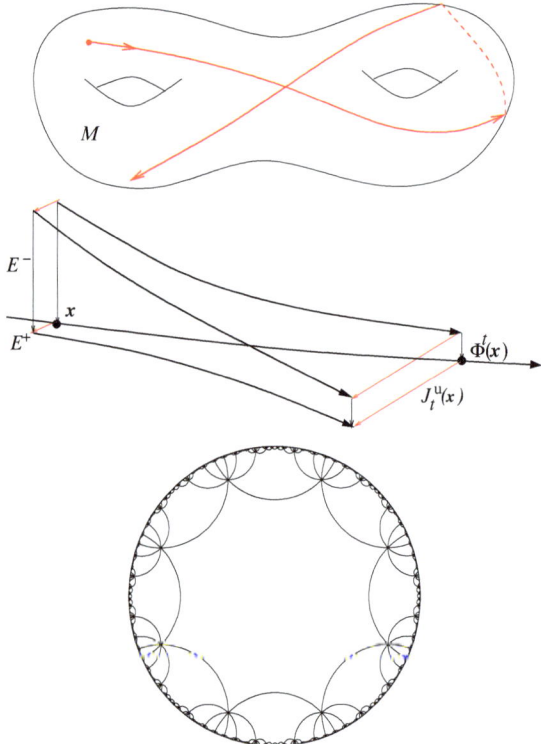

FIGURE 3. Top: Geodesic flow on a surface of negative curvature (such a surface have a genus ≥ 2). Middle: A phase space trajectory and two nearby trajectories approaching it in the future or past. The stable/unstable directions at $\mathbf{x}$ and $\Phi^t(\mathbf{x})$ are shown. The red lines feature the expansion along the unstable direction, measured by the unstable Jacobian $J_t^u(\mathbf{x}) = \det(d\Phi^t \restriction_{E_\mathbf{x}^+})$. Botton: Fundamental domain for an "octagon" surface $\Gamma\backslash\mathbb{H}$ of constant negative curvature. With permission from © C. McMullen.

defined by the property that the flow contracts vectors exponentially in the future (resp. in the past), see Figure 3:
$$\forall v \in E_\mathbf{x}^\pm, \forall t > 0, \quad \|d\Phi_\mathbf{x}^{\mp t} \cdot v\| \leq C\, e^{-\lambda t} \|v\|. \tag{10}$$
Anosov systems present the strongest form of chaos, but their ergodic properties are (paradoxically) better understood than for the billiards of the previous section. The flow has a positive *complexity*, reflected in the exponential proliferation of long periodic geodesics.

For this reason, this geometric model has been at the center of the mathematical investigations of quantum chaos, in spite of its minor physical relevance.

Generalizing the case of the Euclidean billiards, the quantization of the geodesic flow on $(M - g)$ is given by the (semiclassical) Laplace–Beltrami operator,

$$\hat{H}_\hbar = -\frac{\hbar^2 \Delta_g}{2}, \tag{11}$$

acting on the Hilbert space is $L^2(M, dx)$ associated with the Lebesgue measure. The eigenstates of $\hat{H}_\hbar$ with eigenvalues $\approx 1/2$ (equivalently, the high-frequency eigenstates of $-\Delta_g$) constitute a class of quantum chaotic eigenstates, whose study is not impeded by boundary problems present in billiards.

The spectral properties of this Laplacian have interested mathematicians working in Riemannian geometry, PDEs, analytic number theory, representation theory, for at least a century, while the specific "quantum chaotic" aspects have emerged only in the last 30 years.

The first example of a manifold with negative curvature is the Poincaré half-space (or disk) $\mathbb{H}$ with its hyperbolic metric $\frac{dx^2+dy^2}{y^2}$, on which the group $SL_2(\mathbb{R})$ acts isometrically by Moebius transformations. For certain discrete subgroups Γ of $PSL_2(\mathbb{R})$ (called co-compact lattices), the quotient $M = \Gamma \backslash \mathbb{H}$ is a smooth compact surface. This group structure provides detailed information on the spectrum of the Laplacian (for instance, the Selberg trace formula explicitly connects the spectrum with the periodic geodesics of the geodesic flow).

Furthermore, for some of these discrete subgroups Γ, (called *arithmetic*), one can construct a commutative algebra of Hecke operators on $L^2(M)$, which also commute with the Laplacian; it then make sense to study in priority the joint eigenstates of Δ and of these Hecke operators, which we will call the Hecke eigenstates. This arithmetic structure provides nontrivial information on these eigenstates (see §3.3.3), so these eigenstates will appear several times along these notes. Their study composes a part of *arithmetic quantum chaos*, a lively field of research.

2.5. Classical and quantum chaotic maps

Beside the Hamiltonian or geodesic flows, another model system has attracted much attention in the dynamical systems community: chaotic maps on some compact phase space $\mathcal{P}$. Instead of a flow, the dynamics is given by a discrete time transformation $\kappa : \mathcal{P} \to \mathcal{P}$. Because we want to quantize these maps, we require the phase space $\mathcal{P}$ to have a symplectic structure, and the map κ to preserve this structure (in other words, κ is an invertible canonical transformation on $\mathcal{P}$).

The advantage of studying maps instead of flows is multifold. Firstly, a map can be easily constructed from a flow by considering a *Poincaré section* Σ transversal to the flow; the induced return map $\kappa_\Sigma : \Sigma \to \Sigma$, together with the return time, contain all the dynamical information on the flow. Ergodic properties of chaotic maps are usually easier to study than their flow counterpart. For billiards, the natural Poincaré map to consider is the boundary map κ_Σ defined on the phase space associated with the boundary, $T^*\partial\Omega$. The ergodic of this boundary map were understood, and used to address the case of the billiard flow itself [14].

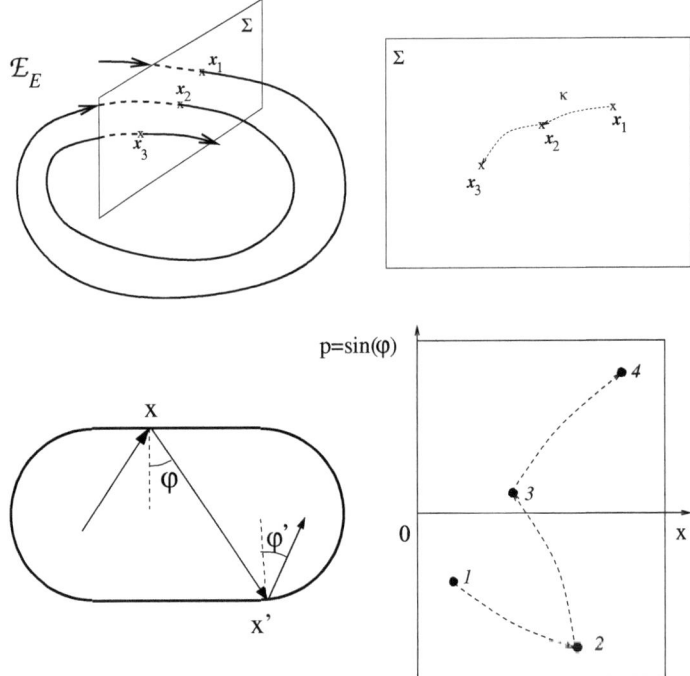

FIGURE 4. Top: Poincaré section and the associated return map constructed from a Hamiltonian flow on $\mathcal{E}_E$. Bottom: Boundary map associated with the stadium billiard.

Secondly, simple chaotic maps can be defined on low-dimensional phase spaces, the most famous ones being the hyperbolic symplectomorphisms on the two-dimensional torus. These are defined by the action of a matrix $S = \begin{pmatrix} a & b \\ c & d \end{pmatrix}$ with integer entries, determinant unity and trace $a + d > 2$ (equivalently, S is unimodular and hyperbolic). Such a matrix obviously acts on $\mathbf{x} = (x,p) \in \mathbb{T}^2$ linearly, through

$$\kappa_S(\mathbf{x}) = (ax + bp, cx + dp) \bmod 1. \tag{12}$$

A schematic view of κ_S for the famous *Arnold's cat map* $S_{\text{cat}} = \begin{pmatrix} 1 & 1 \\ 1 & 2 \end{pmatrix}$ is shows in Figure 5. The hyperbolicity condition implies that the eigenvalues of S are of the form $\{e^{\pm\lambda}\}$ for some $\lambda > 0$. As a result, κ_S has the Anosov property: at each point $\mathbf{x}$, the tangent space $T_{\mathbf{x}}\mathbb{T}^2$ splits into stable and unstable subspaces, identified with the eigenspaces of S, and $\pm\lambda$ are the Lyapunov exponents. Many dynamical properties of κ_S can be explicitly computed. For instance, every rational point $\mathbf{x} \in \mathbb{T}^2$ is periodic, and the number of periodic orbits of period $\leq n$ grows like $e^{\lambda n}$ (thus λ also measures the complexity of the map). This linearity also results in the

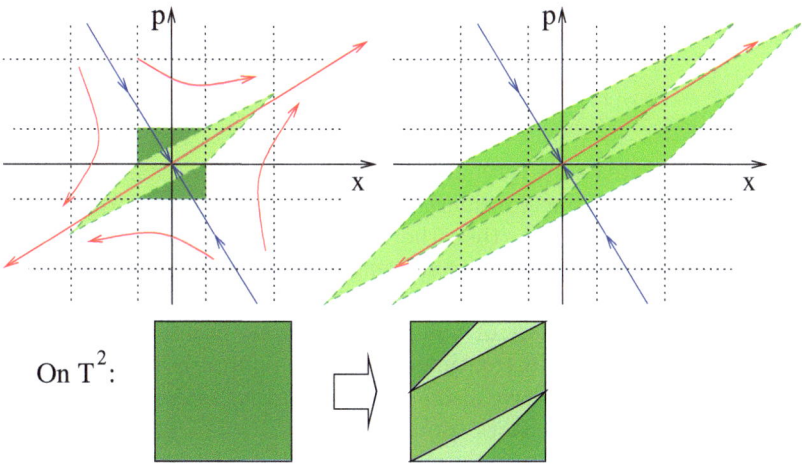

FIGURE 5. Construction of Arnold's cat map $\kappa_{S_{cat}}$ on the 2-torus, obtained by periodizing the linear transformation on $\mathbb{R}^2$. The stable/unstable directions are shown. With permission from © F. Faure.

fact that the decay of correlations (for smooth observables) is superexponential instead of exponential for a generic Anosov diffeomorphism. More generic Anosov diffeomorphisms of the 2-torus can be obtained by smoothly perturbing the linear map κ_S. Namely, for a given Hamiltonian $H \in C^\infty(\mathbb{T}^2)$, the composed map $\Phi_H^\epsilon \circ \kappa_S$ remains Anosov if ϵ is small enough, due to the structural stability of Anosov diffeomorphisms.

Another family of canonical maps on the torus was also much investigated, namely the so-called *baker's maps*, which are piecewise linear. The simplest (sym-

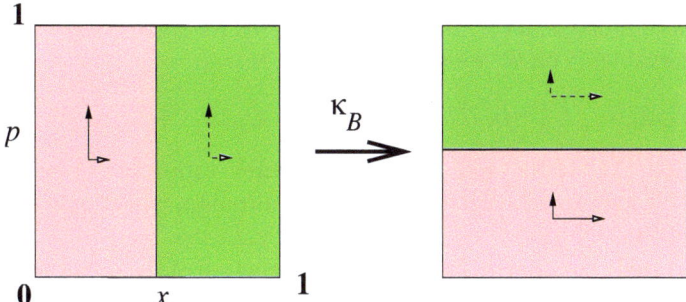

FIGURE 6. Schematic view of the baker's map (13). The arrows show the contraction/expansion directions.

metric) baker's map is defined by

$$\kappa_B(x,p) = \begin{cases} (2x \bmod 1, \frac{p}{2}), & 0 \leq x < 1/2, \\ (2x \bmod 1, \frac{p+1}{2}), & 1/2 \leq x < 1. \end{cases} \quad (13)$$

This map is conjugated to a very simple *symbolic dynamics*, namely the shift on two symbols. Indeed, if one considers the binary expansions of the coordinates $x = 0, \alpha_1\alpha_2\cdots$, $p = \beta_1\beta_2\cdots$, then the map $(x,p) \mapsto \kappa_B(x,p)$ equivalent with the shift to the left on the bi-infinite sequence $\cdots \beta_2\beta_1 \cdot \alpha_1\alpha_2 \cdots$. This conjugacy allows to easily prove that the map is ergodic and mixing, identify all periodic orbits, and provide a large set of nontrivial invariant probability measures. All trajectories not meeting the discontinuity lines are uniformly hyperbolic.

Simple canonical maps have also be defined on the 2-sphere phase space (like the kicked top), but their chaotic properties have, to my knowledge, not been rigorously proven. Their quantization has been intensively investigated.

2.5.1. Quantum maps on the two-dimensional torus. As opposed to the case of Hamiltonian flows, there is no natural rule to quantize a canonical map on a compact phase space $\mathcal{P}$. Already, associating a quantum Hilbert space to this phase space is not obvious. Therefore, from the very beginning, quantum maps have been defined through somewhat arbitrary (or rather, *ad hoc*) procedures, often specific to the considered map $\kappa : \mathcal{P} \to \mathcal{P}$. Still these recipes are always required to satisfy a certain number of properties:

- one needs a *sequence* of Hilbert spaces $(\mathcal{H}_N)_{N \in \mathbb{N}}$ of dimensions N. Here N is interpreted as the inverse of Planck's constant, in agreement with the heuristics that *each quantum state occupies a volume $\hbar^d$ in phase space*. We also want to quantize observables $f \in C(\mathcal{P})$ into Hermitian operators $\hat{f}_N$ on $\mathcal{H}_N$.
- For each $N \geq 1$, the quantization of κ is given by a unitary propagator $U_N(\kappa)$ acting on $\mathcal{H}_N$. The whole family $(U_N(\kappa))_{N \geq 1}$ is called the quantum map associated with κ.
- in the semiclassical limit $N \sim \hbar^{-1} \to \infty$, this propagator satisfies some form of quantum-classical correspondence. Namely, for some (large enough) family of observables f on $\mathcal{P}$, we should have

$$\forall n \in \mathbb{Z}, \quad U_N^{-n} \hat{f}_N U_N^n = \widehat{(f \circ \kappa^n)}_N + \mathcal{O}_n(N^{-1}) \quad \text{as } N \to \infty. \quad (14)$$

The condition (14) is the analogue of the Egorov property (4) satisfied by the propagator $U_\hbar$ associated with a quantum Hamiltonian, which quantizes the stroboscopic map $\mathbf{x} \mapsto \Phi_H^1(\mathbf{x})$.

Let us briefly summarize the explicit construction of the quantizations $U_N(\kappa)$, for the maps $\kappa : \mathbb{T}^2 \to \mathbb{T}^2$ presented in the previous section. Let us start by constructing the quantum Hilbert space. One can see $\mathbb{T}^2$ as the quotient of the phase space $T^*\mathbb{R} = \mathbb{R}^2$ by the discrete translations $\mathbf{x} \mapsto \mathbf{x} + \mathbf{n}$, $\mathbf{n} \in \mathbb{Z}^2$. Hence, it is natural to construct quantum states on $\mathbb{T}^2$ by starting from states $\psi \in L^2(\mathbb{R})$,

and requiring the following periodicity properties
$$\psi(x+n_1) = \psi(x), \quad (\mathcal{F}_\hbar \psi)(p+n_2) = (\mathcal{F}_\hbar \psi)(p), \quad n_1, n_2 \in \mathbb{Z}.$$
It turns out that these two conditions can be satisfied only if $\hbar = (2\pi N)^{-1}$, $N \in \mathbb{N}$, and the corresponding states (which are actually distributions) then form a vector space $\mathcal{H}_N$ of dimension N. A basis of this space is given by the Dirac combs
$$e_j(x) = \frac{1}{\sqrt{N}} \sum_{\nu \in \mathbb{Z}} \delta(x - \frac{j}{N} - \nu), \quad j = 0, \ldots, N-1. \tag{15}$$
It is natural to equip $\mathcal{H}_N$ with the Hermitian structure for which the basis $\{e_j, j = 0, \ldots, N-1\}$ is orthonormal.

Let us now explain how the linear symplectomorphisms κ_S are constructed [48]. Given a unimodular matrix S, its action on $\mathbb{R}^2$ can be generated by a quadratic polynomial $H_S(x, p)$; this action can thus be quantized into the unitary operator $U_\hbar(S) = \exp(-i\hat{H}_{S,\hbar}/\hbar)$ on $L^2(\mathbb{R})$. This operator also acts on distributions $\mathcal{S}'(\mathbb{R})$, and in particular on the finite subspace $\mathcal{H}_N$. Provided the matrix S satisfies some "checkerboard condition", one can show (using group theory) that the action of $U_\hbar(S)$ on $\mathcal{H}_N$ *preserves that space*, and acts on it through a unitary matrix $U_N(S)$. The family of matrices $(U_N(S))_{N \geq 1}$ defines the quantization of the map κ_S on $\mathbb{T}^2$. Group theory also implies that an *exact* quantum-classical correspondence holds (that is, the remainder term in (14) vanishes), and has other important consequences regarding the operators $U_N = U_N(\kappa_S)$ (for each N the matrix U_N is periodic, of period $T_N \leq 2N$). Explicit expressions for the coefficients matrices $U_N(S)$ can be worked out, they depends quite sensitively on the arithmetic properties of the dimension N.

The construction of the quantized baker's map (13) proceeds very differently. An Ansatz was proposed by Balasz–Voros [13], with the following form (we assume that N is an even integer):
$$U_N(\kappa_B) = F_N^* \begin{pmatrix} F_{N/2} & \\ & F_{N/2} \end{pmatrix}, \tag{16}$$
where F_N is the N-dimensional discrete Fourier transform. This Ansatz is obviously unitary. It was guided by the fact that the phases of the matrix elements of the block-diagonal matrix can be interpreted as the discretization of the generating function $S(p, x) = 2px$ for the map (13). A proof that the matrices $U_N(\kappa_B)$ satisfy the Egorov property (14) was given in [39].

Once we have constructed the matrices $U_N(\kappa)$ associated with a chaotic map κ, their eigenstates $\{\psi_{N,j}, j = 1, \ldots, N\}$ enjoy the rôle of quantum chaotic eigenstates. They are of quite different nature from the eigenstates of the Laplacian on a manifold or a billiard: while the latter belong to $L^2(M)$ or $L^2(\Omega)$ (and are actually smooth functions), the eigenstates $\psi_{N,j}$ are N-dimensional vectors. Still, part of "quantum chaos" has consisted in developing common tools to analyze these eigenstates, in spite of the different functional settings.

3. Macroscopic description of the eigenstates

In this section we will study the *macroscopic* localization properties of chaotic eigenstates. Most of the results are mathematically rigorous.

In the case of the semiclassical Laplacian (8) on a billiard Ω we will ask the following question:

> Consider a sequence $(\psi_\hbar)_{\hbar \to 0}$ of normalized eigenstates of $\hat{H}_\hbar$, with energies $E_\hbar \approx 1/2$. For $A \subset \Omega$ a fixed subdomain, what is the probability that the particle described by the stationary state $\psi_\hbar$ lies inside A? How do the probability weights $\int_A |\psi_\hbar(x)|^2\, dx$ behave when $\hbar \to 0$?

This question is quite natural, when contemplating eigenstate plots like in Figure 1. Here, by macroscopic we mean that the domain A is kept fixed while $\hbar \to 0$.

One can obviously generalize the question to integrals of the type

$$\int_\Omega f(x) |\psi_\hbar(x)|^2\, dx,$$

with $f(x)$ a continuous test function on Ω. This integral can be interpreted as the matrix element $\langle \psi_\hbar | \hat{f}_\hbar | \psi_\hbar \rangle$, where the quantum observable $\hat{f}_\hbar$ is just the multiplication operator by $f(x)$. It proves useful to extend the question to phase space observables $f(x,p)$[3]: what is the behaviour of the diagonal matrix elements

$$\mu_{\psi_\hbar}^W(f) \stackrel{\text{def}}{=} \langle \psi_\hbar, \hat{f}_\hbar \psi_\hbar \rangle, \qquad f \in C^\infty(T^*\Omega), \quad \text{in the limit } \hbar \to 0? \qquad (17)$$

Since the quantization procedure $f \mapsto \hat{f}_\hbar$ is linear, these matrix elements define a distribution $\mu_{\psi_\hbar}^W = W_{\psi_\hbar}(\mathbf{x})\, d\mathbf{x}$ in $T^*\Omega$, called the Wigner distribution of the state $\psi_\hbar$ (the density $W_{\psi_\hbar}(\mathbf{x})$ called the Wigner function). Although this function is generally not positive, it is interpreted as a quasi-probability density describing the state $\psi_\hbar$ in *phase space*.

On the Euclidean space, one can define a *nonnegative* phase space density associated to the state $\psi_\hbar$: the *Husimi* measure (and function):

$$\mu_{\psi_\hbar}^H = \mathcal{H}_{\psi_\hbar}(\mathbf{x})\, d\mathbf{x}, \qquad \mathcal{H}_{\psi_\hbar}(\mathbf{x}) = (2\pi\hbar)^{-d/2} |\langle \varphi_{\mathbf{x}}, \psi_\hbar \rangle|^2, \qquad (18)$$

where the Gaussian wavepackets $\varphi_{\mathbf{x}} \in L^2(\mathbb{R}^d)$ are defined in (21). This measure can also be obtained by convolution of $\mu_{\psi_\hbar}^W$ with the Gaussian kernel $e^{-|\mathbf{x}-\mathbf{y}|^2/\hbar}$. In these notes our phase space plots show the Husimi measures.

These questions lead us to the notion of phase space localization, or *microlocalization*[4]. We will say that the family of states $(\psi_\hbar)$ is microlocalized inside a set $B \subset T^*\Omega$ if, for *any* smooth observable $f(x,p)$ vanishing near B, the matrix elements $\langle \psi_\hbar, \hat{f}_\hbar \psi_\hbar \rangle$ decrease faster than any power of $\hbar$ when $\hbar \to 0$.

Microlocal properties are not easy to guess from plots of the spatial density $|\psi_j(x)|^2$ like Figure 1, but they are more natural to study if we want to connect

---

[3] If $\Omega \subset \mathbb{R}^d$ is replaced by a compact Riemannian manifold M, we assume that some quantization scheme $f \mapsto \hat{f}_\hbar$ has been developed on $L^2(M)$.
[4] The prefix *micro* mustn't mislead us: we are still dealing with *macroscopic* localization properties of $\psi_\hbar$!

quantum to classical mechanics, since the latter takes place in phase space rather than position space. Indeed, the major tool we will use is the quantum-classical correspondence (4); for all the flows we will consider, any initial spatial test function $f(x)$ evolves into a genuine phase space function $f_t(x,p)$, so a purely spatial formalism is not very helpful. These microlocal properties are easier to visualize on two-dimensional phase spaces (see below the figures on the 2-torus).

3.1. The case of completely integrable systems

In order to motivate our further discussion of chaotic eigenstates, let us first recall a few facts about the opposite systems, namely completely integrable Hamiltonian flows. For such systems, the energy shell $\mathcal{E}_E$ is foliated by d-dimensional invariant Lagrangian tori. Each such torus is characterized by the values of d independent invariant actions $I_1, \ldots, I_d$, so let us call such a torus $T_{\vec{I}}$. The WKB theory allows one to explicitly construct, in the semiclassical limit, precise *quasimodes* of $\hat{H}_\hbar$ associated with some of these tori[5], that is normalized states $\psi_{\vec{I}} = \psi_{\hbar,\vec{I}}$ satisfying

$$\hat{H}_\hbar \psi_{\vec{I}} = E_{\vec{I}} \psi_{\vec{I}} + \mathcal{O}(\hbar^\infty), \tag{19}$$

with energies $E_{\vec{I}} \approx E$. Such a Lagrangian (or WKB) state $\psi_{\vec{I}}$ takes the following form (away from caustics):

$$\psi_{\vec{I}}(x) = \sum_{\ell=1}^{L} A_\ell(x;\hbar) \exp(iS_\ell(x)/\hbar). \tag{20}$$

Here the functions $S_\ell(x)$ are (local) generating functions[6] for $T_{\vec{I}}$, and each $A_\ell(x;\hbar) = A_\ell^0(x) + \hbar A_\ell^1(x) + \cdots$ is a smooth amplitude.

From this very explicit expression, one can easily check that the state $\psi_{\vec{I}}$ is microlocalized on $T_{\vec{I}}$. On the other hand, our knowledge of $\psi_\hbar$ is much more precise than the latter fact. One can easily construct other states microlocalized on $T_{\vec{I}}$, which are very different from the Lagrangian states $\psi_{\vec{I}}$. For instance, for any point $\mathbf{x}_0 = (x_0, p_0) \in T_{\vec{I}}$ the Gaussian wavepacket (or coherent state)

$$\varphi_{\mathbf{x}_0}(x) = (\pi\hbar)^{-1/4} e^{-|x-x_0|^2/2\hbar} e^{ip_0 \cdot x/\hbar} \tag{21}$$

is microlocalized on the single point $\mathbf{x}_0$, and therefore also on $T_{\vec{I}}$. This example just reflects the fact that a statement about microlocalization of a sequence of states provides much less information on than a formula like (20).

3.2. Quantum ergodicity

In the case of a fully chaotic system, we generally don't have any explicit formula describing the eigenstates, or even quasimodes of $\hat{H}_\hbar$. However, macroscopic informations on the eigenstates can be obtained indirectly, using the quantum-classical correspondence. The main result on this question is a quantum analogue of the

---

[5] The "quantizable" tori $T_{\vec{I}}$ satisfy Bohr–Sommerfeld conditions $I_i = 2\pi\hbar(n_i + \alpha_i)$, with $n_i \in \mathbb{Z}$ and $\alpha_i \in [0,1]$ fixed indices.

[6] Above some neighbourhood $U \in \mathbb{R}^d$, the torus $T_{\vec{I}}$ is the union of L Lagrangian leaves $\{(x, \nabla S_\ell(x)), x \in U\}$, $\ell = 1, \ldots, L$

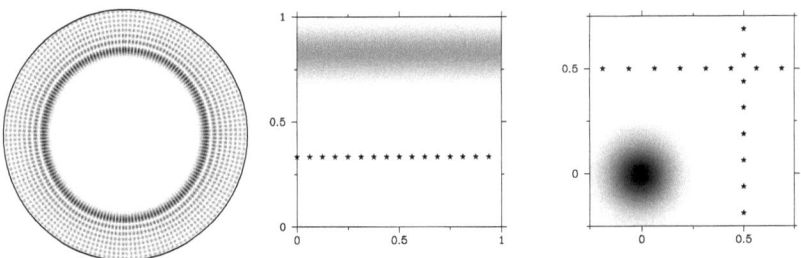

FIGURE 7. Left: One eigenmode of the circle billiard, microlocalized on a certain torus $T_{\vec{I}}$ (I_1, I_2 can be taken to be the energy and the angular momentum). Center: Husimi density of a simple Lagrangian state on $\mathbb{T}^2$, namely a momentum eigenstate microlocalized on the Lagrangian $\{\xi = \xi_0\}$. Right: Husimi density of the Gaussian wavepacket (21). Stars denote the zeros of the Husimi density. From [79] with permission from © 1998 Springer Science+Business Media.

ergodicity property (6) of the classical flow, and it is a consequence of this property. For this reason, it has been named *Quantum Ergodicity* by Zelditch. Loosely speaking, this property states that *almost* all eigenstates $\psi_\hbar$ with $E_\hbar \sim E$ will become *equidistributed* on the energy shell $\mathcal{E}_E$, in the semiclassical limit, provided the classical flow on $\mathcal{E}_E$ is ergodic. We give below the version of the theorem in the case of the Laplacian on a compact Riemannian manifold, using the notations of (9).

Theorem 1. [**Quantum ergodicity**] *Assume that the geodesic flow on (M, g) is ergodic with respect to the Liouville measure. Then, for any orthonormal eigenbasis $(\psi_j)_{j\geq 0}$ of the Laplacian, there exists a subsequence $S \subset \mathbb{N}$ of density 1 (that is, $\lim_{J\to\infty} \frac{\#(S\cap[1,J])}{J} = 1$), such that*

$$\forall f \in C_c^\infty(M), \quad \lim_{j\in S, j\to\infty} \langle \psi_j, \hat{f}_{\hbar_j} \psi_j \rangle = \int_\mathcal{E} f(x,p)\, d\mu_L(x,p),$$

where μ_L is the Liouville measure on $\mathcal{E} = \mathcal{E}_{1/2}$, and $\hbar_j = k_j^{-1}$.

The statement of this theorem was first given by Schnirelman (using test functions $f(x)$) [87], the complete proof was obtained by Zelditch in the case of manifolds of constant negative curvature $\Gamma\backslash\mathbb{H}$ [97], and the general case was then proved by Colin de Verdière [35]. This theorem is "robust": it has been extended to

- quantum ergodic billiards [45, 102]
- quantum Hamiltonians $\hat{H}_\hbar$, such that the flow Φ_H^t is ergodic on $\mathcal{E}_E$ in some energy interval [50]
- quantized ergodic diffeomorphisms on the torus [28] or on more general compact phase spaces [99]
- a general framework of C^* dynamical systems [98]

- a family of quantized ergodic maps with discontinuities [72] and the baker's map [39]
- certain quantum graphs [17]

Let us give the ideas used to prove the above theorem. We want to study the statistical distribution of the matrix elements $\mu_j^W(f) = \langle \psi_j, \hat{f}_{\hbar_j} \psi_j \rangle$ in the range $\{k_j \leq K\}$, with K large. The first step is to estimate the average of this distribution. It is estimated by the generalized Weyl law:

$$\sum_{k_j \leq K} \mu_j(f) \sim \frac{\mathrm{Vol}(M)\sigma_d}{(2\pi)^d} K^d \int_{\mathcal{E}} f \, d\mu_L, \quad K \to \infty, \tag{22}$$

where σ_d is the volume of the unit ball in $\mathbb{R}^d$ [53]. In particular, this asymptotics allows to count the number of eigenvalues $k_j \leq K$:

$$\#\{k_j \leq K\} \sim \frac{\mathrm{Vol}(M)\sigma_d}{(2\pi)^d} K^d, \quad K \to \infty, \tag{23}$$

and shows that the average of the distribution $\{\mu_j(f), k_j \leq K\}$ converges to the phase space average $\mu_L(f)$ when $K \to \infty$.

Now, we want to show that the distribution is concentrated around its average. This will be done by estimating its variance

$$\mathrm{Var}_K(f) \stackrel{\mathrm{def}}{=} \frac{1}{\#\{k_j \leq K\}} \sum_{k_j \leq K} |\langle \psi_j, (\hat{f}_{\hbar_j} - \mu_L(f))\psi_j\rangle|^2,$$

which has been called the *quantum variance*. Because the ψ_j are eigenstates of $U_\hbar$, we may replace $\hat{f}_{\hbar_j}$ by its quantum time average up to some large time T,

$$\hat{f}_{T,\hbar_j} = \frac{1}{2T} \int_{-T}^{T} U_{\hbar_j}^{-t} \hat{f}_{\hbar_j} U_{\hbar_j}^{-t} \, dt,$$

without modifying the matrix elements. At this step, we use the simple inequality

$$|\langle \psi_j, A\psi_j\rangle|^2 \leq \langle \psi_j, A^* A\psi\rangle, \quad \text{for any bounded operator } A,$$

to get the following upper bound for the variance:

$$\mathrm{Var}_K(f) \leq \frac{1}{\#\{k_j \leq K\}} \sum_{k_j \leq K} \langle \psi_j, (\hat{f}_{T,\hbar_j} - \mu_L(f))^*(\hat{f}_{T,\hbar_j} - \mu_L(f))\psi_j\rangle.$$

The Egorov theorem (4) shows that the product operator on the right-hand side is approximately equal to the quantization of the function $|f_T - \mu_L(f)|^2$, where f_T is the classical time average of f. Applying the generalized Weyl law (22) to this function, we get the bound

$$\mathrm{Var}_K(f) \leq \mu_L(|f_T - \mu_L(f)|^2) + \mathcal{O}_T(K^{-1}).$$

Finally, the ergodicity of the classical flow implies that the function $\mu_L(|f_T - \mu_L(f)|^2)$ converges to zero when $T \to \infty$. By taking T, and then K large enough, the above right-hand side can be made arbitrary small. This proves that the quantum variance $Var_K(f)$ converges to zero when $K \to \infty$. A standard Chebychev argument is then used to extract a dense converging subsequence. □

A more detailed discussion on the quantum variance and the distribution of matrix elements $\{\mu_j(f),\ k_j \leq K\}$ will be given in §4.1.3.

3.3. Beyond QE: Quantum Unique Ergodicity vs. strong scarring

3.3.1. Semiclassical measures. Quantum ergodicity can be conveniently expressed by using the concept of *semiclassical measure*. Remember that, using a duality argument, we associate to each eigenstates ψ_j a Wigner distribution μ_j^W on phase space. The quantum ergodicity theorem can be rephrased as follows:

> There exists a density-1 subsequence $S \subset \mathbb{N}$, such that the sequences of Wigner distributions $(\mu_j^W)_{j \in S}$ weak-∗ (or vaguely) converges to the Liouville measure on $\mathcal{E}$.

For any compact Riemannian manifold (M,g), the sequence of Wigner distributions $(\mu_j^W)_{j \in \mathbb{N}}$ remains in a compact set in the weak-∗ topology, so it is always possible to extract an infinite subsequence $(\mu_j)_{j \in S}$ vaguely converging to a limit distribution μ_{sc}, that is

$$\forall f \in C_c^\infty(T^*M), \quad \lim_{j \in S, j \to \infty} \int f\, d\mu_j^W - \int f\, d\mu_{sc}.$$

Such a limit distribution is necessarily a probability measure on $\mathcal{E}$, and is called a *semiclassical measure* of the manifold M. The quantum-classical correspondence implies that μ_{sc} is *invariant* through the geodesic flow: $(\Phi^t)^* \mu_{sc} = \mu_{sc}$. The semiclassical measure μ_{sc} represents the asymptotic (macroscopic) phase space distribution of the eigenstates $(\psi_j)_{j \in S}$. It is the major tool used in the mathematical literature on chaotic eigenstates (see below). The definition can be obviously generalized to any quantized Hamiltonian flow or canonical map.

3.3.2. Quantum Unique Ergodicity conjecture. The quantum ergodicity theorem provides an incomplete information, which leads to the following question:

Question 1. *Do all eigenstates become equidistributed in the semiclassical limit? Equivalently, is the Liouville measure the* unique *semiclassical measure for the manifold M? On the opposite, are there exceptional subsequences converging to a semiclassical measure $\mu_{sc} \neq \mu_L$?*

This question makes sense if the geodesic flow admits invariant measures different from μ_L (that is, the flow is not uniquely ergodic). Our central example, manifolds of negative curvature, admit many different invariant measures, e.g., the singular measures μ_γ supported on each of the (countably many) periodic geodesics.

This question was already raised in [35], where the author conjectured that no subsequence of eigenstates can concentrate along a single periodic geodesic, that is, μ_γ cannot be a semiclassical measure. Such an unlikely subsequence was later called a *strong scar* by Rudnick and Sarnak [84], in reference to the *scars* discovered by Heller on the stadium billiard (see §4.2). In the same paper the authors formulated a stronger conjecture:

Conjecture 1 (Quantum unique ergodicity). *Let (M, g) be a compact Riemannian manifold with negative sectional curvature. Then all high-frequency eigenstates become equidistributed with respect to the Liouville measure.*

The term *quantum unique ergodicity* refers to the notion of *unique ergodicity* in ergodic theory: a system is uniquely ergodic system if it admits unique invariant probability measure. The geodesic flows we are considering admit many invariant measures, but the conjecture states that the corresponding quantum system selects only one of them.

3.3.3. Arithmetic quantum unique ergodicity. This conjecture was motivated by the following result proved in the cited paper. The authors specifically considered arithmetic surfaces, obtained by quotienting the Poincaré disk $\mathbb{H}$ by certain congruent co-compact groups Γ. As explained in §2.4.2, on such a surface it is natural to consider Hecke eigenstates, which are joint eigenstates of the Laplacian and the (countably many) Hecke operators[7]. It was shown in [84] that any semiclassical measure μ_{sc} associated with Hecke eigenstates does not contain any periodic orbit component μ_γ. The methods of [84] were refined by Bourgain and Lindenstrauss [27], who showed that the measure μ_{sc} of an ϵ-thin tube along any geodesic segment is bounded from above by $C\,\epsilon^{2/9}$. This bound implies that the *Kolmogorov–Sinai entropy* of μ_{sc} is bounded from below by $2/9$[8]. Finally, using advanced ergodic theory methods, Lindenstrauss completed the proof of the QUE conjecture in the arithmetic context.

Theorem 2 (Arithmetic QUE [69]). *Let $M = \Gamma\backslash\mathbb{H}$ be an arithmetic[9] surface of constant negative curvature. Consider an eigenbasis $(\psi_j)_{j\in\mathbb{N}}$ of Hecke eigenstates of $-\Delta_M$. Then, the only semiclassical measure associated with this sequence is the Liouville measure.*

Lindenstrauss and Brooks recently announced an improvement of this theorem: the result holds true, assuming the (ψ_j) are joint eigenstates of Δ_M and of a single Hecke operator T_{n_0}. Their proof uses a new delocalization estimate for regular graphs [30], which replaces the entropy bounds of [27].

This positive QUE result was preceded by a similar statement for the hyperbolic symplectomorphisms on $\mathbb{T}^2$ introduced in §2.5.1. These quantum maps $U_N(S_0)$ have the nongeneric property to be periodic, so one has an explicit expression for their eigenstates. It was shown in [38] that for a certain family hyperbolic matrices S_0 and certain sequences of *prime* values of N, the eigenstates of $U_N(S_0)$

---

[7]The spectrum of the Laplacian on such a surface is believed to be simple; if this is the case, then an eigenstate of Δ is automatically a Hecke eigenstate.
[8]The notion of KS entropy will be further explained in §3.4.
[9]For the precise definition of these surfaces, see [69].

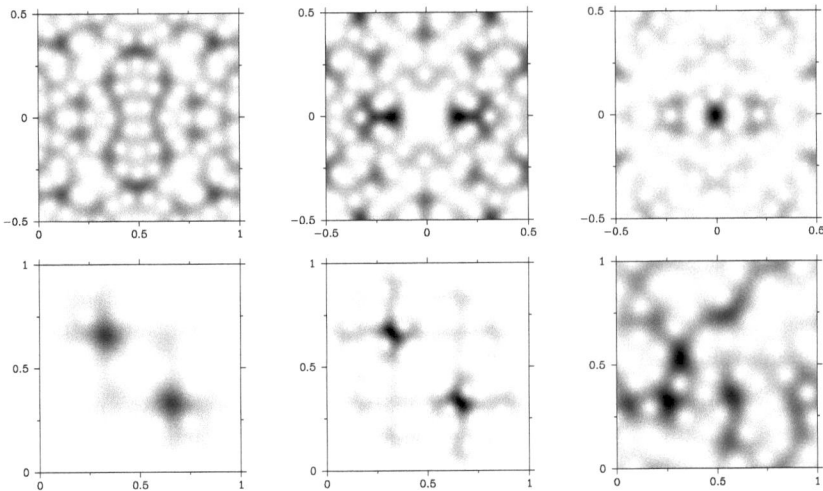

FIGURE 8. Top: Husimi densities of various states in $\mathcal{H}_N$ (large values=dark regions). Top: 3 (Hecke) eigenstates of the quantum cat map $U_N(S_{DEGI})$, for $N = 107$. Bottom left, center: 2 eigenstates of the quantum baker $U_N(\kappa_D)$ scarred on the period-2 orbit, for $N = 48$ and $N = 128$. Bottom right: Random state (35) for $N = 56$. From [79] with permission from © 1998 Springer Science + Business Media.

become equidistributed in the semiclassical limit, with an explicit bound on the remainder. Some eigenstates, corresponding to the matrix $S_{DEGI} = \begin{pmatrix} 2 & 1 \\ 3 & 2 \end{pmatrix}$ are plotted in Figure 8, in the Husimi representation.

A few years later, Kurlberg and Rudnick [61] were able to construct, attached to any matrix S_0 and any value of N, a finite commutative family of operators $\{U_N(S'),\ S' \in \mathcal{C}(S_0, N)\}$ including $U_N(S_0)$, which they called "Hecke operators" by analogy with the case of arithmetic surfaces. They then considered specifically the joint ("Hecke") eigenbases of this family, and proved QUE in this framework:

Theorem 3 ([61]). *Let $S_0 \in SL_2(\mathbb{Z})$ be a quantizable symplectic matrix. For each $N > 0$, consider a Hecke eigenbasis $(\psi_{N,j})_{j=1,\ldots,N}$ of the quantum map $U_N(S_0)$. Then, for any observable $f \in C^\infty(\mathbb{T}^2)$ and any $\epsilon > 0$, we have*

$$\langle \psi_{N,j}, \hat{f}_N \psi_{N,j} \rangle = \int f\, d\mu_L + \mathcal{O}_{f,\epsilon}(N^{-1/4+\epsilon}),$$

where μ_L is the Liouville (or Lebesgue) measure on $\mathbb{T}^2$.

As a result, the Wigner distributions of the eigenstates $\psi_{N,j}$ become uniformly equidistributed on the torus, as $N \to \infty$. The eigenstates considered in [38] were instances of Hecke eigenstates.

In view of these positive results, it is tempting to generalize the QUE conjecture to other chaotic systems.

Conjecture 2 (Generalized QUE). *Let Φ_H^t be an ergodic Hamiltonian flow on some energy shell $\mathcal{E}_E$. Then, all eigenstates $\psi_{\hbar,j}$ of $\hat{H}_\hbar$ of energies $E_{\hbar,j} \approx E$ become equidistributed when $\hbar \to 0$.*

Let κ be a canonical ergodic map on $\mathbb{T}^2$. Then, the eigenstates $\psi_{N,j}$ of $U_N(\kappa)$ become equidistributed when $N \to \infty$.

An intensive numerical study for eigenstates of a Sinai-like billiard was carried on by Barnett [15]. It seems to confirm QUE for this system.

In the next subsection we will exhibit particular systems for which this conjecture **fails**.

3.3.4. Counterexamples to QUE: half-scarred eigenstates.
In this section we will exhibit sequences of eigenstates converging to semiclassical measures different from μ_L, thus disproving the above conjecture.

Let us continue our discussion of symplectomorphisms on $\mathbb{T}^2$. For any $N \geq 1$, the quantum symplectomorphism $U_N(S_0)$ are periodic (up to a global phase) of period $T_N \leq 3N$, so that its eigenvalues are essentially T_N-roots of unity. For values of N such that $T_N \ll N$, the spectrum of $U_N(S_0)$ is very degenerate, in which case imposing the eigenstates to be of Hecke type becomes a strong requirement. In [62] it is shown that, provided the period is not too small (namely, $T_N \gg N^{1/2-\epsilon}$, which is the case for almost all values of N), then QUE holds for any eigenbasis.

On the opposite, there exist (sparse) values of N, for which the period can be as small as $T_N \sim C \log N$, so that the eigenspaces of huge dimensions $\sim C^{-1} N / \log N$. This freedom allowed Faure, De Bièvre and the author to explicitly construct eigenstates with different localization properties [42, 43].

Theorem 4. *Take $S_0 \in SL_2(\mathbb{Z})$ a (quantizable) hyperbolic matrix. Then, there exists an infinite (sparse) sequence $\mathcal{S} \subset \mathbb{N}$ such that, for any periodic orbit γ of κ_{S_0}, one can construct a sequence of eigenstates $(\psi_N)_{N \in \mathcal{S}}$ of $U_N(S_0)$ associated with the semiclassical measure*

$$\mu_{sc} = \frac{1}{2}\mu_\gamma + \frac{1}{2}\mu_L. \qquad (24)$$

More generally, for any κ_{S_0}-invariant measure μ_{inv}, one can construct sequences of eigenstates associated with the semiclassical measure

$$\mu_{sc} = \frac{1}{2}\mu_{inv} + \frac{1}{2}\mu_L.$$

This result provided the first counterexample to the generalized QUE conjecture. The eigenstates associated with $\frac{1}{2}\mu_\gamma + \frac{1}{2}\mu_L$ can be called *half-localized*. The coefficient $1/2$ in front of the singular component of μ_γ was shown to be optimal [43], a phenomenon which was then generalized using entropy methods (see §3.4).

Let us briefly explain the construction of eigenstates half-localized on a fixed point $\mathbf{x}_0$. They are obtained by projecting on any eigenspace the Gaussian

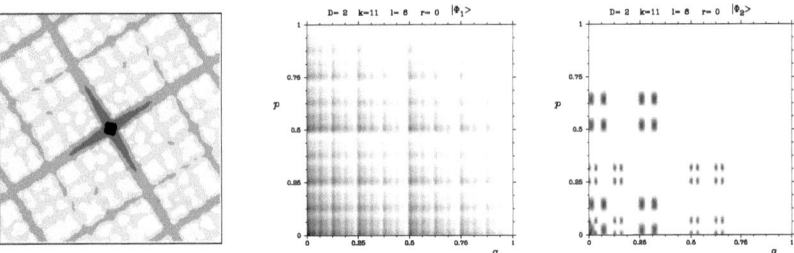

FIGURE 9. Left: Husimi density of an eigenstate of $U_N(S_{\text{cat}})$, strongly scarred at the origin (from [42]). Notice the hyperbolic structure around the fixed point. Center, right: Two eigenstates of the Walsh-quantized baker's map (plotted using a "Walsh–Husimi measure"); the corresponding semiclassical measures are fractal. From [2] with permission from © 2007 Springer Basel AG.

wavepacket $\varphi_{\mathbf{x}_0}$ (see (21)). Each spectral projection can be expressed as a linear combination of the evolved states $U_N(S_0)^n \varphi_{\mathbf{x}_0}$, for $n \in [-T_N/2, T_N/2 - 1]$. Now, we use the fact that, for N in an infinite subsequence $\mathcal{S} \subset \mathbb{N}$, the period T_N of the operator $U_N(S_0)$ is close to twice the *Ehrenfest time*

$$T_E = \frac{\log h^{-1}}{\lambda}, \qquad (25)$$

(here λ is the positive Lyapunov exponent). The above linear combination can be split in two components: during the time range $n \in [-T_E/2, T_E/2]$ the states $U_N(S_0)^n \varphi_{\mathbf{x}_0}$ remain microlocalized at the origin; on the opposite, for times $T_E/2 < |n| \leq T_E$, these states expand along long stretches of stable/unstable manifolds, and densely fill the torus. As a result, the sum of these two components is half-localized, half-equidistributed. □

In Figure 9 (left) we plot the Husimi density associated with one half-localized eigenstate of the quantum cat map $U_N(S_{\text{cat}})$.

A nonstandard (Walsh-) quantization of the 2-baker's map was constructed in [2], with properties similar to the above quantum cat map. It allows to exhibit semiclassical measures $\frac{1}{2}\mu_\gamma + \frac{1}{2}\mu_L$ as in the above case, but also purely fractal semiclassical measures void of any Liouville component (see Figure 9).

Studying hyperbolic toral symplectomorphisms on $\mathbb{T}^{2d}$ for $d \geq 2$, Kelmer [60] identified eigenstates microlocalized on a proper subspace of dimension $\geq d$. He extended his analysis to certain nonlinear perturbations of κ_S. Other very explicit counterexamples to generalized QUE were constructed in [33], based on interval-exchange maps of the interval (such maps are ergodic, but have zero Lyapunov exponents).

3.3.5. Counterexamples to QUE for the stadium billiard. The only counterexample to (generalized) QUE in the case of a chaotic flow concerns the stadium billiard and similar surfaces. This billiard admits a one-dimensional family of marginally stable periodic orbits, the so-called *bouncing-ball* orbits hitting the horizontal sides of the stadium orthogonally (these orbits form a set of Liouville measure zero, so they do not prevent the flow from being ergodic). In 1984 Heller [51] had observed that some eigenstates are concentrated in the rectangular region (see Figure 10). These states were baptized *bouncing-ball modes*, and studied quite thoroughly, both numerically and theoretically [52, 9]. In particular, the relative number of these modes becomes negligible in the limit $K \to \infty$, so they are still compatible with quantum ergodicity.

Hassell recently proved [49] that some high-frequency eigenstates of some stadia indeed fail to equidistribute. To state his result, let us parametrize the shape of a stadium billiard is by the ratio β between the length and the height of the rectangle.

Theorem 5 ([49]). *For any $\epsilon > 0$, there exists a subset $B_\epsilon \subset [1, 2]$ of measure $\geq 1 - 4\epsilon$ and a number $m(\epsilon) > 0$ such that, for any $\beta \in B_\epsilon$, the β-stadium admits a semiclassical measure with a weight $\geq m(\epsilon)$ on the bouncing-ball orbits.*

Although the theorem only guarantees that a fraction $m(\epsilon)$ of the semiclassical measure is localized along the bouncing-ball orbits, numerical studies suggest that the modes are asymptotically fully concentrated on these orbits. Besides, such modes are expected to exist for all ratios $\beta > 0$.

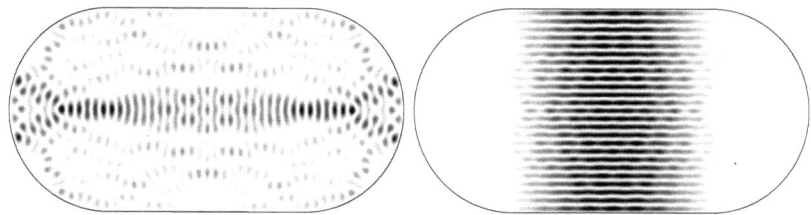

FIGURE 10. Two eigenstates of the stadium billiard ($\beta = 2$). Left ($k = 39.045$): The mode has a *scar* along the unstable horizontal orbit. Right ($k = 39.292$): The mode is localized in the *bouncing-ball region*.

3.4. Entropy of the semiclassical measures

To end this section on the macroscopic properties, let us mention a recent approach to contrain the possible semiclassical measures occurring in a chaotic system. This approach, initiated by Anantharaman [1], consists in proving nontrivial lower bounds for the *Kolmogorov–Sinai entropy* of the semiclassical measures. The KS (or *metric*) entropy is a common tool in classical dynamical systems, flows or maps [57]. To be brief, the entropy $H_{KS}(\mu)$ of an invariant probability measure μ is a nonnegative number, which quantifies the information-theoretic complexity of μ-typical trajectories. It does not directly measure the localization of μ, but gives some information about it. Here are some relevant properties:

- the delta measure μ_γ on a periodic orbit has entropy zero.
- for an Anosov system (flow or diffeomorphism), the entropy is connected to the unstable Jacobian $J^u(\mathbf{x})$ (see Figure 3) through the Ruelle–Pesin formula:

$$\text{invariant}, \quad H_{KS}(\mu) \leq \int \log J^u(\rho) \, d\mu,$$

  with equality iff μ is the Liouville measure.
- the entropy is an affine function on the set of probability measures:
  $H_{KS}(\alpha\mu_1 + (1-\alpha)\mu_2) = \alpha H_{KS}(\mu_1) + (1-\alpha)H_{KS}(\mu_2)$.

In particular, the invariant measure $\alpha\mu_\gamma + (1-\alpha)\mu_L$ of a hyperbolic symplectomorphism S has entropy $(1-\alpha)\lambda$, where λ is the positive Lyapunov exponent.

Anantharaman considered the case of geodesic flows on manifolds M of negative curvature, see §2.4.2. She proved the following constraints on semiclassical measures of M:

Theorem 6 ([1]). *Let (M, g) be a smooth compact Riemannian manifold of negative sectional curvature. Then there exists $c > 0$ such that any semiclassical measure μ_{sc} of (M, g) satisfies $H_{KS}(\mu_{sc}) \geq c$.*

In particular, this result forbids semiclassical measures from being supported on unions of periodic geodesics. A more quantitative lower bound was obtained in [2, 4], related with the instability of the flow.

Theorem 7 ([4]). *Under the same assumptions as above, any semiclassical measure must satisfy*

$$H_{KS}(\mu_{sc}) \geq \int \log J^u \, d\mu_{sc} - \frac{(d-1)\lambda_{\max}}{2}, \tag{26}$$

where $d = \dim M$ and $\lambda_{\max}$ is the maximal expansion rate of the flow.

This lower bound was generalized to the case of the Walsh-quantized baker's map [2], and the hyperbolic symplectomorphisms on $\mathbb{T}^2$ [29, 78], where it takes the form $H_{KS}(\mu_{sc}) \geq \frac{\lambda}{2}$. For these maps, the bound is *saturated* by the half-localized semiclassical measures $\frac{1}{2}(\mu_\gamma + \mu_L)$.

The lower bound (26) is certainly not optimal in cases of variable curvature. Indeed, the right-hand side may become negative when the curvature varies too much. A more natural lower bound has been obtained by Rivière in two dimensions:

Theorem 8 ([82, 83]). *Let (M, g) be a compact Riemannian surface of nonpositive sectional curvature. Then any semiclassical measure satisfies*

$$H_{KS}(\mu_{sc}) \geq \frac{1}{2} \int \lambda_+ \, d\mu_{sc}, \tag{27}$$

where λ_+ is the positive Lyapunov exponent.

The same lower bound was obtained by Gutkin for a family of nonsymmetric baker's map [46]; he also showed that the bound is optimal for that system.

The lower bound (27) is also expected to hold for ergodic billiards, like the stadium; in particular, it would not contradict the existence of semiclassical measures supported on the bouncing ball orbits.

In higher dimension, one expects the lower bound $H_{KS}(\mu_{sc}) \geq \frac{1}{2}\int \log J^u \, d\mu_{sc}$ to hold for Anosov systems. Kelmer's counterexamples [60] show that this bound may be saturated for certain Anosov diffeomorphisms on $\mathbb{T}^{2d}$.

To close this section, we notice that the QUE conjecture (which remains open) amounts to improving the entropic lower bound (26) to $H_{KS}(\mu_{sc}) \geq \int \log J^u \, d\mu_{sc}$.

4. Statistical description

The macroscopic distribution properties described in the previous section give a poor description of the eigenstates, compared with our knowledge of eigenmodes of integrable systems. At the practical level, one is interested in quantitative properties of the eigenmodes at finite values of $\hbar$. It is also desirable to understand their structure at the microscopic scale (the scale of the wavelength $R \sim \hbar$), or at least some *mesoscopic* scale ($\hbar \ll R \ll 1$).

The results we will present are of two types. On the one hand, *individual* eigenfunctions will be analyzed statistically, e.g., by computing correlation functions or value distributions of various representations (position density, Husimi). On the other hand, one can also perform a statistical study of a whole bunch of eigenfunctions (around some large wavevector K), for instance by studying how global indicators of localization (e.g., the norms $\|\psi_j\|_{L^p}$) are distributed. We will not attempt to review all possible statistical indicators, but only some "popular" ones.

4.1. Chaotic eigenstates as random states?

It has realized quite early that the statistical data of chaotic eigenstates (obtained numerically) could be reproduced by considering instead ensembles of *random states*. The latter are, so far, the best Ansatz we can find to describe chaotic eigenstates. Yet, one should keep in mind that this Ansatz is of a different type from the WKB Ansatz pointwise describing individual eigenstates of integrable systems. By definition, random states only have a chance to capture the *statistical* properties of the chaotic eigenstates. This "typicality" of chaotic eigenstates should of course be put in parallel with the typicality of spectral correlations, embodied by the Random matrix conjecture (see J. Keating's lecture).

A major open problem in quantum chaos is to prove this "typicality" of chaotic eigenstates. The question seems as difficult as the Random Matrix conjecture.

4.1.1. Spatial correlations.
Let us now introduce in more detail the ensembles of random states. For simplicity we consider the Laplacian on a Euclidean planar domain Ω with chaotic geodesic flow (say, the stadium billiard). As in (9), we

denote by k_j^2 the eigenvalue of $-\Delta$ corresponding to the eigenmode ψ_j. Let us recall some history.

Facing the absence of explicit expression for the eigenstates, Voros [95, §7] and Berry [18] proposed to (brutally) approximate the Wigner measures μ_j^W of high-frequency eigenstates ψ_j by the Liouville measure μ_L on $\mathcal{E}$. This approximation is justified by the quantum ergodicity theorem, as long as one investigates macroscopic properties of ψ_j. However, the game consisted in also extract some *microscopic* information on ψ_j, so erasing all small-scale fluctuations of the Wigner function could be a dangerous approximation.

Berry [18] showed that this approximation provides nontrivial predictions for the *microscopic correlations* of the eigenstates. Indeed, a partial Fourier transform of the Wigner function leads to the *autocorrelation function* describing the short-distance oscillations of ψ. He defined the correlation function by averaging over some distance R:

$$C_{\psi,R}(x,r) = \overline{\psi^*(x-r/2)\psi(x+r/2)}^R \stackrel{\text{def}}{=} \frac{1}{\pi R^2} \int_{|y-x|\leq R} \psi^*(y-r/2)\psi(y+r/2)\,dy,$$

taking R to be a *mesoscopic scale* $k_j^{-1} \ll R \leq 1$ in order to average over many oscillations of ψ. Inserting μ_L in the place of μ_ψ^W then provides a simple expression for this function in the range $0 \leq |r| \ll 1$:

$$C_{\psi,R}(x,r) \approx \frac{J_0(k|r|)}{\text{Vol}(\Omega)}. \tag{28}$$

Such a homogeneous and isotropic expression could be expected from our approximation. Replacing the Wigner distribution by μ_L suggests that, near each point $x \in \Omega$, the eigenstate ψ is an equal mixture of particles of energy k^2 travelling in all possible directions.

4.1.2. A random state Ansatz. Yet, the approximation μ_L for the Wigner distributions μ_ψ^W is NOT the Wigner distribution of any quantum state[10]. The next question is thus [95]: can one exhibit a family of quantum states whose Wigner measures resemble μ_L? Or equivalently, whose microscopic correlations behave like (28)?

Berry proposed a *random superposition of plane waves* Ansatz to account for these isotropic correlations. One form of this Ansatz reads

$$\psi_{\text{rand},k}(x) = \left(\frac{2}{N\,\text{Vol}(\Omega)}\right)^{1/2} \Re\left(\sum_{j=1}^N a_j \exp(k\hat{n}_j \cdot x)\right), \tag{29}$$

where $(\hat{n}_j)_{j=1,\ldots,N}$ are unit vectors distributed on the unit circle, and the coefficients $(a_j)_{j=1,\ldots,N}$ are independent identically distributed (i.i.d.) complex normal Gaussian random variables. In order to span all possible velocity directions (within the uncertainty principle), one should include $N \approx k$ directions $\hat{n}_j$ The normalization ensures that $\|\psi_{\text{rand},k}\|_{L^2(\Omega)} \approx 1$ with high probability when $k \gg 1$.

---

[10]Characterizing the function on $T^*\mathbb{R}^d$ which are Wigner functions of individual quantum states is a nontrivial question.

Alternatively, one can replace the plane waves in (29) by circular-symmetric waves, namely Bessel functions. In circular coordinates, the random state reads

$$\psi_{\mathrm{rand},k}(r,\theta) = (\mathrm{Vol}(\Omega))^{-1/2} \sum_{m=-M}^{M} b_m J_{|m|}(kr)\, e^{im\theta}, \qquad (30)$$

where the coefficients i.i.d. complex Gaussian satisfying the symmetry $b_m = b^*_{-m}$, and $M \approx k$. Both random ensembles asymptotically produce the same statistical results.

The random state $\psi_{\mathrm{rand},k}$ satisfies the equation $(\Delta + k^2)\psi = 0$ in the interior of Ω. Furthermore, $\psi_{\mathrm{rand},k}$ satisfies a "local quantum ergodicity" property: for any observable $f(x,p)$ supported in the interior of $T^*\Omega$, the matrix elements $\langle \psi_{\mathrm{rand},k}, \hat{f}_{k^{-1}} \psi_{\mathrm{rand},k} \rangle \approx \mu_L(f)$ with high probability (more is known about these elements, see §4.1.3).

The stronger claim is that, in the interior of Ω, the *local statistical properties* of $\psi_{\mathrm{rand},k}$, including its microscopic ones, should be similar with those of the eigenstates ψ_j with wavevectors $k_j \approx k$.

The correlation function of eigenstates of chaotic planar billiards has been numerically studied, and compared with this random models, see, e.g., [71, 11]. The agreement with (28) is fair for some eigenmodes, but not so good for others; in particular the authors observe some anisotropy in the experimental correlation function, which may be related to some form of *scarring* (see §4.2), or to the bouncing-ball modes of the stadium billiard.

The *value distribution* of the random wavefunction (29) is Gaussian, and compares very well with numerical studies of eigenmodes of chaotic billiards [71]. A similar analysis has been performed for eigenstates of the Laplacian on a compact surface of constant negative curvature [6]. In this geometry the random Ansatz was defined in terms of adapted circular hyperbolic waves. The authors checked that the coefficients of the individual eigenfunctions in this expansion were indeed Gaussian distributed; they also checked that the value distribution of individual eigenstates $\psi_j(x)$ is Gaussian to a good accuracy, without any exceptions.

4.1.3. On the distribution of quantum averages. The random state model also predicts the statistical distribution of diagonal matrix elements $\langle \psi_j, \hat{f}_\hbar \psi_j \rangle$, equivalently the average of the observable f w.r.t. the Wigner distributions, $\mu_j^W(f)$. The quantum variance estimate in the proof of Thm. 1 shows that the distribution of these averages becomes semiclassically concentrated around the classical value $\mu_L(f)$. Using a mixture of semiclassical and random matrix theory arguments, Feingold and Peres [44] conjectured that, in the semiclassical limit, the matrix elements of eigenstates in a small energy window $\hbar k_j \in [1-\epsilon, 1+\epsilon]$ should be Gaussian distributed, with the mean $\mu_L(f)$ and the (quantum) variance related with the *classical variance* of f. The latter is defined as the integral of the

autocorrelation function $C_{f,f}(t)$ (see (7)):
$$\mathrm{Var}_{cl}(f) = \int_{\mathbb{R}} C_{f,f}(t)\,dt\,.$$

A more precise semiclassical derivation [41], using the Gutzwiller trace formula, and supported by numerical computations on several chaotic systems, confirmed both the Gaussian distribution of the matrix elements, and the showed the following connection between quantum and classical variances (expressed in semiclassical notations):
$$\mathrm{Var}_\hbar(f) = g\,\frac{\mathrm{Var}_{cl}(f)}{T_H}\,. \tag{31}$$

Here g is a symmetry factor ($g=2$ in presence of time reversal symmetry, $g=1$ otherwise), and $T_H = 2\pi\hbar\bar{\rho}$ is the Heisenberg time, where $\bar{\rho}$ is the smoothed density of states. In the case of the semiclassical Laplacian $-\hbar^2\Delta/2$ on a compact surface or a planar domain, the above right-hand side reads $\mathrm{Var}_\hbar(f) = g\hbar\,\mathrm{Var}_{cl}(f)/\mathrm{Vol}(\Omega)$. Equivalently, the quantum variance corresponding to wavevectors $k_j \in [K, K+1]$ is predicted to take the value
$$\mathrm{Var}_K(f) \sim \frac{g}{K}\,\frac{\mathrm{Var}_{cl}(f)}{\mathrm{Vol}(\Omega)}\,. \tag{32}$$

Successive numerical studies on chaotic Euclidean billiards [10, 15] and manifolds or billiards of negative curvature [7] globally confirmed this prediction for the quantum variance, as well as the Gaussian distribution for the matrix elements at high frequency. Still, the convergence to this law can be slowed down for billiards admitting bouncing-ball eigenmodes, like the stadium billiard [10].

For a generic chaotic system, rigorous semiclassical methods could only prove logarithmic upper bounds for the quantum variance [88], $\mathrm{Var}_\hbar(f) \leq C/|\log\hbar|$. Schubert showed that this slow decay can be sharp for certain eigenbases of the quantum cat map, in the case of large spectral degeneracies [89] (as we have seen in §3.3.4, such degeneracies are also responsible for the existence exceptionally localized eigenstates, so a large variance is not surprising).

The only systems for which an algebraic decay is known are of arithmetic nature. Luo and Sarnak [70] proved that, in the case of on the modular domain $M = SL_2(\mathbb{Z})\backslash\mathbb{H}$ (a noncompact, finite volume arithmetic surface), the quantum variance corresponding to high-frequency Hecke eigenfunctions[11] is of the form $\mathrm{Var}_K(f) = \frac{B(f)}{K}$: the polynomial decay is the same as in (32), but the coefficient $B(f)$ is equal the classical variance, "decorated" by an extra factor of arithmetic nature.

More precise results were obtained for quantum symplectomorphisms on the 2-torus. Kurlberg and Rudnick [64] studied the distribution of matrix elements $\{\sqrt{N}\langle\psi_{N,j}, \hat{f}_N\psi_{N,j}\rangle,\ j=1,\ldots,N\}$, where the $(\psi_{N,j})$ form a Hecke eigenbasis of $U_N(S)$ (see §3.3.3). They computed the variance, which is asymptotically of the

---

[11] The proof is fully written for the holomorphic cusp forms, but the authors claim that it adapts easily to the Hecke eigenfunctions.

form $\frac{B(f)}{N}$, with $B(f)$ a "decorated" classical variance. They also computed the fourth moment of the distribution, which suggests that the latter is not Gaussian, but given by a combination of several semi-circle laws on $[-2, 2]$ (or Sato-Tate distributions). The same semi-circle law had been shown in [63] to correspond to the asymptotic value distribution of the Hecke eigenstates, at least for N along a subsequence of "split primes".

4.1.4. Maxima of eigenfunctions.
Another interesting quantity is the statistics of the maximal values of eigenfunctions, that is their L^∞ norms, or more generally their L^p norms for $p \in (2, \infty]$ (we always assume the eigenfunctions to be L^2-normalized). The maxima belong to the far tail of the value distribution, so their behaviour is a priori uncorrelated with the Gaussian nature of the latter.

The random wave model gives the following estimate [8]: for $C > 0$ large enough,

$$\frac{\|\psi_{\mathrm{rand},k}\|_\infty}{\|\psi_{\mathrm{rand},k}\|_2} \leq C\sqrt{\log k} \quad \text{with high probability when } k \to \infty. \tag{33}$$

Numerical tests on some Euclidean chaotic billiards and a surface of negative curvature show that this order of magnitude is correct for chaotic eigenstates [8]. Small variations were observed between arithmetic/non-arithmetic surfaces of constant negative curvature, the sup-norms appearing slightly larger in the arithmetic case, but still compatible with (33). For the planar billiards, the largest maxima occured for states *scarred* along a periodic orbit (see §4.2).

Mathematical results concerning the maxima of eigenstates of generic manifolds of negative curvature are scarce. A general upper bound

$$\|\psi_j\|_\infty \leq C\, k_j^{(d-1)/2} \tag{34}$$

holds for arbitrary compact manifolds [53], and is saturated in the case of the standard spheres. On a manifold of negative curvature, this upper bound can be improved by a logarithmic factor $(\log k_j)^{-1}$, taking into account a better bound on the remainder in Weyl's law.

Once again, more precise results have been obtained only in the case of Hecke eigenstates on arithmetic manifolds. Iwaniec and Sarnak [54] showed that, for some arithmetic surfaces, the Hecke eigenstates satisfy the bound

$$\|\psi_j\|_\infty \leq C_\epsilon\, k_j^{5/12+\epsilon},$$

and conjecture a bound $C_\epsilon\, k_j^\epsilon$, compatible with the random wave model. More recently, Milićević [74] showed that, on certain arithmetic surfaces, a subsequence of Hecke eigenstates satisfies a *lower* bound

$$\|\psi_j\|_\infty \geq C \exp\left\{\left(\frac{\log k_j}{\log \log k_j}\right)^{1/2}(1+o(1))\right\},$$

thereby violating the random wave result. The large values are reached on specific CM-points of the surface, of arithmetic nature.

On higher-dimensional arithmetic manifolds, Rudnick and Sarnak [84] had already identified some Hecke eigenstates with larger values, namely
$$\|\psi_j\|_\infty \geq C\, k_j^{1/2}.$$
A general discussion of this phenomenon appears in the recent work of Milićević [75]; the author presents a larger family of arithmetic 3-manifolds featuring eigenstate with abnormally large values, and conjectures that his list is exhaustive.

4.1.5. Random states on the torus. In the case of quantized chaotic maps on $\mathbb{T}^2$, one can easily setup a model of random states mimicking the statistics of eigenstates. The choice is particularly when the map does not possess any particular symmetry: the ensemble of random states in $\mathcal{H}_N$ is then given by

$$\psi_{\mathrm{rand},N} = \frac{1}{\sqrt{N}} \sum_{\ell=1}^{N} a_\ell\, e_\ell, \qquad (35)$$

where $(e_\ell)_{\ell=1,\ldots,N}$ is the orthonormal basis (15) of $\mathcal{H}_N$, and the (a_ℓ) are i.i.d. normal complex Gaussian variables. This random ensemble is $U(N)$-invariant, so it can be defined w.r.t. any orthonormal basis of $\mathcal{H}_N$.

This random model, and some variants taking into account symmetries, have been used to describe the spatial, but also the phase space distributions of eigenstates of quantized chaotic maps. In [79], various indicators of localization of the Husimi densities (18) have been computed for this random model, and compared with numerical results for the eigenstates of the quantized "cat" and baker's maps (see Figure 11). The distributions seem compatible with the random state model, except for the large deviations of the sup-norms of the Husimi densities, due to eigenstates "scarred" at the fixed point $(0,0)$ (see Figure 8).

In [63] the Hecke eigenstates of $U_N(S)$, expressed as N-vectors in the position basis (e_ℓ), were shown to satisfy nontrivial L^∞ bounds[12]:

$$\|\psi_{N,j}\|_\infty \leq C_\epsilon\, N^{3/8+\epsilon}.$$

Besides, for N along a subsequence of "split primes", the description of the Hecke eigenstates is much more precise. Their position vectors are uniformly bounded, $\|\psi_{N,j}\|_\infty \leq 2$, and the value distribution of individual eigenstates $\{|\psi_{N,j}(\ell/N)|,\ \ell = 0,\ldots,N-1|\}$ is asymptotically given by the semicircle law on $[0,2]$, showing that these eigenstates are very different from Gaussian random states.

In spite of this fact, the value distribution of their Husimi function $\mathbf{x} \to |\langle \varphi_{\mathbf{x}}, \psi_{N,j}\rangle|^2$, which combines $\approx \sqrt{N}$ position coefficients, seems to be exponential.

4.2. Scars of periodic orbits

Around the time the quantum ergodicity theorem was proved, an interesting phenomenon was observed by Heller in numerical studies on the stadium billiard [51]. Heller noticed that for certain eigenfunctions, the spatial density $|\psi_j(x)|^2$ is *abnormally enhanced* along one or several *unstable* periodic geodesics. He called such

---

[12]In the chosen normalization, the trivial bound reads $\|\psi\|_\infty \leq \|e_\ell\|_\infty = N^{1/2}$.

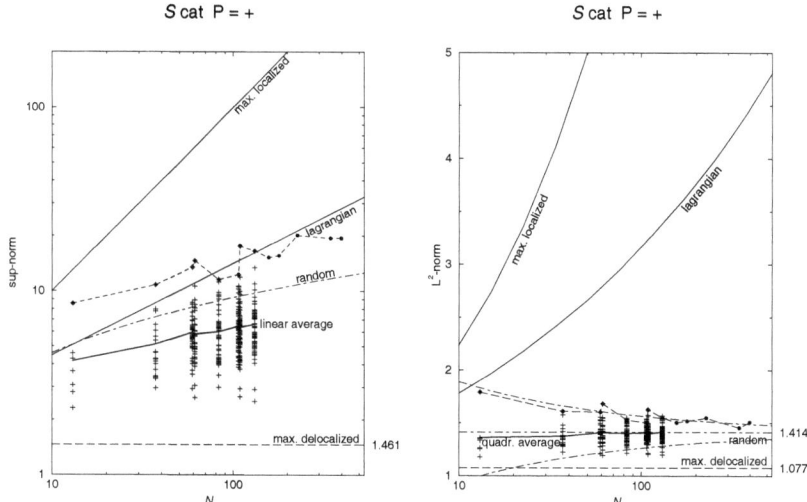

FIGURE 11. L^∞ (left) and L^2 (right) norms of the Husimi densities for eigenstates of the quantum symplectomorphism $U_N(S_{DEGI})$ (crosses; the dots indicate the states maximally scarred on $(0,0)$). The data are compared with the values for maximally localized states, Lagrangian states, random states and the maximally delocalized state. From [79] with permission from © 1998 Springer Science + Business Media.

an enhancement a *scar* of the periodic orbit on the eigenstate ψ_j. See Figure 10 (left) and Figure 12 for scars at low and relatively high frequencies.

This phenomenon was observed to persist at higher and higher frequencies for various Euclidean billiards [94, 8], but was not detected on manifolds of negative curvature [6]. Could scarred states represent counterexamples to quantum ergodicity? More precise numerics [15] showed that that the probability weights of these enhancements near the periodic orbits decay in the high-frequency limit,

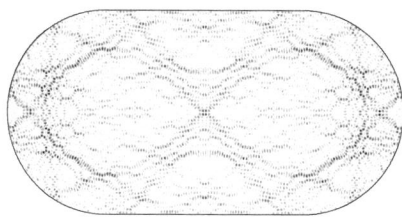

FIGURE 12. A high-energy eigenmode of the stadium billiard ($k \approx 130$). Do you see any scar?

because the *areas* covered by these enhancements decay faster than their *intensities*. As a result, a sequence of scarred states may still become equidistributed in the classical limit [55, 15].

In order to quantitatively characterize the scarring phenomenon, it turned more convenient to switch to phase space representations, in particular the Husimi density of the boundary function $\Psi(q) = \partial_\nu \psi(q)$, which lives on the phase space $T^*\partial\Omega$ of the billiard map [37, 93]. Periodic orbits are then represented by discrete points. Scars were then detected as enhancements of the Husimi density $\mathcal{H}_{\Psi_j}(\mathbf{x})$ on periodic phase space points $\mathbf{x}$.

Similar studies were performed in the case of quantum chaotic maps on the torus, like the baker's map [86] or hyperbolic symplectomorphisms [79]. The scarred state showed in Figure 8 (left) has the largest value of the Husimi density among all eigenstates of $U_N(S_{DEGI})$, but it is nevertheless a Hecke eigenstate, so that its Husimi measure should be (macroscopically) close to μ_L. This example unambiguously shows that the scarring phenomenon is a microscopic phenomenon, compatible with quantum unique ergodicity.

4.2.1. A statistical theory of scars. Heller first tried to explain the scarring phenomenon using the *smoothed local density of states*

$$S_{\chi,\mathbf{x}_0}(E) = \sum_j \chi(E - E_j)|\langle \varphi_{\mathbf{x}_0}, \psi_j \rangle|^2 = \langle \varphi_{\mathbf{x}_0}, \chi(E - \hat{H}_\hbar)\varphi_{\mathbf{x}_0} \rangle, \qquad (36)$$

where $\varphi_{\mathbf{x}_0}$ is a Gaussian wavepacket (21) sitting on a point of the periodic orbit; the energy cutoff χ is constrained by the fact that this expression is estimated through its Fourier transform, that is the time autocorrelation function

$$t \mapsto \langle \varphi_{\mathbf{x}_0}, U_\hbar^t \varphi_{\mathbf{x}_0} \rangle \tilde{\chi}(t), \qquad (37)$$

where $\tilde{\chi}$ is the $\hbar$-Fourier transform of χ. Because we can control the evolution of $\varphi_{\mathbf{x}_0}$ only up to the Ehrenfest time (25), we must take $\tilde{\chi}$ supported on the interval $[-T_E/2, T_E/2]$, so that χ has width $\gtrsim \hbar/|\log \hbar|$[13].

Since $U_\hbar^t \varphi_{\mathbf{x}_0}$ comes back to the point $\mathbf{x}_0$ at each period T, $S_{\chi,\mathbf{x}}(E)$ has peaks at the *Bohr–Sommerfeld energies* of the orbit, separated by $2\pi\hbar/T$ from one another; however, due to the hyperbolic spreading of the wavepacket, these peaks have widths $\sim \lambda\hbar/T$, where λ is the Lyapunov exponent of the orbit. Hence, the peaks can only be significant for small enough λ, that is weakly unstable orbits. Even if λ is small, the width $\lambda\hbar/T$ becomes much larger than the mean level spacing $1/\bar{\rho} \sim C\hbar^2$ in the semiclassical limit, so that $S_{\chi,\mathbf{x}}(E)$ is a mixture of many eigenstates. In particular, this mechanism can not predict which individual eigenstate will show an enhancement at $\mathbf{x}_0$, nor can it predict the value of the enhancements.

Following Heller's work, Bogomolny [23] and Berry [19] showed that certain linear combinations of eigenstates show some "extra density" in the spatial density

---

[13] We stick here to two-dimensional billiards, or maps on $\mathbb{T}^2$, so that the unstable subspaces are one dimensional.

(resp. "oscillatory corrections" in the Wigner density) around a certain number of closed geodesics. In the semiclassical limit, these combinations also involve many eigenstates in some energy window.

A decade later, Heller and Kaplan developed a "nonlinear" theory of scarring, which proposes a *statistical* definition of the scarring phenomenon [55]. They noticed that, given an energy interval I of width $\hbar^2 \ll |I| \ll \hbar$, the distribution of the overlaps $\{|\langle \varphi_\mathbf{x}, \psi_j \rangle|^2, E_j j \in I\}$ depends on the phase point $\mathbf{x}$: if $\mathbf{x}$ lies on a (mildly unstable) periodic orbit, the distribution is spread between some large values (scarred states) and some low values (antiscarred states). On the opposite, if $\mathbf{x}$ is a "generic" point, the distribution of the overlaps is narrower.

This remark was made quantitative by defining a *stochastic model* for the unsmoothed local density of states $S_\mathbf{x}(E)$ (that is, taking χ in (36) to be a delta function), as an effective way to take into account the (uncontrolled) long time recurrences in the autocorrelation function (37). According to this model, the overlaps $\langle \varphi_\mathbf{x}, \psi_j \rangle$ in an energy window $I \ni E$ should behave like random Gaussian variables, of *variance given by the smoothed local density* $S_{\chi,\mathbf{x}}(E)$. Hence, if $\mathbf{x}$ lies on a short periodic orbit, the states in energy windows close to the Bohr–Sommerfeld energies (where $S_{\chi,\mathbf{x}}(E)$ is maximal) statistically have larger overlaps with $\varphi_\mathbf{x}$, while states with energies E_j close to the anti-Bohr–Sommerfeld energies statistically have smaller overlaps. The concatenation of these Gaussian random variables with smoothly-varying variances produces a non-Gaussian distribution, with a tail larger than the one predicted by Berry's random model. On the opposite, if $\mathbf{x}$ is a "generic" point, the variance should not depend on the energy E_j and the full distribution of the $\langle \varphi_\mathbf{x}, \psi_j \rangle$ remains Gaussian.

Although not rigorously justified, this statistical definition of scarring gives quantitative predictions, and can be viewed as an interesting dynamical correction of the random state model (29).

5. Nodal structures

After having described the "macroscopic skeleton" of the eigenfunctions ψ_j, namely their semiclassical measures, and the possible large values taken by the ψ_j near a periodic orbit or elsewhere, we now focus on the opposite feature of the ψ_j, namely their *nodal sets*[14] $\mathcal{N}_{\psi_j} = \{x \in \Omega, \psi_j(x) = 0\}$. Since the eigenfunctions ψ_j can be chosen real, their nodal set is a union of hypersurfaces (in 2 dimensions, nodal lines), which sometimes intersect each other, or intersect the boundary. This set separates connected domains where ψ_j has a definite sign, called *nodal domains*. The nodal set can be viewed as a *microscopic skeleton* of the eigenfunction ψ_j: it fully determines the function (up to a global factor), and the typical scale separating two nearby hypersurfaces is the wavelength k_j^{-1} (or $\hbar_j$ in the semiclassical formalism).

---

[14] For a moment we will focus on eigenstates of chaotic billiards.

The study of the nodal patterns of eigenfunctions has a long history in mathematical physics and Riemannian geometry. Except for integrable systems[15], we have no explicit knowledge of these sets. However, some global properties are known, independently on any assumption on the geometry. In 1923 Courant [36] showed that, for the Dirichlet Laplacian on a plane domain Ω, the number of nodal domains ν_j of the jth eigenstate (counted with multiplicities) satisfies $\nu_j \leq j$, an inequality to be compared with the equality valid in 1 dimension. This upper bound was sharpened by Pleijel [81] for high-frequency eigenstates:

$$\limsup_{j\to\infty} \frac{\nu_j}{j} \leq 0.692.$$

5.1. Nodal count statistics for chaotic eigenstates

The specific study of nodal structures of eigenstates for *chaotic* billiards is more recent. Blum, Gnutzmann and Smilansky [22] seem to be the first authors to propose using nodal statistics to differentiate regular from chaotic wavefunctions – at least in 2 dimensions. They compared the *nodal count sequence* $(\xi_j = \frac{\nu_j}{j})_{j\geq 1}$ for separable vs. chaotic planar domains, and observed different statistical behaviours. In the separable case the nodal lines can be explicitly computed: they form a "grid",

---

[15] Actually, nodal domains are well identified only for *separable* systems, a stronger assumption than integrability.

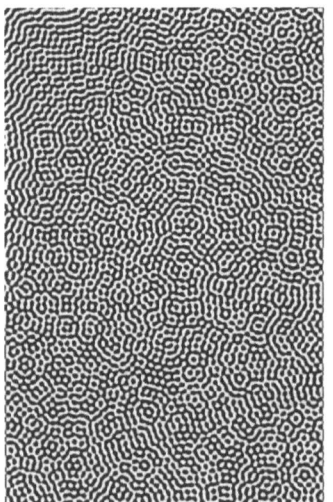 

FIGURE 13. Left: Nodal domains of an eigenfunction of the quarter-stadium with $k \approx 100.5$ (do you see the boundary of the stadium?). Right: Nodal domains of a random state (30) with $k = 100$. From [26] with permission from © 2007 IOP Publishing.

defined in terms of the two independent invariant actions I_1, I_2. The distribution of the sequence is peaked near some value ξ_m depending on the geometry.

In the chaotic case, the authors found that very few nodal lines intersect each other, and conjectured that the sequence should have the same statistics than in the case of the random models (29), (30), therefore showing some *universality*. The numerical plot of Figure 13 perfectly illustrates this assertion. In the random model, the random variable corresponding to $\xi_j = \frac{\nu_j}{j}$ should be

$$\frac{\nu(k)}{\bar{N}(k)}, \quad k \gg 1, \tag{38}$$

where $\nu(k)$ is the number of nodal domains in Ω for the random state (29), and $\bar{N}(k) = \frac{\text{Vol}(\Omega)k^2}{4\pi}$ is the integrated density of states in Ω.

Motivated by these results, Bogomolny and Schmit proposed a *percolation model* to predict the nodal count statistics in random or chaotic wavefunctions [25]. Their model starts from a separable eigenfunction, of the form $\cos(kx/\sqrt{2})\cos(ky/\sqrt{2})$, for which nodal lines form a grid; they perturb this function near each intersection, so that each crossing becomes an avoided crossing (see Figure 14). Although the length of the nodal set is almost unchanged, the structure of the *nodal domains* is drastically modified by this perturbation. Assuming these local perturbations are *uncorrelated*, they obtain a representation of the nodal domains as *clusters* of a critical bond percolation model, a well-known model in two-dimensional statistical mechanics.

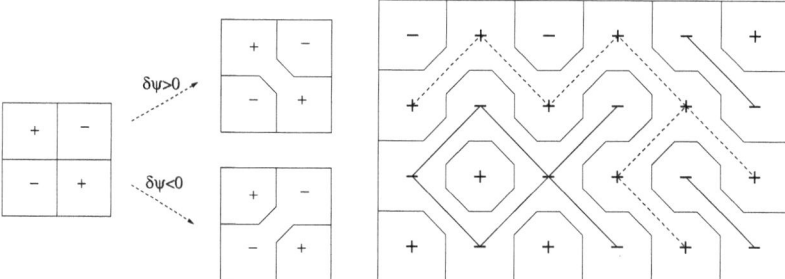

FIGURE 14. Construction of the random-bond percolation model. Left: Starting from a square grid where $\psi(x)$ alternatively takes positive $(+)$ and negative $(-)$ values, a small perturbation $\delta\psi(x)$ near each crossing creates an avoided crossing. Right: The resulting positive and negative nodal domains can be described by setting up bonds (thick/dashed lines) between adjacent sites. From [25] with permission from © 2002 The American Physical Society.

The (somewhat amazing) claim made in [25] is that such a perturbation of the separable wavefunction has the same nodal count statistics as a random function, eventhough the latter is very different from the former in may ways. This fact can be attributed to the *instability* of the nodal domains of the separable wavefunction (due to the large number of crossings), as opposed to the relative stability of the domains of random states (with generically no crossing). The uncorrelated local perturbations instantaneously transform microscopic square domains into mesoscopic (sometimes macroscopic) "fractal" domains.

The high-frequency limit ($k \to \infty$) for the random state (29), (30) corresponds to the thermodynamics limit of the percolation model, a limit in which the statistical properties of percolation clusters have been much investigated. The nodal count ratio (39) counts the number of clusters on a lattice of $N_{tot} = \frac{2}{\pi}\bar{N}(k)$ sites. The distribution of the number of domains/clusters $\nu(k)$ was computed in this limit [25]: it is a Gaussian with properties

$$\frac{\langle \nu(k) \rangle}{\bar{N}(k)} \to 0.0624, \qquad \frac{\sigma^2(\nu(k))}{\bar{N}(k)} \to 0.0502. \tag{39}$$

Nazarov and Sodin [76] have considered random spherical harmonics on the 2-sphere (namely, Gaussian random states within each $2n+1$-dimensional eigenspace, $n \geq 0$), and proved that the number of nodal domains on the eigenspace associated with the eigenvalue $n(n+1)$ statistically behaves like $(a+o(1))n^2$, for some constant $a > 0$. Although they compute neither the constant a, nor the variance of the distribution, this result indicates that the percolation model may indeed correctly predict the nodal count statistics for Gaussian random states.

Remark 2. We now have *two* levels of modelization. First, the chaotic eigenstates are statistically modelled by the random states (29), (30). Second, the nodal structure of random states is modelled by critical percolation. These two conjectures appeal to different methods: the second one is a purely statistical problem, while the first one belongs to the "chaotic=random" meta-conjecture.

5.2. Other nodal observables

The percolation model also predicts the statistical distribution of the *areas* of nodal domains: this distribution has the form $\mathcal{P}(s) \sim s^{-187/91}$, where s is the area. Of course, this scaling can only hold in the mesoscopic range $k^{-2} \ll s \ll 1$, since any domain has an area $\geq C/k^2$.

The number $\tilde{\nu}_j$ of nodal lines of ψ_j touching the boundary $\partial \Omega$ is also an intersting nodal observable. The random state model (30) was used in [22] to predict the following distribution:

$$\frac{\langle \tilde{\nu}(k) \rangle}{k} \xrightarrow{k \to \infty} \frac{\text{Vol}(\partial\Omega)}{2\pi}, \qquad \frac{\sigma^2(\tilde{\nu}(k))}{k} \xrightarrow{k \to \infty} 0.0769\, \text{Vol}(\partial\Omega).$$

The same expectation value was rigorously obtained by Toth and Wigman [92], when considering a different random model, namely random superpositions of eigenstates ψ_j of the Laplacian in frequency intervals $k_j \in [K, K+1]$:

$$\psi_{\text{rand},K} = \sum_{k_j \in [K,K+1]} a_j\, \psi_j, \tag{40}$$

with the a_j are i.i.d. normal Gaussians.

5.3. Macroscopic distribution of the nodal set

The volume of the nodal set of eigenfunctions is another interesting quantity. A priori, this volume should be less sensitive to perturbations than the nodal count ν_j. Several rigorous results have been obtained on this matter. Donnelly and Fefferman [40] showed that, for any d-dimensional compact real-analytic manifold, the $(d-1)$-dimensional volume of the nodal set of the Laplacian eigenstates satisfies the bounds

$$C^{-1} k_j \operatorname{Vol}_{d-1} \mathcal{N}(\psi_j) \leq C k_j,$$

for some $C > 0$ depending on the manifold. The statistics of this volume has been investigated for various ensembles of random states [16, 20, 85]. The average length grows like $c_M k$ with a constant $c_M > 0$ depending on the manifold; estimates for the variance are more difficult to obtain. Berry [20] argued that for the two-dimensional random model (29), the variance should be of order $\log(k)$, showing an unusually strong concentration property for this random variable. Such a logarithmic variance was recently proved by Wigman in the case of random spherical harmonics of the 2-sphere [96].

Counting or volume estimates do not provide any information on the spatial localization of the nodal set. At the microscopic level, Brüning [31] showed that for any compact Riemannian manifold, the nodal set $\mathcal{N}(\psi_j)$ is "dense" at the scale of the wavelength: for some constant $C > 0$, any ball $B(x, C/k_j)$ intersects $\mathcal{N}(\psi_j)$.

One can also consider the "macroscopic distribution" of the zero set, by integrating weight functions over the $(d-1)$-dimensional Riemannian measure on $\mathcal{N}(\psi_j)$:

$$\forall f \in C^0(M), \qquad \tilde{\mu}^Z_{\psi_j}(f) \overset{\text{def}}{=} \int_{\mathcal{N}(\psi_j)} f(x) \, d\operatorname{Vol}_{d-1}(x). \tag{41}$$

Similarly with the case of the density $|\psi_j(x)|^2$ or its phase space cousins, the spatial distribution of the nodal set can then be described by the weak-$*$ limits of the renormalized measures $\mu^Z_{\psi_j} = \frac{\tilde{\mu}^Z_{\psi_j}}{\tilde{\mu}^Z_{\psi_j}(M)}$ in the high-frequency limit. In the case of chaotic eigenstates, the following conjecture[16] seems a reasonable "dual" to the QUE conjecture:

Conjecture 3. *Let (M, g) be a compact smooth Riemannian manifold, with an ergodic geodesic flow. The, for any orthonormal basis $(\psi_j)_{j \geq 1}$, the probability measures $\mathcal{Z}_{\psi_j}$ weak-$*$ converge to the Lebesgue measure on M, in the limit $j \to \infty$.*

This conjecture is completely open. One slight weakening would be to request that the convergence holds on a density 1 subsequence, as in Thm. 1. Indeed, a similar property can be proved in the complex analytic setting (see §5.5). For M a real analytic manifold, eigenfunctions can be analytically continued in some complex neighbourhood of M, into holomorphic functions $\psi_j^{\mathbb{C}}$. For $(\psi_j)_{j \in S}$ a sequence of ergodic eigenfunctions, Zelditch has obtained the asymptotic distribution of the

---
[16] probably first mentioned by Zelditch

(complex) nodal set of $\psi_j^{\mathbb{C}}$ [100]; however, his result says nothing about the real zeros (that is, $\mathcal{N}(\psi_j^{\mathbb{C}}) \cap M$).

Once more, it is easier to deal with some class of randoms states, than the true eigenstates. In the case of the random spherical harmonics on the sphere, the random ensemble is rotation invariant, so for each level n the expectation of the measures $\mu_{\psi(n)}^Z$ is equal to the (normalized) Lebesgue measure. Zelditch [101] generalized this result to arbitrary compact Riemannian manifolds M, by considering random superpositions of eigenstates of the type (40), showing that the expectation of $\mu_{\psi_{\text{rand},K}}^Z$ converges to the Legesgue measure in the semiclassical limit.

The study of nodal sets of random wavefunctions represents lively field of research in probability theory [77].

5.4. Nodal sets for eigenstates of quantum maps on the torus

So far we have only considered the nodal set for the eigenfunctions ψ_j of the Laplacian, viewed in their spatial representation $\psi \in L^2(\Omega)$. These eigenfunctions can be taken real, due to the fact that Δ is a real operator. At the classical level, this reality corresponds to the fact that the classical flow is time reversal invariant.

In the case of quantized maps on the torus with time reversal symmetry, the eigenvectors ψ_j are real elements of $\mathcal{H}_N \equiv \mathbb{C}^N$, with components $\psi_j(\ell/N) = \langle e_\ell, \psi_j \rangle$, $\ell = 0, \ldots, N-1$. Keating, Mezzadri and Monastra [58] defined the nodal domains of such states as the intervals $\{\ell_1, \ell_1 + 1, \ldots, \ell_2 - 1\}$ on which $\psi_j(\ell/N)$ has a constant sign. Using this definition, they computed the exact nodal statistics for a random state (35) (with real Gaussian coefficients a_j), and numerically show that these statistics are satisfied by eigenstates of a generic quantized Anosov map (namely, a perturbation of the symplectomorphism S_{DEGI}).

In a further publication [59], these ideas are extended to canonical maps on the four-dimensional torus. In this setting, they show that the random state Ansatz directly leads to a percolation model on a triangular lattice, with the same thermodynamical behaviour as the model of [25]. From this remark, they conjecture that, in the appropriate scaling limit, the boundaries of the nodal domains for chaotic eigenstates belong to the universality class of SLE_6 curves, and support this claim by some numerical evidence.

5.5. Husimi nodal sets

The position representation $\psi(x)$ is physically natural, but it is not always the most appropriate to investigate semiclassical properties: phase space representations of the quantum states offer valuable informations, and are more easily compared with invariant sets of the classical flow. We have singled out two such phase space representations: the Wigner function and the Bargmann–Husimi representation. Both can be defined on $T^*\mathbb{R}^d$, but also on the tori $\mathbb{T}^{2d}$. The Wigner function is real and changes sign, so its nodal set is an interesting observable. However, in this section we will focus on the nodal sets of the Husimi (or Bargmann) functions.

The Gaussian wavepackets (21) can be appropriately renormalized into states $\tilde{\varphi}_{\mathbf{x}} \propto \varphi_{\mathbf{x}}$ depending *antiholomorphically* on the complex variable $z = x - ip$. As a result, the Bargmann function associated with $\psi \in L^2(\mathbb{R}^d)$,

$$z \mapsto \mathcal{B}\psi(z) = \langle \varphi_z, \psi \rangle,$$

is an entire function of z. For $d = 1$, the nodal set of $\mathcal{B}\psi(z)$ is thus a discrete set of points in $\mathbb{C}$, which we will denote[17] by $\mathcal{Z}_\psi$. More interestingly, through Hadamard's factorization one can essentially recover the Bargmann function $\mathcal{B}\psi$ from its nodal set, and therefore the quantum state ψ. Assuming $0 \notin \mathcal{Z}_\psi$, we have

$$\mathcal{B}\psi(z) = e^{\alpha z^2 + \beta z + \gamma} \prod_{0 \neq z_i \in \mathcal{Z}_\psi} (1 - z/z_i)\, e^{\frac{z}{z_i} + \frac{1}{2}\frac{z^2}{z_i^2}},$$

leaving only 3 undetermined parameters. Leboeuf and Voros [67] called the set $\mathcal{Z}_\psi$ the *stellar representation* of ψ, and proposed to characterize the chaotic eigenstates using this representation. This idea is especially appealing in the case of a compact phase space like the 2-torus: in that case the Bargmann function $\mathcal{B}\psi$ of a state $\psi_N \in \mathcal{H}_N$ is an entire function on $\mathbb{C}$ satisfying quasiperiodicity conditions, so that its nodal set is $\mathbb{Z}^2$-periodic, and contains exactly N zeros in each fundamental cell. One can then reconstruct the state ψ_N from this set of N points on $\mathbb{T}^2$ (which we denote by $\mathcal{Z}_{\psi_N}^{\mathbb{T}^2}$):

$$\mathcal{B}\psi_N(z) = e^\gamma \prod_{z_i \in \mathcal{Z}_{\psi_N}^{\mathbb{T}^2}} \chi(z - z_i),$$

where $\chi(z)$ is a fixed Jacobi theta function vanishing on $\mathbb{Z} + i\mathbb{Z}$. This stellar representation is *exact*, *minimal* (N complex points represent $\psi \in \mathcal{H}_N \equiv \mathbb{C}^N$) and lives in *phase space*. The conjugation of these three properties makes it interesting from a semiclassical point of view.

In [67] the authors noticed a stark difference between the nodal patterns of integrable vs. chaotic eigenstates. In the integrable case, zeros are regularly *aligned along certain curves*, which were identified as anti-Stokes lines in a complex WKB formalism. Namely, the Bargmann function can be approximated by a WKB Ansatz similar with (20) with phase functions $S_j(z)$, and anti-Stokes lines are defined by equations $\Im(S_j(z) - S_k(z)) = 0$ in regions where $e^{iS_j(z)/\hbar}$ and $e^{iS_k(z)/\hbar}$ dominate the other terms. These curves of zeros are sitting at the "antipodes" of the Lagrangian torus where the Husimi density is concentrated (see Figure 7 (center)).

On the opposite, the zeros of chaotic eigenstates appear more or less equidistributed across the whole torus (see Figure 15), like the Husimi density itself. This fact was checked on other systems, e.g., planar billiards, for which the stellar representations of the boundary functions $\partial_\nu \psi_j(x)$ were investigated in [93], leading to similar conclusions. This observation was followed by a rigorous statement, which

---

[17] No confusion will appear between this set and the real nodal sets of the previous section.

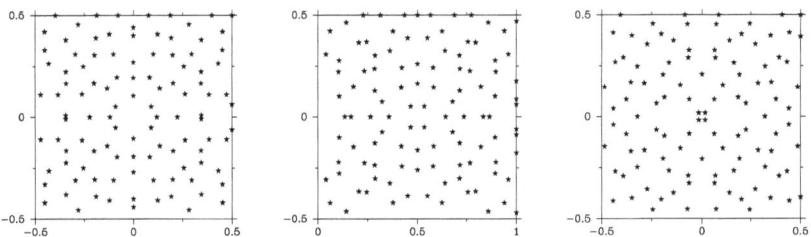

FIGURE 15. Stellar representation for 3 eigenstates of the quantum cat map $U_N(S_{DEGI})$, the Husimi densities of which were shown in Figure 8, top, in a *different order*. Can you guess the correspondence?

we express by defining (using the same notation as in the previous section) the "stellar measure" of a state $\psi_N \in \mathcal{H}_N$.

$$\mu_{\psi_N}^Z \stackrel{\text{def}}{=} N^{-1} \sum_{z_i \in \mathcal{Z}_{\psi_N}^{\mathbb{T}^2}} \delta_{z_i}.$$

Theorem 9 ([79]). *Assume that a sequence of normalized states $(\psi_N \in \mathcal{H}_N)_{N>1}$ becomes equidistributed on $\mathbb{T}^2$ in the limit $N \to \infty$ (that is, their Husimi measures $\mu_{\psi_N}^H$ weak-$*$ converge to the Liouville measure μ_L).*
Then, the corresponding stellar measures $\mu_{\psi_N}^Z$ also weak-$*$ converge to μ_L.

Using the quantum ergodicity theorem (or quantum unique ergodicity when available), one deduces that (almost) all sequences of chaotic eigenstates have asymptotically equidistributed Husimi nodal sets.

This result was proved independently (and in greater generality) by Shiffman and Zelditch [90]. The strategy is to first show that the *electrostatic potential*

$$u_{\psi_N}(\mathbf{x}) = N^{-1} \log \mathcal{H}_{\psi_N}(\mathbf{x}) = 2N^{-1} \log |\mathcal{B}\psi_N(z)| - \pi|z|^2$$

decays (in L^1) in the semiclassical limit, and then use the fact that $\mu_{\psi_N}^Z = 4\pi \Delta u_{\psi_N}$. This use of potential theory (specific to the holomorphic setting) explains why such a corresponding statement has not been proved yet for the nodal set of real eigenfunctions (see the discussion at the end of §5.3). To my knowledge, this result is the only rigorous one concerning the stellar representation of chaotic eigenstates.

In parallel, many studies have been devoted to the statistical properties of stellar representation of random states (35), which can then be compared with those of chaotic eigenstates. Zeros of random holomorphic functions (e.g., polynomials) have a long history in probability theory, see, e.g., the recent review [77], which mentions the works of Kac, Littlewood, Offord, Rice. The topic has been revived in the years 1990 through questions appearing in quantum chaos [66, 24, 47].

For a Gaussian ensemble like (35), one can explicitly compute the n-point correlation functions of the zeros: besides being equidistributed, the zeros statistically *repel* each other quadratically on the microscopic scale $N^{-1/2}$ (the typical distance between nearby zeros), but are uncorrelated at larger distances. Such a local repulsion (which imposes a certain *rigidity* of random nodal sets) holds in great generality, showing a form of universality at the microscopic scale [21, 77].

The study of [79] suggested that the localization properties of the Husimi measure (e.g., a scar on a periodic orbit) could not be directly visualized in the distribution of the few zeros near the scarring orbit, but rather in the *collective* distribution of all zeros. We thus studied in detail the Fourier coefficients of the stellar measures,
$$\mu_\psi^Z(e^{2i\pi \mathbf{k}\cdot\mathbf{x}}), \qquad 0 \neq \mathbf{k} \in \mathbb{Z}^2.$$
In the case of the random model (35), the variance of the Fourier coefficients could be explicitly computed: for fixed $\mathbf{k} \neq 0$ and $N \gg 1$, the variance is $\sim \pi^2 \zeta(3)|\mathbf{k}|^4/N^3$, showing that typical coefficients are of size $\sim N^{-3/2}$. In the case of chaotic eigenstates, we conjectured an absolute upper bound $o_\mathbf{k}(N^{-1})$ for the $\mathbf{k}$th (the equidistribution of Thm. 9 only forces these coefficients to be $o(1)$). We also argued that the presence of scars in individual eigenfunctions could be detected through the abnormally large values of a few low Fourier modes.

To finish this section, let us mention a few recent rigorous results concerning zeros of random holomorphic functions, after [77]:

- central limit theorems and large deviations of the "linear statistics" $\mu^Z(f)$, with f a fixed test function (including the characteristic function on a bounded domain)
- comparison with other point processes, e.g., the Ginibre ensemble of ensembles of randomly deformed lattices. Here comes the question of the "best matching" of a random zero set (of mean density N) with a square lattice of cell area N^{-1}.
- how to use zero sets to partition the plane into cells.

To my knowledge these properties have not been compared with chaotic eigenstates.

To summarize, the stellar representation provides complementary (*dual*) information to the macroscopic features of the Husimi (or Wigner) measures. The very nonlinear relation with the wavefunction makes its study difficult, but at the same time interesting.

References

[1] N. Anantharaman, *Entropy and the localization of eigenfunctions*, Ann. Math. **168** (2008), 435–475.

[2] N. Anantharaman and S. Nonnenmacher, *Entropy of Semiclassical Measures of the Walsh-Quantized Baker's Map*, Ann. Henri Poincaré **8** (2007), 37–74.

[3] N. Anantharaman and S. Nonnenmacher, *Half-delocalization of eigenfunctions of the laplacian on an Anosov manifold*, Ann. Inst. Fourier **57** (2007), 2465–2523.

[4] N. Anantharaman, H. Koch and S. Nonnenmacher, *Entropy of eigenfunctions*, in *New Trends in Mathematical Physics*, 1–22, V. Sidoravičius (ed.), Springer, Dordrecht, 2009.

[5] D.V. Anosov, *Geodesic flows on closed Riemannian manifolds of negative curvature*, Trudy Mat. Inst. Steklov. **90** (1967).

[6] R. Aurich and F. Steiner, *Statistical properties of highly excited quantum eigenstates of a strongly chaotic system*, Physica D **64** (1993), 185–214.

[7] R. Aurich and P. Stifter, *On the rate of quantum ergodicity on hyperbolic surfaces and for billiards*, Physica D **118** (1998), 84–102.

[8] R. Aurich, A. Bäcker, R. Schubert and M. Taglieber, *Maximum norms of chaotic quantum eigenstates and random waves*, Physica D **129** (1999) 1–14.

[9] A. Bäcker, R. Schubert and P. Stifter, *On the number of bouncing ball modes in billiards*, J. Phys. A **30** (1997), 6783–6795.

[10] A. Bäcker, R. Schubert and P. Stifter, *Rate of quantum ergodicity in Euclidean billiards*, Phys. Rev. E **57** (1998) 5425–5447. Erratum: Phys. Rev. E **58** (1998), 5192.

[11] A. Bäcker and R. Schubert, *Autocorrelation function for eigenstates in chaotic and mixed systems*, J. Phys. A **35** (2002), 539–564.

[12] N.L. Balasz and A. Voros, *Chaos on the pseudosphere*, Phys. Rep. **143** (1986), 109–240.

[13] N.L. Balasz and A. Voros, *The quantized baker's transformation*, Ann. Phys. (NY) **190** (1989), 1–31.

[14] P. Bàlint and I. Melbourne, *Decay of correlations and invariance principles for dispersing billiards with cusps, and related planar billiard flows*, J. Stat. Phys. **133** (2008), 435–447.

[15] A.H. Barnett, *Asymptotic rate of quantum ergodicity in chaotic Euclidean billiards*, Comm. Pure Appl. Math. **59** (2006), 1457–1488.

[16] P. Bérard, *Volume des ensembles nodaux des fonctions propres du laplacien*, Bony-Sjöstrand–Meyer seminar, 1984–1985, Exp. No. 14, Ecole Polytech., Palaiseau, 1985.

[17] G. Berkolaiko, J.P. Keating and U. Smilansky, *Quantum ergodicity for graphs related to interval maps*, Commun. Math. Phys. **273** (2007), 137–159.

[18] M.V. Berry, *Regular and irregular semiclassical wave functions*, J. Phys. **A, 10** (1977), 2083–91.

[19] M.V. Berry, *Quantum Scars of Classical Closed Orbits in Phase Space*, Proc. R. Soc. Lond. A **423** (1989), 219–231.

[20] M.V. Berry, *Statistics of nodal lines and points in chaotic quantum billiards: perimeter corrections, fluctuations, curvature*, J. Phys. A **35** (2002), 3025–3038.

[21] P. Bleher, B. Shiffman and S. Zelditch, *Universality and scaling of correlations between zeros on complex manifolds*, Invent. Math. **142** (2000), 351–395.

[22] G. Blum, S. Gnutzmann and U. Smilansky, *Nodal Domains Statistics: a criterium for quantum chaos*, Phys. Rev. Lett. **88** (2002), 114101.

[23] E.B. Bogomolny, *Smoothed wave functions of chaotic quantum systems*, Physica D **31** (1988), 169–189.

[24] E. Bogomolny, O. Bohigas and P. Leboeuf, *Quantum chaotic dynamics and random polynomials*, J. Stat. Phys. **85** (1996), 639–679.

[25] E. Bogomolny and C. Schmit, *Percolation model for nodal domains of chaotic wave functions*, Phys. Rev. Lett. **88** (2002), 114102.

[26] E. Bogomolny and C. Schmit, *Random wave functions and percolation*, J. Phys. A **40** (2007), 14033–14043.

[27] J. Bourgain and E. Lindenstrauss, *Entropy of quantum limits*, Comm. Math. Phys. **233** (2003), 153–171; corrigendum available at http://www.math.princeton.edu/ elonl/Publications/.

[28] A. Bouzouina et S. De Bièvre, *Equipartition of the eigenfunctions of quantized ergodic maps on the torus*, Commun. Math. Phys. **178** (1996), 83–105.

[29] S. Brooks, *On the entropy of quantum limits for 2-dimensional cat maps*, Commun. Math. Phys. **293** (2010), 231–255.

[30] S. Brooks and E. Lindenstrauss, *Non-localization of eigenfunctions on large regular graphs*, preprint arXiv:0912.3239.

[31] J. Brüning, *Über Knoten von Eigenfunktionen des Laplace–Beltrami-Operators*, Math. Z. **158** (1978), 15–21.

[32] L.A. Bunimovich, *On the ergodic properties of nowhere dispersing billiards*, Commun. Math. Phys. **65** (1979), 295–312.

[33] C.-H. Chang, T. Krüger, R. Schubert and S. Troubetzkoy, *Quantisations of Piecewise Parabolic Maps on the Torus and their Quantum Limits*, Commun. Math. Phys. **282** (2008), 395–418.

[34] N. Chernov, *A stretched exponential bound on time correlations for billiard flows*, J. Stat. Phys. **127** (2007) 21–50

[35] Y. Colin de Verdière, *Ergodicité et fonctions propres du Laplacien*, Commun. Math. Phys. **102** (1985), 597–502.

[36] R. Courant and D. Hilbert, *Methoden der mathematischen Physik, Vol. I*, Springer, Berlin, 1931.

[37] B. Crespi, G. Perez and S.-J. Chang, *Quantum Poincaré sections for two-dimensional billiards*, Phys. Rev. E **47** (1993), 986–991.

[38] M. Degli Esposti, S. Graffi and S. Isola, *Classical limit of the quantized hyperbolic toral automorphisms*, Comm. Math. Phys. **167** (1995), 471–507.

[39] M. Degli Esposti, S. Nonnenmacher and B. Winn, *Quantum variance and ergodicity for the baker's map*, Commun. Math. Phys. **263** (2006), 325–352.

[40] H. Donnelly and C. Fefferman, *Nodal sets of eigenfunctions on Riemannian manifolds*, Invent. Math. **93** (1988), 161–183.

[41] B. Ekhardt et al., *Approach to ergodicity in quantum wave functions*, Phys. Rev. E **52** (1995), 5893–5903.

[42] F. Faure, S. Nonnenmacher and S. De Bièvre, *Scarred eigenstates for quantum cat maps of minimal periods*, Commun. Math. Phys. **239**, 449–492 (2003).

[43] F. Faure and S. Nonnenmacher, *On the maximal scarring for quantum cat map eigenstates*, Commun. Math. Phys. **245** (2004), 201–214.

[44] M. Feingold and A. Peres, *Distribution of matrix elements of chaotic systems*, Phys. Rev. **A 34** (1986), 591–595.

[45] P. Gérard et G. Leichtnam, *Ergodic properties of eigenfunctions for the Dirichlet problem*, Duke Math. J. **71** (1993), 559–607.

[46] B. Gutkin, *Entropic bounds on semiclassical measures for quantized one-dimensional maps*, Commun. Math. Phys. **294** (2010), 303–342.

[47] J.H. Hannay, *Chaotic analytic zero points: exact statistics for those of a random spin state*, J. Phys. **A 29** (1996), L101–L105.

[48] J.H. Hannay and M.V. Berry, *Quantization of linear maps – Fresnel diffraction by a periodic grating*, Physica **D 1** (1980), 267–290.

[49] A. Hassell, *Ergodic billiards that are not quantum unique ergodic*, with an appendix by A. Hassell and L. Hillairet. Ann. of Math. **171** (2010), 605–618.

[50] B. Helffer, A. Martinez and D. Robert, *Ergodicité et limite semi-classique*, Commun. Math. Phys. **109** (1987), 313–326.

[51] E.J. Heller, *Bound-state eigenfunctions of classically chaotic hamiltonian systems: scars of periodic orbits*, Phys. Rev. Lett. **53** (1984), 1515–1518.

[52] E.J. Heller and P. O'Connor, *Quantum localization for a strongly classically chaotic system*, Phys. Rev. Lett. **61** (1988), 2288–2291.

[53] L. Hörmander, *The spectral function for an elliptic operator*, Acta Math. **127** (1968), 193–218.

[54] H. Iwaniec and P. Sarnak, L^∞ *norms of eigenfunctions of arithmetic surfaces*, Ann. of Math. **141** (1995), 301–320.

[55] L. Kaplan and E.J. Heller, *Linear and nonlinear theory of eigenfunction scars*, Ann. Phys. (NY) **264** (1998), 171–206.

[56] L. Kaplan, *Scars in quantum chaotic wavefunctions*, Nonlinearity **12** (1999), R1–R40.

[57] A. Katok and B. Hasselblatt, *Introduction to the modern theory of dynamical systems*, Cambridge Univ. Press, Cambridge, 1995.

[58] J.P. Keating, F. Mezzadri, and A.G. Monastra, *Nodal domain distributions for quantum maps*, J. Phys. **A 36** (2003), L53–L59.

[59] J.P. Keating, J. Marklof and I.G. Williams, *Nodal domain statistics for quantum maps, percolation, and stochastic Loewner evolution*, Phys. Rev. Lett. **97** (2006), 034101.

[60] D. Kelmer, *Arithmetic quantum unique ergodicity for symplectic linear maps of the multidimensional torus*, Ann. of Math. **171** (2010), 815–879.

[61] P. Kurlberg and Z. Rudnick, *Hecke theory and equidistribution for the quantization of linear maps of the torus*, Duke Math. J. **103** (2000), 47–77.

[62] P. Kurlberg and Z. Rudnick *On quantum ergodicity for linear maps of the torus* Commun. Math. Phys. **222** (2001), 201–227.

[63] P. Kurlberg and Z. Rudnick, *Value distribution for eigenfunctions of desymmetrized quantum maps*, Int. Math. Res. Not. **18** (2001), 985–1002.

[64] P. Kurlberg and Z. Rudnick, *On the distribution of matrix elements for the quantum cat map*, Ann. of Math. **161** (2005), 489–507.

[65] V.F. Lazutkin, *KAM theory and semiclassical approximations to eigenfunctions* (Addendum by A. Shnirelman), Springer, 1993.

[66] P. Leboeuf and P. Shukla, *Universal fluctuations of zeros of chaotic wavefunctions*, J. Phys. **A 29** (1996), 4827–4835.

[67] P. Leboeuf and A. Voros, *Chaos-revealing multiplicative representation of quantum eigenstates*, J. Phys. **A 23** (1990), 1765–1774.

[68] A.J. Lichtenberg and M.A. Lieberman, *Regular and chaotic dynamics*, 2d edition, Springer, 1992.

[69] E. Lindenstrauss, *Invariant measures and arithmetic quantum unique ergodicity*, Annals of Math. **163** (2006), 165–219.

[70] W. Luo and P. Sarnak, *Quantum variance for Hecke eigenforms*, Ann. Sci. ENS. **37** (2004), 769–799.

[71] S.W. McDonald and A.N. Kaufmann, *Wave chaos in the stadium: statistical properties of short-wave solutions of the Helmholtz equation* Phys. Rev. **A 37** (1988), 3067–3086.

[72] J. Marklof and S. O'Keefe, *Weyl's law and quantum ergodicity for maps with divided phase space*; appendix by S. Zelditch *Converse quantum ergodicity*, Nonlinearity **18** (2005), 277–304.

[73] I. Melbourne, *Decay of correlations for slowly mixing flows*, Proc. London Math. Soc. **98** (2009), 163–190.

[74] D. Milićević, *Large values of eigenfunctions on arithmetic hyperbolic surfaces*, to appear in Duke Math. J.

[75] D. Milićević, *Large values of eigenfunctions on arithmetic hyperbolic 3-manifolds*, preprint.

[76] F. Nazarov and M. Sodin, *On the number of nodal domains of random spherical harmonics*, Amer. J. Math. **131** (2009), 1337–1357.

[77] F. Nazarov and M. Sodin, *Random Complex Zeroes and Random Nodal Lines*, preprint, arXiv:1003.4237.

[78] S. Nonnenmacher, *Entropy of chaotic eigenstates*, CRM Proceedings and Lecture Notes **52** (2010), arXiv:1004.4964.

[79] S. Nonnenmacher and A. Voros, *Chaotic eigenfunctions in phase space*, J. Stat. Phys. **92** (1998), 431–518.

[80] I.C. Percival, *Regular and irregular spectra*, J. Phys. **B 6** (1973), L229–232.

[81] Å. Pleijel, *Remarks on Courant's nodal line theorem*, Comm. Pure Appl. Math. **8** (1956), 553–550.

[82] G. Rivière, *Entropy of semiclassical measures in dimension 2*, Duke Math. J. (in press), arXiv:0809.0230.

[83] G. Rivière, *Entropy of semiclassical measures for nonpositively curved surfaces*, preprint, arXiv:0911.1840.

[84] Z. Rudnick and P. Sarnak, *The behaviour of eigenstates of arithmetic hyperbolic manifolds*, Commun. Math. Phys. **161** (1994), 195–213.

[85] Z. Rudnick and I. Wigman, *On the volume of nodal sets for eigenfunctions of the Laplacian on the torus*, Ann. H. Poincaré **9** (2008), 109–130.

[86] M. Saraceno *Classical structures in the quantized baker transformation*, Ann. Phys. (NY) **199** (1990), 37–60.

[87] A. Schnirelman, *Ergodic properties of eigenfunctions*, Uspekhi Mat. Nauk **29** (1974), 181–182.

[88] R. Schubert, *Upper bounds on the rate of quantum ergodicity*, Ann. H. Poincaré **7** (2006), 1085–1098.

[89] R. Schubert, *On the rate of quantum ergodicity for quantised maps*, Ann. H. Poincaré **9** (2008), 1455–1477.

[90] B. Shiffman and S. Zelditch, *Distribution of zeros of random and quantum chaotic sections of positive line bundles*, Commun. Math. Phys. **200** (1999), 661–683.

[91] Ja.G. Sinai, *Dynamical systems with elastic reflections. Ergodic properties of dispersing billiards*, Uspehi Mat. Nauk **25** (1970) no. 2 (152), 141–192.

[92] J.A. Toth and I. Wigman, *Title: Counting open nodal lines of random waves on planar domains*, preprint arXiv:0810.1276.

[93] J.-M. Tualle and A. Voros, *Normal modes of billiards portrayed in the stellar (or nodal) representation*, Chaos, Solitons and Fractals **5** (1995), 1085–1102.

[94] E. Vergini and M. Saraceno, *Calculation by scaling of highly excited states of billiards*, Phys. Rev. E **52** (1995), 2204–2207.

[95] A. Voros, *Asymptotic ℏ-expansions of stationary quantum states*, Ann. Inst. H. Poincaré A **26** (1977), 343–403.

[96] I. Wigman, *Fluctuations of the nodal length of random spherical harmonics*, preprint 0907.1648.

[97] S. Zelditch, *Uniform distribution of the eigenfunctions on compact hyperbolic surfaces*, Duke Math. J. **55** (1987), 919–941.

[98] S. Zelditch, *Quantum ergodicity of C^* dynamical systems*, Commun. Math. Phys. **177** (1996), 507–528.

[99] S. Zelditch, *Index and dynamics of quantized contact transformations*, Ann. Inst. Fourier, **47** (1997), 305–363.

[100] S. Zelditch, *Complex zeros of real ergodic eigenfunctions*, Invent. Math. **167** (2007), 419–443.

[101] S. Zelditch, *Real and complex zeros of Riemannian random waves*, Proceedings of the conference *Spectral analysis in geometry and number theory*, Contemp. Math. **484** 321–342, AMS, Providence, 2009.

[102] S. Zelditch et M. Zworski, *Ergodicity of eigenfunctions for ergodic billiards*, Commun. Math. Phys. **175** (1996), 673–682.

Stéphane Nonnenmacher
Institut de Physique Théorique
CEA-Saclay
91191 Gif-sur-Yvette, France
e-mail: stephane.nonnenmacher@cea.fr

Is the Solar System Stable?

Jacques Laskar

> **Abstract.** Since the formulation of the problem by Newton, and during three centuries, astronomers and mathematicians have sought to demonstrate the stability of the Solar System. As mentioned by Poincaré, several demonstrations of the stability of the Solar System have been published. By Laplace and Lagrange in the first place, then by Poisson, and more recently by Arnold. Others came after again. *Were the old demonstrations insufficient, or are the new ones unnecessary?* These rigorous demonstrations are in fact various approximations of idealized systems, but thanks to the numerical experiments of the last two decades, we know now that the motion of the planets in the Solar System is chaotic. This prohibits any accurate prediction of the planetary trajectories beyond a few tens of millions of years. The recent simulations even show that planetary collisions or ejections are possible on a period of less than 5 billion years, before the end of the life of the Sun.

1. Historical introduction[1]

Despite the fundamental results of Henri Poincaré about the non-integrability of the three-body problem in the late 19th century, the discovery of the non-regularity of the Solar System's motion is very recent. It indeed required the possibility of calculating the trajectories of the planets with a realistic model of the Solar System over very long periods of time, corresponding to the age of the Solar System. This was only made possible in the past few years. Until then, and this for three centuries, the efforts of astronomers and mathematicians were devoted to demonstrate the stability of the Solar System.

1.1. Solar System stability

The problem of the Solar System stability dates back to Newton's statement concerning the law of gravitation. If we consider a unique planet around the Sun, we retrieve the elliptic motion of Kepler, but as soon as several planets orbit around

---

[1]This part is adapted from a lecture of the author on October 19th 2006, at Lombardo Institute (Milan) in honor of Lagrange: *Lagrange et la stabilité du Système solaire* (Laskar, 2006).

the Sun, they are subjected to their mutual attraction which disrupts their Kleperian motion. At the end of the volume of Opticks (1717,1730), Newton himself expresses his doubts on this stability which he believes can be compromised by the perturbations of other planets and also of the comets, as it was not known at the time that their masses were very small.

> And to show that I do not take Gravity for an essential Property of Bodies, I have added one Question concerning its Cause, chosing to propose it by way of a Question, because I am not yet satisfied about it for want of Experiments.
> ...
> For while comets move in very excentrick orbs in all manner of positions, blind fate could never make all the planets move one and the same way in orbs concentrick, some inconsiderable irregularities excepted, which may have risen from the mutual actions of comets and planets upon one another, and which will be apt to increase, till this system wants a reformation.

These planetary perturbations are weak because the masses of the planets in the Solar System are much smaller than the mass of the Sun (Jupiter's mass is about 1/1000 of the mass of the Sun). Nevertheless, one may wonder as Newton whether their perturbations could accumulate over very long periods of time and destroy the system. Indeed, one of the fundamental scientific questions of the 18th century was to first determine if Newton's law does account in totality for the motion of celestial bodies, and then to know if the stability of the Solar System was granted in spite of the mutual perturbations of planets resulting from this gravitation law. This problem was even more important as observations actually showed that Jupiter was getting closer to the Sun while Saturn was receding from it. In a chapter devoted to the *secular terms*, De la Lande reports in the first edition of his "Abrégé d'Astronomie" (1774) the problems that arose from these observations[2].

> Kepler écrivait en 1625 qu'ayant examiné les observations de Régiomontanus et de Waltherus, faites vers 1460 et 1500, il avait trouvé constamment les lieux de Jupiter & de Saturne plus ou moins avancés qu'ils ne devaient l'être selon les moyens mouvements déterminés par les anciennes observations de Ptolémée & celles de Tycho faites vers 1600.
>
> *Kepler wrote in 1625, after having considered the observations of Regiomontanus and Waltherus made in 1460 and 1500, that he found consistently that the locations of Jupiter & Saturn were more or less advanced as they should be when their mean motions was determined according to ancient observations of Ptolemy & those of Tycho made around 1600.*

---

[2] Translating French language of the XVIIIth century is not an easy matter, and the translations in English are provided here only to give a rapid view of the original text. The reader is welcome to propose some better translations to the author.

Following the work of Le Monnier (1746a,b) which, according to De La Lande[3]

a démontré le premier, d'une manière suivie et détaillée, après un travail immense sur les oppositions de Saturne (Mémoire de l'Académie 1746), que non seulement il y a dans cette planète des inégalités périodiques dépendantes de la situation par rapport à Jupiter, mais que dans les mêmes configurations qui reviennent après cinquante-neuf ans, l'erreur des Tables va toujours en croissant.

demonstrated for the first time, in a detailed manner, after great work on the oppositions of Saturn (Mémoire de l'Académie 1746), that not only there are some periodic inequalities in this planet that depends on its position relative to Jupiter, but in the same configurations returning after fifty-nine years, the error in the Tables is always growing.

These observations led Halley to introduce a quadratic secular term in the mean longitudes of Jupiter and Saturn. The Tables of Halley became an authority during several decades, and were reproduced in various forms. In particular, by the Royal Academy of Prussia (1776) (Figures 1, 2) during the period when Lagrange lived in Berlin. These apparent irregularities of Jupiter's and Saturn's motions constituted one of the most important scientific problems of the 18th century because it was a question of knowing if Newton's law do account for the motion of planets, and also of deciding on the stability of the Solar System. This led Paris Academy of Sciences to propose several prizes for the resolution of this problem. Euler was twice awarded a Prize of the Academy for these questions, in 1748 and 1752. In his last memoir (Euler, 1752), which laid the foundations of the methods of perturbations, Euler believed that he had demonstrated that Newton's law induces secular variations in the mean motion of Jupiter and Saturn, variations he found to be of the same sign, contrary to the observations. In reality, we know now that these results from Euler were wrong.

1.2. The 1766 memoir

It is still this important question that Lagrange is trying to solve in his 1766 memoir *Solution de différents problèmes de calcul intégral*, which appeared in Turin's memoirs:

je me bornerai à examiner ici, d'après les formules données ci-dessus, les inégalités des mouvements de Jupiter et Saturne qui font varier l'excentricité et la position de l'aphélie de ces deux planètes, aussi bien que l'inclinaison et le lieu du nœud de leur orbites, et qui produisent surtout une altération apparente dans leurs moyens mouvements, inégalités que les observations ont fait connaître depuis longtemps, mais que personne jusqu'ici n'a encore entrepris de déterminer avec toute l'exactitude qu'on peut exiger dans un sujet si important.

---

[3] De la Lande, Tables Astronomiques de M. Halley pour les planètes et les comètes, Paris, 1759

				Table II.		
		Mouvements moyens de Saturne pendant les Années Juliennes.				
Années	Longitude de ♄	Équat. sécul.	Aphélie	Anomalie moyenne	Nœud	Argument de Latitude
	S. D. M. S	D. M.	S. D. M. S.	S. D. M. S.	S. D. M. S.	S. D. M. S.
1	0 12 13 21	0 0	0 0 1 20	0 12 13 1	0 0 0 18	0 12 13 3
2	0 24 26 43	0 0	0 0 2 40	0 24 24 3	0 0 0 36	0 24 26 7
3	1 6 40 4	0 0	0 0 4 0	1 6 36 4	0 0 0 54	1 6 39 10
B. 4	1 18 55 26	0 0	0 0 5 20	1 18 50 6	0 0 1 12	1 18 54 14
5	2 1 8 48	0 0	0 0 6 40	2 1 2 8	0 0 1 30	2 1 7 18
6	2 13 22 9	0 0	0 0 8 0	2 13 14 9	0 0 1 48	2 13 20 21
7	2 25 35 30	0 0	0 0 9 20	2 25 26 10	0 0 2 6	2 25 33 24
B. 8	3 7 50 53	0 0	0 0 10 40	3 7 40 13	0 0 2 24	3 7 48 29
9	3 20 4 14	0 0	0 0 12 0	3 19 52 14	0 0 2 42	3 20 1 32
10	4 2 17 35	0 0	0 0 13 20	4 2 4 15	0 0 3 0	4 2 14 35
11	4 14 30 57	0 0	0 0 14 40	4 14 16 17	0 0 3 18	4 14 27 39
B. 12	4 26 46 19	0 0	0 0 16 0	4 26 30 19	0 0 3 36	4 26 42 43
13	5 8 59 41	0 0	0 0 17 20	5 8 42 21	0 0 3 54	5 8 55 47
14	5 21 13 2	0 0	0 0 18 40	5 20 54 22	0 0 4 12	5 21 8 50
15	6 3 26 23	0 0	0 0 20 0	6 3 6 23	0 0 4 30	6 3 21 53
B. 16	6 15 41 46	0 0	0 0 21 20	6 15 20 26	0 0 4 48	6 15 36 58
17	6 27 55 7	0 0	0 0 22 40	6 27 32 27	0 0 5 6	6 27 50 1
18	7 10 8 28	0 0	0 0 23 0	7 9 44 28	0 0 5 24	7 10 3 4
19	7 22 21 50	0 0,1	0 0 25 20	7 21 56 30	0 0 5 42	7 22 16 8
B. 20	8 4 37 12	0 0,1	0 0 26 40	8 4 10 32	0 0 6 0	8 4 31 12
40	4 9 14 24	0 0,2	0 0 53 20	4 8 21 4	0 0 12 0	4 9 2 24
60	0 13 51 36	0 0,5	0 1 20 0	0 12 31 36	0 0 18 0	0 13 33 36
80	8 18 28 48	0 0,9	0 1 46 40	8 16 42 8	0 0 24 0	8 18 4 48
100	4 23 6 0	0 1,4	0 2 13 20	4 20 52 40	0 0 30 0	4 22 36 0
200	9 16 12 0	0 5,6	0 4 26 40	9 11 45 20	0 1 0 0	9 15 12 0
300	2 9 18 0	0 12,5	0 6 40 0	2 2 38 0	0 1 30 0	2 7 48 0
400	7 2 24 0	0 22,2	0 8 53 20	6 23 30 40	0 2 0 0	7 0 24 0
500	11 25 30 0	0 34,8	0 11 6 40	11 14 23 20	0 2 30 0	11 23 0 0
600	4 18 36 0	0 50,0	0 13 20 0	4 5 16 0	0 3 0 0	4 15 36 0
70	9 11 42 0	1 8,1	0 15 33 20	8 26 8 40	0 3 30 0	9 8 12 0
800	2 4 48 0	1 29,0	0 17 46 40	1 17 1 20	0 4 0 0	2 0 48 0
900	6 27 54 0	1 52,6	0 20 0 0	6 7 54 0	0 4 30 0	6 23 24 0
1000	11 21 0 0	2 19,0	0 22 13 20	10 28 46 40	0 5 0 0	11 16 0 0
2000	11 12 0 0	9 16,1	1 14 26 40	9 27 33 20	0 10 0 0	11 2 0 0
3000	11 3 0 0	20 51,1	2 6 40 0	8 26 20 0	0 15 0 0	10 18 0 0
4000	10 24 0 0	37 4,4	2 28 53 20	7 25 6 40	0 20 0 0	10 4 0 0
5000	10 15 0 0	57 55,6	3 21 6 40	6 23 53 20	0 25 0 0	9 20 0 0
6000	10 6 0 0	83 24,4	4 13 20 0	5 22 40 0	1 0 0 0	9 6 0 0

FIGURE 1. Reproduction of the Tables of Halley *Recueil de Tables Astronomiques publié sous la direction de l'Académie Royale des Sciences et Belles-Lettres de Prusse, Vol. II,* 1776.

I will only consider here, according to the above formulas, the inequalities in the motions of Jupiter and Saturn which induce some variations in the eccentricity and the position of aphelion of these two planets, as well as the inclination and location of the node of their orbits, and mostly which induce some apparent alteration in their means motions,

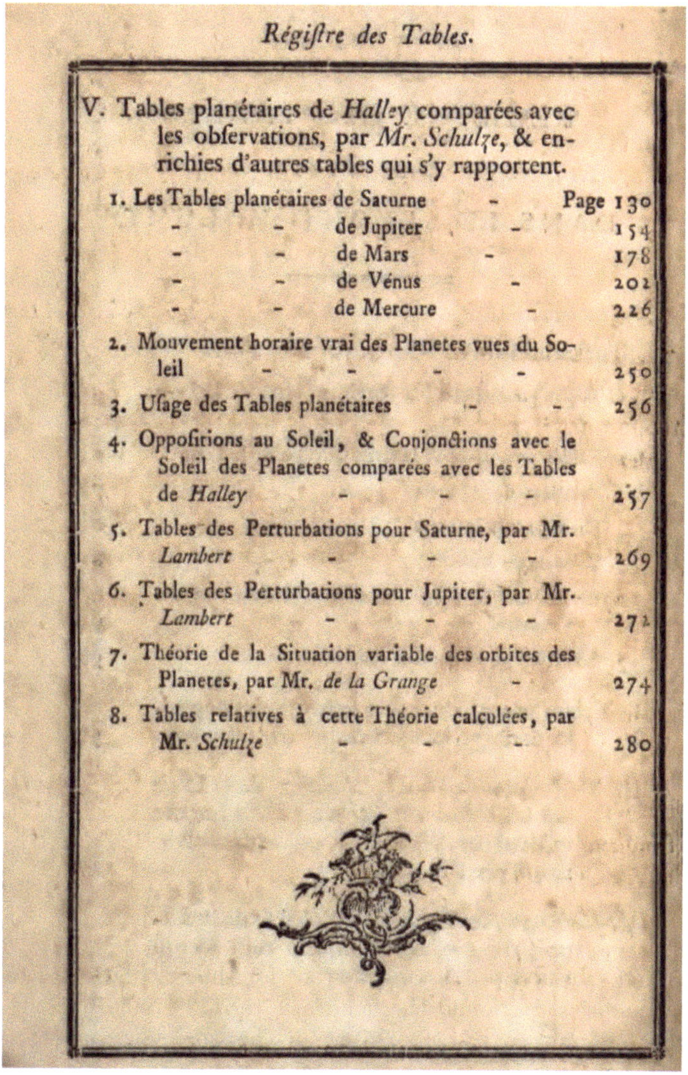

Régiſtre des Tables.

V. Tables planétaires de *Halley* comparées avec les obſervations, par *Mr. Schulze*, & enrichies d'autres tables qui s'y rapportent.

FIGURE 2. Table of content of *Recueil de Tables Astronomiques publié sous la direction de l'Académie Royale des Sciences et Belles-Lettres de Prusse, Vol. II,* 1776. It includes a paper from Lagrange where he summarizes the results of his memoir on the nodes and inclinations of the planets (1774–1778).

inequalities that the observations have been made long known, but that so far nobody has yet undertaken to determine with the accuracy that is required for such an important subject.

TABLE DE LA VARIATION DES ÉLÉMENTS DE JUPITER ET DE SATURNE, SUIVANT LA THÉORIE.

	JUPITER.	SATURNE.
Variation de la plus grande équation du centre.	$+ 7'',4254\, n$	$- 32'',6086\, n$
Variation de l'inclinaison à l'écliptique.	$- 1'',0030\, n$	$+ 2'',7449\, n$
Mouvement moyen de l'aphélie par rapport aux étoiles fixes.	$+86'',6311\, n$	$+471'',8632\, n$
Inégalité croissante dans le mouvement de l'aphélie.	$+ 0'',0262\, n^2$	$+ 0'',1141\, n^2$
Mouvement moyen des nœuds par rapport aux étoiles fixes.	$+86'',1075\, n$	$-256'',4655\, n$
Inégalité croissante dans le mouvement des nœuds.	$+ 0'',0513\, n^2$	$- 0'',0405\, n^2$
Inégalité croissante dans le mouvement en longitude.	$+ 2'',7402\, n^2$	$- 14'',2218\, n^2$

FIGURE 3. Results of Lagrange (1766) for the calculation of the secular inequalities. He found the quadratic term $2''.7402 n^2$ in the mean longitude of Jupiter, and $-14''.2218 n^2$ in the mean longitude of Saturn, where n is the number of revolutions of each planet.

It is obvious that Lagrange does not believe in Euler's results. However, the care with which Lagrange conducted his own study of the same problem will not be sufficient. Although Lagrange's results are in agreement with the behavior of the observations (he actually found that Jupiter accelerates while Saturn is slowing down (Figure 3), his calculations are still incorrect. However, this memoir remains a milestone for the development of new methods of resolutions of differential equations (see the more detailed work of F. Brechenmacher 2007).

1.3. The invariance of the semi-major axis

It is finally Laplace, who will first demonstrate the secular invariance of the semi-major axes of planets, results he publishes in the *Memoires de l'Académie des Sciences de Paris* in 1776. On the secular inequality of the semi-major axes, he writes:

> Elle ne paraît pas cependant avoir été déterminée avec toute la précision qu'exige son importance. M. Euler, dans sa seconde pièce sur les irrégularités de Jupiter et de Saturne, la trouve égale pour l'une et l'autre de ces planètes. Suivant M. de Lagrange, au contraire, dans le troisième Volume des *Mémoires de Turin*, elle est fort différente pour ces deux corps. ... j'ai lieu de croire, cependant, que la formule n'est pas encore exacte. Celle à laquelle je parviens est fort différente. ... en substituant ces valeurs dans la formule de l'équation séculaire, je l'ai trouvée absolument nulle ; d'où je conclus que l'altération du mouvement moyen de Jupiter, si elle existe, n'est point due à l'action de Saturne.
>
> *It does not seem, however, to have been determined with all the precision required by its importance. Mr. Euler, in his second piece on the irregularities of Jupiter and Saturn, find it equal for both these planets. According to Mr. de Lagrange, on the contrary, the third volume of Mémoires de Turin, it is very different for these two bodies ... I have some reasons to believe, however, that the formula is still not accurate. The one which I obtain is quite different ... by substituting these values in the formula of the secular equation, I found absolutely zero, from which I conclude the alteration of the mean motion of Jupiter, if it exists, does not result from the action of Saturn.*

Laplace's result is admirable, because he succeeds where the most outstanding intellects of the century, Euler and Lagrange, have failed, although they set up (with d'Alembert) the components which have permitted this discovery. Laplace's result is all the more striking as it runs counter to the observations, which, it is necessary to underline it, had not bothered Euler either. However Laplace does not call into question Newton's law of gravitation, but makes it necessary to find another cause for the irregularities of Jupiter and of Saturn. Luckily, there is another suitable culprit. Next to the planets, the movement of which seems regular and well ordered, other bodies exist, the comets, of which one had already noticed the very diverse trajectories. As their masses remained unknown at this time, one could evoke their attraction to explain any irregularity in the Solar System.

> Il résulte de la théorie précédente que ces variations ne peuvent être attribuées à l'action mutuelle de ces deux planètes ; mais, si l'on considère le grand nombre de comètes qui se meuvent autour du Soleil, si l'on fait ensuite réflexion qu'il est très possible que quelques-unes d'entre elles aient passé assez près de Jupiter et de Saturne pour altérer leurs mouvements, ... il serait donc fort à désirer que le nombre des comètes,

leurs masses et leurs mouvements fussent assez connus pour que l'on pût déterminer l'effet de leur action sur les planètes (Laplace, 1776a).

It follows from the above theory that these variations cannot be attributed to the mutual action of these two planets, but if we consider the large number of comets that move around the Sun, if we then imagine that it is very possible that some of them have passed close enough to Jupiter and Saturn to alter their motions, ... it would be very desirable that the number of comets, their masses and their movements were quite known so that we could determine the effect of their action on the planets (Laplace, 1776a).

The importance of the analysis of the comets' trajectories will also be fundamental for the interest that Laplace will take in the study of probability theory. He indeed had to discriminate whether the variety of the trajectories of comets are the result of chance or not (Laplace, 1776b)

1.4. Inclinations and eccentricities

Laplace had presented his results concerning the invariance of semi-major axes to the Academy in 1773. The following year, in October 1774, Lagrange, then in Berlin, submitted to the Paris Academy of Sciences a new memoir about the secular motions of inclinations and nodes of the planets. In this memoir, appear for the first time the linear differential equations with constant coefficients that represent to the first order the averaged motion of the planetary orbits.

Ce Mémoire contient une nouvelle Théorie des mouvements des nœuds et des variations des inclinaisons des orbites des planètes, et l'application de cette Théorie à l'orbite de chacune des six planètes principales. On y trouvera des formules générales, par lesquelles on pourra déterminer dans un temps quelconque la position absolue de ces orbites, et connaître par conséquent les véritables lois des changements auxquels les plans de ces orbites sont sujets.

This Memoir contains a new theory for the motion of the nodes and variations of the inclinations of the orbits of the planets, and the application of this theory to the orbit of each of the six main planets. It contains general formulas by which one can determine, for any time value, the absolute position of any such orbits, and therefore know the true laws to which are subject the planes of these orbits.

One of the important elements in the resolution of these equations is the use of the Cartesian variables

$$s = \tan i \sin \Omega \; ; \quad u = \tan i \cos \Omega , \tag{1}$$

where i is the inclination, and Ω is the longitude of the node. These variables are almost the same as those which are still used today for the study of planetary motions. Lagrange provided here for the first time a quasi-periodic expression for the motion of the orbital plane of the planets which we can now write in a more synthetic way thanks to complex notation

$$u(t) + \sqrt{-1}s(t) = \sum_{k=1}^{6} \beta_k \exp(\sqrt{-1}s_k t) \,. \qquad (2)$$

The s_k are the eigenvalues of the matrix with constant coefficients of the linear secular System. Of course, Lagrange did not use the matrices formalism which will only be put in place much later (see Brechenmacher 2007), but he had to carry out the same computation of the eigenvalues of a 6 x 6 matrix in an equivalent way. In order to do this, he will proceed by iteration, beginning by the resolution of the Sun-Jupiter-Saturn system. It is impressive to see that in spite of the uncertainties concerning the values of the masses of the inner planets (Mercury, Venus and Mars)[4], Lagrange obtained values of the fundamental frequencies of the secular system (s_k) that are very close to the present ones (Table 1).

k	Lagrange (1774)	Laskar et al., 2004
s_1	5.980	5.59
s_2	6.311	7.05
s_3	19.798	18.850
s_4	18.308	17.755
s_5	0	0
s_6	25.337	26.347

TABLE 1. Secular frequencies s_k for the motion of the nodes and inclinations of the planetary orbits. The values of Lagrange (1774) and modern values of (Laskar et al., 2004) are given in arcseconds per year. It may be surprising that modern values give less significant digits than those of Lagrange, but the chaotic diffusion in the Solar System causes a significant change in these frequencies, making it vain for a precise determination of the latter. The secular zero frequency s_5 results from the invariance of the angular momentum.

At the Academy of Sciences in Paris, Laplace was very impressed by the results of Lagrange. He himself had temporarily left aside his own studies concerning the secular motion of the planetary orbits. He understood immediately the originality and the interest of Lagrange's work and submitted without delay a new memoir to the Academy, concerning the application of Lagrange's method to the motion of eccentricities and aphelions of planetary orbits (Laplace 1775).

---

[4]Mercury and Venus do not possess satellites which could provide a good determination of the masses of the planet by applying the third law of Kepler. The satellites of Mars, Phobos and Deimos will only be discovered many years later, in 1877.

Je m'étais proposé depuis longtemps de les intégrer ; mais le peu d'utilité de ce calcul pour les besoins de l'Astronomie, joint aux difficultés qu'il présentait, m'avait fait abandonner cette idée, et j'avoue que je ne l'aurais pas reprise, sans la lecture d'une excellent mémoire *Sur les inégalités séculaires du mouvement des nœuds et de l'inclinaison des orbites des planètes*, que M. de Lagrange vient d'envoyer à l'Académie, et qui paraîtra dans un des volumes suivants. (*Laplace, œuvres t. VIII, p. 355*)

I had proposed myself, since a long time to incorporate them, but the little utility of these calculations for the purposes of Astronomy, added to the difficulties that it presented, made me abandon this idea, and I confess that I would not have returned to it, without the lecture of the excellent paper "Sur les inégalités séculaires du mouvement des nœuds et de l'inclinaison des orbites des planètes" that M. Lagrange has just send to the Academy, and which will appear in one of the following volumes. (Laplace, œuvres t. VIII, p. 355)

What is surprising, is that Laplace's memoir, submitted in December of 1774, is very quickly published, in 1775, with the Academy's memoirs of 1772, while the original memoir of Lagrange will have to wait until 1778 to be published with the other memoirs of the year 1774. The application to eccentricities and to aphelion is in fact immediate, using the variables

$$l = e \cos \varpi \; ; \qquad h = e \sin \varpi \; . \tag{3}$$

J'ai de plus cherché si l'on ne pourrait pas déterminer d'une manière analogue les inégalités séculaires de l'excentricité et du mouvement de l'aphélie, et j'y suis heureusement parvenu ; en sorte que je puis déterminer, non seulement les inégalités séculaires du mouvement des nœuds et de l'inclinaison des orbites des planètes, les seules que M. de Lagrange ait considérées, mais encore celles de l'excentricité et du mouvement des aphélies, et comme j'ai fait valoir que les inégalités du moyen mouvement et de la distance moyenne sont nulles, on aura ainsi une théorie complète et rigoureuse de toutes les inégalités séculaires des orbites des planètes. (*Laplace, œuvres t. VIII, p. 355*)

In addition, I have searched if one could determine similarly the secular inequalities of eccentricity and motion of the aphelion, and I happily succeeded, so that I can determine not only the secular inequalities of the motion of nodes and the inclination of the orbits of the planets, the only ones that were considered by Mr. Lagrange, but also those of the eccentricity and motion of aphelion, and as I have argued that the inequalities of the mean motion and the average distance is zero, we will thus have a complete and rigorous theory of all secular inequalities of the orbits of the planets. (Laplace, œuvres t. VIII, p. 355)

One may be amazed that Laplace's memoir was published before Lagrange's, and Laplace himself feels obliged to add a note upon this

> J'aurai dû naturellement attendre que les recherches de M. de Lagrange fussent publiées avant que de donner les miennes ; mais, venant de faire paraître dans les *Savants étrangers*, année 1773, un Mémoire sur cette matière, j'ai cru pouvoir communiquer ici aux géomètres, en forme de supplément, ce qui lui manquait encore pour être complet, en rendant d'ailleurs au Mémoire de M. de Lagrange toute la justice qu'il mérite ; je m'y suis d'autant plus volontiers déterminé, que j'espère qu'ils me sauront gré de leur présenter d'avance l'esquisse de cet excellent Ouvrage. (*Laplace, œuvres t. VIII, p.* 355)

> *I should have naturally waited that the research of Mr. Lagrange were published before to give mines, but as I just published in "Savants étrangers", year 1773, a Memoire on this matter, I thought I could communicate here to the geometers, in the form of a supplement, which was still lacking for completeness, giving back to the memoire of Mr. Lagrange all the justice it deserves; I was even more resolute to do this, as I hope they would be grateful to me to present them in advance the sketch of this excellent work. (Laplace, œuvres t. VIII, p. 355)*

Laplace sends his memoir to Lagrange who sends him back a long letter from Berlin on April 10, 1775:

> Monsieur et très illustre Confrère, j'ai reçu vos Mémoires, et je vous suis obligé de m'avoir anticipé le plaisir de les lire. Je me hâte de vous en remercier, et de vous marquer la satisfaction que leur lecture m'a donnée. Ce qui m'a le plus intéressé, ce sont vos recherches sur les inégalités séculaires. Je m'étais proposé depuis longtemps de reprendre mon ancien travail sur la théorie de Jupiter et de Saturne, de le pousser plus loin et de l'appliquer aux autres planètes ; j'avais même dessein d'envoyer à l'Académie un deuxième Mémoire sur les inégalités séculaires du mouvement de l'aphélie et de l'excentricité des planètes, dans lequel cette matière serait traitée d'une manière analogue à celle dont j'ai déterminé les inégalités du mouvement du nœud et des inclinaisons, et j'en avais déjà préparé les matériaux ; mais, comme je vois que vous avez entrepris vous-même cette recherche, j'y renonce volontiers, et je vous sais même très bon gré de me dispenser de ce travail, persuadé que les sciences ne pourront qu'y gagner beaucoup.

> *Mr. and illustrious Colleague, I received your memoirs, and I am obliged to you to have anticipated the pleasure of reading it. I look forward to thank you, and to mark the satisfaction their reading has given me. What I was most interested in, are your research on the secular inequalities. I thought long ago to take back my old work on the theory of Jupiter and Saturn, to push it further and apply it to other planets. I even planned to send a second memoir Academy on the inequalities of the secular motion*

of the aphelion and eccentricity of the planets, in which the material is treated in a similar manner as what I have determined for the motion of the inequalities of the node and inclinations, and I had already prepared the materials, but as I see that you have undertaken yourself this research, I happily renounce to it, and I even thanks you for dispensing me of this work, convinced that science will largely gain from this.

So Lagrange specifies that he also had understood that the problem of the eccentricities could be treated in the same way, and because Laplace now deals with this question, Lagrange proposes to give up this subject to him. In fact, this "promise" will not last, and he sends back to d'Alembert a letter dated from May 29, 1775 which shows that he cannot resist continuing his research on this fascinating subject.

Je suis près à donner une théorie complète des variations des éléments des planètes en vertu de leur action mutuelle. Ce que M. de la Place a fait sur cette matière m'a beaucoup plu, et je me flatte qu'il ne me saura pas mauvais gré de ne pas tenir l'espèce de promesse que j'avais faite de la lui abandonner entièrement ; je n'ai pas pu résister à l'envie de m'en occuper de nouveau, mais je ne suis pas moins charmé qu'il y travaille aussi de son côté ; je suis même fort empressé de lire ses recherches ultérieures sur ce sujet, mais je le prie de ne m'en rien communiquer en manuscrit et de ne me les envoyer qu'imprimées ; je vous prie de bien vouloir le lui dire, en lui faisant en même temps mille compliments de ma part.

I am ready to give a complete theory for the variations of the elements of the planets under their mutual action. That Mr. de la Place did on this subject I liked, and I flatter myself that he will not be offended if I do not hold the kind of promise that I made to completely abandon this subject to him; I could not resist to the desire to look into it again, but I am no less charmed that he is also working on it on his side; I am even very eager to read his subsequent research on this topic, but I do ask him not to send me any manuscript and send them to me only in printed form; I would be obliged that you tell him, with a thousand compliments from my side.

Indeed, Lagrange resumed his work and published his results in several memoirs in 1781, 1782, 1783a, b, and 1784 in which he gives the first complete solution of the motion of the six main planets. Perhaps due to the deception he felt following the submission of his article of 1774 to Paris Academy of Sciences, he chose this time to publish his works in the Memoirs of the Academy of Berlin.

1.5. The great inequality of Jupiter and Saturn[5]

Laplace had demonstrated the invariance of secular variations of the semi-major axes, considering only the first terms of the expansion of their average perturba-

---

[5] A more detailed account of this quest can be found in (Wilson, 1985), and (Laskar, 1992).

tions, but the problem of the accordance with observations remained. He resumed his search with his theory of Jupiter and Saturn as a base. A first element put him on the right track: the observation of the energy's conservation in the Sun-Jupiter-Saturn system. If Newton's law is correct, the conservation of the system's energy implies that when one of the mean motion increases, the other must decrease. This is clearly observed. By neglecting the terms of order two compared to the masses, he found out that the quantity

$$\frac{m_J}{a_J} + \frac{m_S}{a_S} \qquad (4)$$

must remain constant. With Kepler's law ($n^2 a^3 = Cte$) it gives:

$$\frac{dn_S}{dn_J} = -\frac{m_J}{m_S}\sqrt{\frac{a_J}{a_S}} \qquad (5)$$

where for every planet Jupiter (J) or Saturn (S), m is the mass, a the semi-major axis and n the mean motion. Using the observations available at the time (Table 2), one finds $dn_S/dn_J = -2.32$, which is translated by Laplace as "Saturn deceleration must be compared to the acceleration of Jupiter, roughly, as 7 is with 3". Using the values obtained by Halley by comparison with the observations, one obtains $dn_S/dn_J = -2.42$, allowing Laplace to think with great confidence that "the variations observed in the motions of Jupiter and of Saturn result from their mutual action". Newton's law does thus not seem to be challenged, but it remained necessary to find the reason for these variations from Newton's equations. As Laplace demonstrated that there are no secular terms in the first-order equations of semi major axes, he inferred that these changes of the average motion of the planets are probably due to short period terms (periodic terms with frequencies that are integer combinations of the mean motions of Jupiter and Saturn) which would be of period that is long enough to look like a secular term. A good candidate for this is the term associated to the combination of longitudes $2\lambda_J - 5\lambda_S$, with period of about 900 years.

planète	1/m	a (UA)	n ("/365j)
Jupiter	1067.195	5.20098	109182
Saturne	3358.40	9.54007	43966.5

TABLE 2. Values of the parameters of Jupiter and Saturn used in Laplace's work (Laplace, 1785).

This research led Laplace to undertake the construction of a more complete theory of the motion of the Jupiter-Saturn couple. After very long calculations, because to obtain these terms it is necessary to develop the perturbations to a high degree with respect to the eccentricities of Jupiter and Saturn, he obtained the following formulas (reduced here to their dominant terms) for the mean longitudes

of Jupiter and Saturn:
$$\lambda_J = n_J t + \epsilon_J + 20' \quad \sin(5n_S t - 2n_S t + 49°8'40'')$$
$$\lambda_S = n_S t + \epsilon_S + 46'50'' \sin(5n_S t - 2n_S t + 49°8'40'') \tag{6}$$

ϵ_J and ϵ_S being the initial conditions for 1700 after J.C. Laplace then corrected the values of the mean motions of Jupiter and Saturn n_J and n_S with respect to Halley's tables. He was then able to compare his new theory, without secular terms in the mean motions (or what is equivalent, without quadratic terms in the mean longitude) to modern and ancient observations. The differences in longitude between his theory and those of new observations (from 1582 to 1786) were all less than 2', while the differences with Halley's tables reached more than 20'. He also compared his theory with the Chaldean observations of Saturn in 228 BC and of Jupiter at 240 BC transmitted by Ptolemy in the Almageste. These observations are of particularly good quality, because they identify precisely the planet positions in comparison with known stars. Laplace found a difference with his formulas only 55'' for the first and 5'' for the second.

The new theory of the Jupiter-Saturn couple that Laplace completed was therefore in perfect agreement with the observations from 240 BC to 1715 AD, without the necessity of an empirical secular term in the mean motion. The whole theory was entirely derived from Newton's law of gravitation. Laplace saw here *a new proof of the admirable theory of universal gravity*. He also obtained an important side result, that is, the mass of comets is certainly very small, otherwise their perturbations would have disturbed Saturn's orbital motion.

After this work, the secular terms of the mean motions will once and for all disappear from the astronomical tables, and in the second edition of his *Abrégé d'Astronomie*, De la Lande (1795) will reduce the chapter about the secular equations to a simple paragraph, recalling that Laplace's calculations on the great inequality of Jupiter and Saturn *make the acceleration of the one (Jupiter) and the delaying of the other (Saturn) disappear; their effect is only to make the duration of their revolutions to seem more or less long during nine centuries.*

1.6. Back to the semi-major axis

Laplace's demonstration on the secular invariance of the semi-major axes of the planets considered the expansion up to degree two in eccentricities and inclinations of the perturbing potential of the planets. Lagrange went back to this problem in 1776 using his method of variations of constants, which allowed him to redo the demonstration, without expansion in eccentricity, and therefore valid for all eccentricities. His demonstration is also particularly simple and very close to the current demonstration. Lagrange will once again come back over this problem in 1808 after Poisson had presented his famous Memoir of about 80 pages (Poisson 1808), where he showed that the invariance of the semi-major axes of the planets is still valid at the second order with respect to the masses.

Lagrange was in Paris during this time, Member of the Institute, where he had been called by Laplace, in 1787, as a *subsidized veteran of the Academy of*

Sciences. In this memoir of 1808, Lagrange showed that using coordinates referring to the barycenter of the Solar System instead of using, as was previously the case, heliocentric coordinates, he succeeded in giving a more symmetric shape to the equations and considerably simplified Poisson's demonstration.

Indeed, he derived the differential equations of motion from a single function, and this was the beginning of Lagrangian formalism of variations of constants which already begun to express its considerable power in this difficult problem. This study conducted to the general method of Lagrange, described in the *Mémoire sur la théorie générale de la variation des constantes arbitraires dans tous les problèmes de la mécanique* (Lagrange 1808, 1809). This problem of the Solar System's stability and of the calculus of the secular terms observed by the astronomers was therefore fundamental in the development of mechanics and perturbative methods and more generally in the development of science in the XVIIIth century.

1.7. The proof of stability of Lagrange and Laplace and Le Verrier's question

After the work of Lagrange and Laplace, the stability of the Solar System seemed to be acquired. The semi-major axes of the orbits had no long-term variations, and their eccentricities and inclinations showed only small variations which did not allow the orbits to intersect and planets to collide. However, it should be noted that Lagrange and Laplace solutions are very different from Kepler's ellipses: the planetary orbits are no longer fixed. They are subject to a double movement of precession with periods ranging from 45000 years to a few million years: precession of the perihelion, which is the slow rotation of the orbit in its plan, and precession of nodes, which is the rotation of the orbital plane in space.

Lagrange and Laplace have written the first *proof* of the stability of the Solar System. But as Poincaré (1897) emphasizes in a general audience paper about the stability of the Solar System:

> Les personnes qui s'intéressent aux progrès de la Mécanique céleste, ..., doivent éprouver quelque étonnement en voyant combien de fois on a démontré la stabilité du système solaire.
>
> Lagrange l'a établie d'abord, Poisson l'a démontrée de nouveau, d'autres démonstrations sont venues depuis, d'autres viendront encore. Les démonstrations anciennes étaient-elles insuffisantes, ou sont-ce les nouvelles qui sont superflues?
>
> L'étonnement de ces personnes redoublerait sans doute, si on leur disait qu'un jour peut-être un mathématicien fera voir, par un raisonnement rigoureux, que le système planétaire est instable.
>
> *Those who are interested in the progress of celestial mechanics, ... must feel some astonishment at seeing how many times the stability of the Solar System has been demonstrated.*

> *Lagrange established it first, Poisson has demonstrated it again, other demonstrations came afterwards, others will come again. Were the old demonstrations insufficient, or are the new ones unnecessary?*
>
> *The astonishment of those people would probably double, if they would be told that perhaps one day a mathematician will demonstrate, by a rigorous reasoning, that the planetary system is unstable.*

In fact, the work of Lagrange and Laplace concerned only the linear approximation of the average motion of the planets. In modern language, we can say they demonstrated that the origin (equivalent to planar circular motions) is an elliptical fixed point in the secular phase space, obtained after averaging of order one with respect to the mean longitudes. Later on, Le Verrier (1840, 1841) resumed the computations of Lagrange and Laplace. This was before his discovery of Neptune in 1846, from the analysis of the irregularities in the motion of Uranus. In a first paper (1840) he computed the secular system for the planets, following the previous works of Lagrange and Laplace, with the addition of the computation of the change in the solutions resulting from possible new determinations of the planetary masses. Soon after (1841), he reviewed the effects of higher-order terms in the perturbation series. He demonstrated that these terms produce significant corrections to the linear equations, and that the calculations of Laplace and Lagrange could not be used for an indefinite period. Le Verrier (1840, 1841) raised the question of the existence of small divisors in the secular system of the inner planets. This was even more important as some of the values of the planetary masses were very imprecise, and a change of the mass values could lead to a very small divisor, eventually equal to zero. The problem for Le Verrier was then that the terms of third order could be larger than the terms of second order, which in his view compromised the convergence of the solutions.

> Ces termes acquièrent par l'intégration de très petits diviseurs ; et ainsi il en résulte, dans les intégrales, des termes dus à la seconde approximation, et dont les coefficients surpassent ceux de la première approximation. Si l'on pouvait répondre de la valeur absolue de ces petits diviseurs, la conclusion serait simple : la méthode des approximations successives devrait être rejetée.
>
> *Through integration, these terms acquire very small divisors; and thus it results in the integrals, some terms from the second approximation, and which coefficients exceed those of the first approximation. If we could bound the absolute value of these small divisors, the conclusion would be simple: the method of successive approximations should be rejected.*

The indeterminacy of the masses of the inner planets thus did not allow Le Verrier to decide of the stability of the system and he could only ask for the mathematicians' help to solve the problem.

> Il paraît donc impossible, par la méthode des approximations successives, de prononcer si, en vertu des termes de la seconde approximation, le système composé de Mercure, Vénus, la Terre et Mars, jouira d'une

stabilité indéfinie ; et l'on doit désirer que les géomètres, par l'intégration des équations différentielles, donnent les moyens de lever cette difficulté, qui peut très bien ne tenir qu'à la forme.

It seems thus impossible, by the method of successive approximations to decide whether, owing the terms of the second approximation, the system consisting of Mercury, Venus, Earth and Mars will enjoy indefinite stability and we must desire that geometers, by the integration of the differential equations of motion will provide the means to overcome this difficulty, which may well result only from the form.

1.8. Poincaré: the geometer's answer

But Poincaré (1892–99) will give a negative answer to Le Verrier's question. To do this, he completely re-thought the methods of celestial mechanics from the work of Hamilton and Jacobi. Poincaré demonstrated that it is not possible to integrate the equations of the movement of three celestial bodies subject to their mutual gravitational interaction, and that it is impossible to find an analytical solution representing the planetary motion, valid over an infinite time interval. In the same way, he concluded that the perturbations series used by astronomers to calculate the motion of the planets are not converging on an open set of initial conditions.

Poincaré therefore showed that the series of the astronomers are generally divergent. However, he had a high regard for the work of the astronomers of the time, and also pointed out that these divergent series can still be used as a very good approximation for the motion of the planets for some time, which can be long, but not infinite. Poincaré did not seem to think that his results may have great practical importance, if not precisely for the study of the stability of the Solar System.

Les termes de ces séries, en effet, décroissent d'abord très rapidement et se mettent ensuite à croître ; mais, comme les astronomes s'arrêtent aux premiers termes de la série et bien avant que ces termes aient cessé de décroître, l'approximation est suffisante pour les besoins de la pratique. La divergence de ces développements n'aurait d' inconvénients que si l'on voulait s'en servir pour établir rigoureusement certains résultats, par exemple la stabilité du système solaire.

The terms of these series, in fact, decrease first very quickly and then begin to grow, but as the Astronomers stop after the first terms of the series, and well before these terms have stop to decrease, the approximation is sufficient for the practical use. The divergence of these expansions would have some disadvantages only if one wanted to use them to rigorously establish some specific results, as the stability of the Solar System.

It should be noted that Poincaré means here stability on infinite time, which is very different from the practical stability of the Solar System, which only makes sense on a time comparable to its life expectancy time interval. Le Verrier had reformulated the question of the stability of the Solar System by pointing out the need to take into account the terms of higher degree than those considered by Laplace and Lagrange; Poincaré is even more demanding, asking for the convergence of the series:

> Ce résultat aurait été envisagé par Laplace et Lagrange comme établissant complètement la stabilité du système solaire. Nous sommes plus difficiles aujourd'hui parce que la convergence des développements n'est pas démontrée ; ce résultat n'en est pas moins important.
>
> *This result would have been considered by Laplace and Lagrange as establishing completely the stability of the Solar System. We are more demanding today because the convergence of expansions has not been demonstrated; This result is nevertheless important.*

Poincaré demonstrated the divergence of the series used by astronomers in their perturbations computations. As usual, he studied a much larger variety of perturbation series, but apparently not of immediate interest to astronomers, as they required to modify the initial conditions of the planets. However, Poincaré raise some doubts on the divergence of this type of series:

> Les séries ne pourraient-elles pas, par exemple, converger quand x_1^0 et x_2^0 ont été choisis de telle sorte que le rapport n_1/n_2 soit incommensurable, et que son carré soit au contraire commensurable (ou quand le rapport n_1/n_2 est assujetti à une autre condition analogue à celle que je viens d'énoncer un peu au hasard)? Les raisonnements de ce chapitre ne me permettent pas d'affirmer que ce fait ne se présentera pas. Tout ce qu'il m' est permis de dire, c'est qu'il est fort invraisemblable.
>
> *The series could they not, for example, converge when x_1^0 and x_2^0 have been chosen so that the ratio n_1/n_2 is incommensurable, and its square is instead commensurable (or when the ratio n_1/n_2 is subject to another condition similar to that I have enounced somewhat randomly). The arguments of this chapter do not allow me to say that this does not exist. All I am allowed to say is that this is highly unlikely.*

Half a century later, in line with the work of Poincaré, the Russian mathematician A.N. Kolmogorov actually demonstrated that these convergent perturbation series exists.

1.9. Back to stability

Kolmogorov (1954) analyzed again the problem of convergence of the perturbation series of celestial mechanics and demonstrated that for non-degenerated perturbed Hamiltonian systems, close to the non-regular solutions described by Poincaré, there are still regular quasiperiodic trajectories filling tori in the phase space. This result is not in contradiction with the result of non-integrability of Poincaré, because these tori, parameterized by the action variables do not form a continuum. This result has been completed by Arnold (1963a) which demonstrated that, for a sufficiently small perturbation, the set of invariant tori foliated by quasi-periodic

trajectories is of strictly positive measure, measure that tends to unity when the
perturbation tends to zero. Moser (1962) has established the same kind of results
for less strong conditions that do not require the analyticity of the Hamiltonian.
These theorems are generically called KAM theorems, and have been used in various fields. Unfortunately, they do not directly apply to the planetary problem
that present some proper degenerescence (the unperturbed Hamiltonian depends
only on the semi-major axis, and not on the other action variables (related to
eccentricity and inclination). This led Arnold to extend the proof of the existence
of invariant tori, taking into account the phenomenon of degenerescence. He then
applied his theorem explicitly to a planar planetary system with two planets,
for a semi-major axes' ratio close to zero, then demonstrating the existence of
quasiperiodic trajectories for sufficiently small values of the planetary masses and
eccentricities (1963b). This result was later on extended to more general two planets spatial planetary systems (Robutel, 1995). More recently, (Féjoz and Herman,
2004; Chierchia and Pinzari, 2011) have shown the existence of tori of quasiperiodic orbits in a general system of N planets, but this result still requires extremely
small planetary masses.

The results of Arnold brought many discussions, indeed, as the quasiperiodic
KAM tori form a totally discontinuous set, an infinitely small variation of the
initial conditions will turn an infinitely stable quasiperiodic solution into a chaotic,
unstable solution. Furthermore, as the planetary system has more than two degrees
of freedom, none of the KAM tori separates the phase space, leaving the possibility
for the chaotic trajectories to travel great distances in the phase space. This is the
diffusion phenomenon highlighted by Arnold.

In fact, later results showed that in the vicinity of a regular KAM torus,
the diffusion of the trajectories is very slow (Nekhoroshev, 1977, Giorgilli et al.,
1989, Lochak, 1993, Morbidelli and Giorgilli, 1995), and may be negligible for a
very long time, eventually as long as the age of the universe. Finally, although
the masses of the actual planets are much too large for these results to be applied
directly to the Solar System[6], it is generally assumed that the scope of these
mathematical results goes much further than their demonstrated limits, and until
recently it was generally accepted that the Solar System is *probably stable (by
any reasonable definition of the term) over time scales comparable with its age*[7].
Over the last twenty years, the problem of the stability of the Solar System has
considerably progressed, largely through the assistance provided by computers
which allow extensive analytical calculations and numerical integrations of realistic
models of the Solar System on durations that are now equivalent to his age. But
this progress is also due as well to the understanding of the underlying dynamics,
resulting from the development of the theory of dynamical systems since Poincaré.

---

[6]The application of Nekhoroshev theorem for the stability in finite time of the Solar System was
made by Niederman (1996), but required planetary masses of the order of 10^{-13} solar mass.
[7]Quoted from C. Murray (1988) in his review of the proceeding of a meeting devoted to the
Stability of Planetary Systems.

2. Numerical computations

The orbital motion of the planets in the Solar System has a very privileged status. Indeed, it is one of the best modeled problems of physics, and its study may be practically reduced to the study of the behavior of the solutions of the gravitational equations (Newton's equations supplemented by relativistic corrections) by considering point masses, except in the case of the interactions of the Earth-Moon system. The dissipative effects are also very small, and even if we prefer to take into account the dissipation by tide effect in the Earth-Moon system to obtain a solution as precise as possible for the motion of the Earth (e.g., Laskar et al., 2004), we can very well ignore the loss of mass of the Sun. The mathematical complexity of this problem, despite its apparent simplicity (especially if it is limited to the Newtonian interactions between point masses) is daunting, and has been a challenge for mathematicians and astronomers since its formulation three centuries ago. Since the work of Poincaré, it is also well known that the perturbative methods that were used in the planetary calculations for almost two centuries cannot provide precise approximations of solutions on an infinite time. Furthermore, as indicated above, the rigorous results of stability by Arnold (1963ab) do not apply to realistic planetary systems.

Since the apparition of computers, the numerical integration of the planetary equations has emerged as a simple way to overcome this complexity of the solutions, but this approach has always been limited up to the present by computer technology. The first long numerical integrations of the Solar System orbits were limited to the outer planets, from Jupiter to Pluto (Cohen et al., 1973, Kinoshita and Nakai, 1984). Indeed, the more the orbital motion of the planets is fast, the more it is difficult to integrate them numerically, because the required integration step decreases with the period of the planet. Using a conventional numerical method, to integrate the orbit of Jupiter, a integration step size of 40 days is sufficient, whereas a 0.5 days is necessary to integrate the motion of the entire Solar System including Mercury. The first integrations of the outer planets system that were performed over 100 Myr and then 210 Myr (Carpino et al., 1987, Nobili et al., 1989, Applegate et al., 1986) essentially confirmed the stability of the outer planets system, finding quasiperiodic orbits similar to those of Lagrange or Le Verrier. It is only when Sussman and Wisdom (1988) have extended their calculations on 875 Myr that the first signs of instability in the motion of Pluto have appeared, with a Lyapunov time (the inverse of the Lyapunov exponent) of 20 Myr. But as the mass of Pluto (which is no longer considered as a *planet* since the resolution of the International Astronomical Union in 2006) is very low (1/130 000 000 the solar mass), this instability does not manifest itself by macroscopic instabilities in the remaining part of the Solar System, which appeared very stable in all these studies.

2.1. Chaos in the Solar System

Numerical integrations allow to obtain very accurate solutions for the trajectories of planets, but are limited by the short time step, necessary to achieve this precision in the case of the complete Solar System, where it is necessary to take into account the motion of Mercury, and even of the Moon. It should be noted that, until 1991, the only available numerical integration for a realistic model of the whole Solar System was the numerical integration of the Jet Propulsion Laboratory DE102 (Newhall et al., 1983), calculated over only 44 centuries.

I opted then for a different approach, using analytical perturbation methods, in the spirit of the work of Lagrange, Laplace, and Le Verrier. Indeed, since these pioneering works, the *Bureau des Longitudes*[8], always has been the place of development of analytical planetary theories based on the classical perturbation series (Brumberg and Chapront, 1973, Bretagnon, 1974, Duriez, 1979). Implicitly, these studies assumed that the movement of celestial bodies is quasiperiodic and regular. These methods were essentially the same as those which were used by Le Verrier, with the additional help of computers for symbolic calculations. Indeed, these methods can provide very good approximations of the solutions of the planets over thousands of years, but they will not be able to provide answers to the questions of the stability of the Solar System. This difficulty, which has been known since Poincaré is one of the reasons that motivated the direct numerical integration of the equations of motion.

Nevertheless, the results of the KAM theorems suggested the possibility that classic perturbative solutions could be developed using computer algebra, to find quasiperiodic solutions of the orbital motions in the Solar System. However, seeking to build such a solution, I realized that the existence of multiple resonances in the averaged system of the inner planets rendered illusory such an approach (Laskar, 1984). This difficulty led me to proceed in two distinct stages:

The first step is the construction of an average system, similar to the systems studied by Lagrange and Laplace. The equations then do not represent the motion of the planets, but the slow deformation of their orbit. This system of equations, obtained by an averaging of order two over the fast angles (the mean longitudes) thanks to dedicated computer algebra programs, comprises 153824 polynomial terms. Nevertheless, it can be considered as a simplified system of equations, because its main frequencies are now the frequencies of precession of the orbit of the planets, and not their orbital periods. The complete system can therefore be numerically integrated with a very big step size of about 500 years. The averaged contributions of the Moon and of general relativity are added without difficulty and represent just a few additional terms (Laskar, 1985, 1986).

The second step, namely the numerical integration of the average (or secular) system, is then very effective and could be performed over more than 200 Myr in only a few hours of computation time. The main result of this integration was to reveal that the whole Solar System, and specifically the inner Solar System (Mercury, Venus, Earth and Mars), is chaotic, with a Lyapunov time of about 5

---

[8]The *Bureau des Longitudes* was founded on 7 messidor an III (June 25, 1795) to develop astronomy and celestial mechanics. Its founding members were Laplace, Lagrange, Lalande, Delambre, Méchain, Cassini, Bougainville, Borda, Buache, Caroché.

million years (Laskar, 1989). An error of 15 meters in the initial position of the Earth gives rise to an error of about 150 meters after 10 Ma, but the same error becomes 150 million km after 100 Ma. It is therefore possible to construct precise ephemeris over a period of a few tens of Ma (Laskar et al., 2004, 2011), but it becomes practically impossible to predict the movement of the planets beyond 100 million years.

When these results were published, the only possible comparison was a comparison with the planetary ephemeris DE102, over only 44 centuries. This however allowed to be confident about the results, by comparing the derivatives of the averaged solutions at the origin (Laskar, 1986, 1990). At this time, there was no possibility of obtaining similar results by direct numerical integration.

Thanks to the rapid advances in computer industry, just two years later, Quinn et al. (1991) have been able to publish a numerical integration of the whole Solar System, taking into account the effects of general relativity and of the Moon, over 3 Myr in the past (later complemented by an integration from −3 Myr to +3 Myr). The comparison with the secular solution (Laskar, 1990) then showed a very good agreement, and confirmed the existence of secular resonances in the inner Solar System (Laskar *et al.*, 1992a). Later, using a symplectic integrator that allowed them to use a large step size for the numerical integration of 7.2 days, Sussman and Wisdom (1992) obtained an integration of the Solar System over 100 Myr, which confirmed the value of the Lyapunov time of approximately 5 Myr for the Solar System.

2.2. Planetary motions over several million years (Myr)

The variations of eccentricities and inclinations of the planetary orbits are clearly visible on a few Myr (Figure 4). Over one million years, the solutions resulting from the perturbation methods of Lagrange and Le Verrier would already give a good estimate of these variations that are essentially due to the linear coupling in the secular equations. On several hundreds of Myr, the behavior of solutions of the external planets (Jupiter, Saturn, Uranus and Neptune) are very similar to the one of the first Myr, and the motion of these planets appears to be very regular, which has also been shown very accurately by means of frequency analysis (Laskar, 1990).

2.3. Planetary motions over several billion years (Gyr)

Once it is known that the motion of the Solar System is chaotic, with exponential divergence of trajectories that multiplies the error on the initial positions by 10 every 10 Myr, it becomes illusory to try to retrieve, or predict the movement of the planets beyond 100 Myr by the calculation of a single trajectory. However, one can make such a computation to explore the phase space of the system. The calculated trajectory should then only be considered as a possible trajectory among others after 100 Myr. In (Laskar, 1994), such calculations have even been pushed on several billion years to highlight the impact of the chaotic diffusion of the orbits. In Figure 5 we no longer represent the eccentricities of the planets, but their

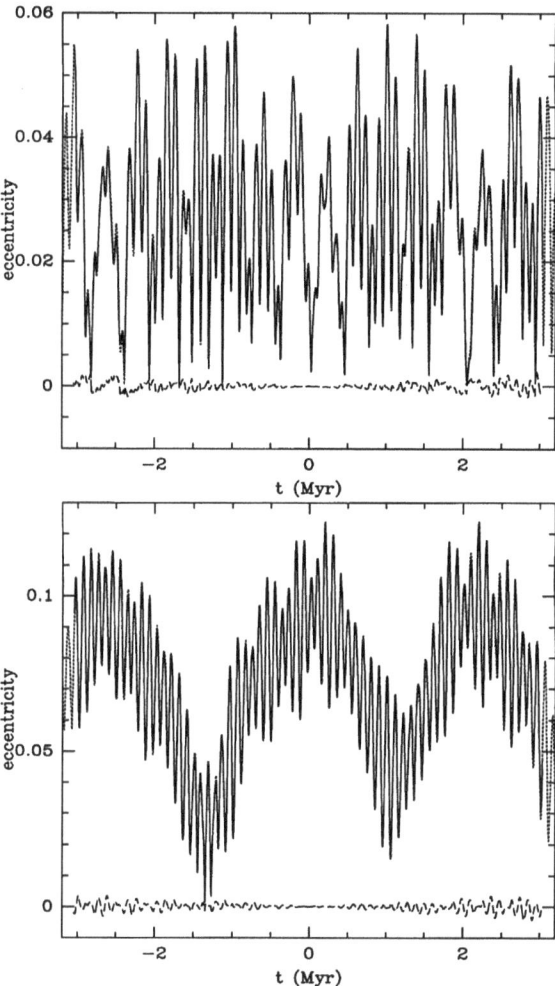

FIGURE 4. The eccentricity of the Earth (top) Mars and (bottom) from −3Myr to +3Myr. The solid line is the numerical solution from (Quinn et al., 1991), and the dotted line, the secular solution (Laskar, 1990). For clarity, the difference between the two solutions is also plotted (Laskar et al., 1992).

maximum value, calculated on slices of 1 Myr. If the trajectory is quasiperiodic or close to quasiperiodic, this maximum will behave as a straight horizontal line, corresponding to the sum of the modulus of the amplitudes of the various periodic terms in the quasiperiodic expansion of the solution.

In this way, we are able to eliminate the oscillation of eccentricities resulting from the linear coupling already present in solutions of Lagrange or Le Verrier.

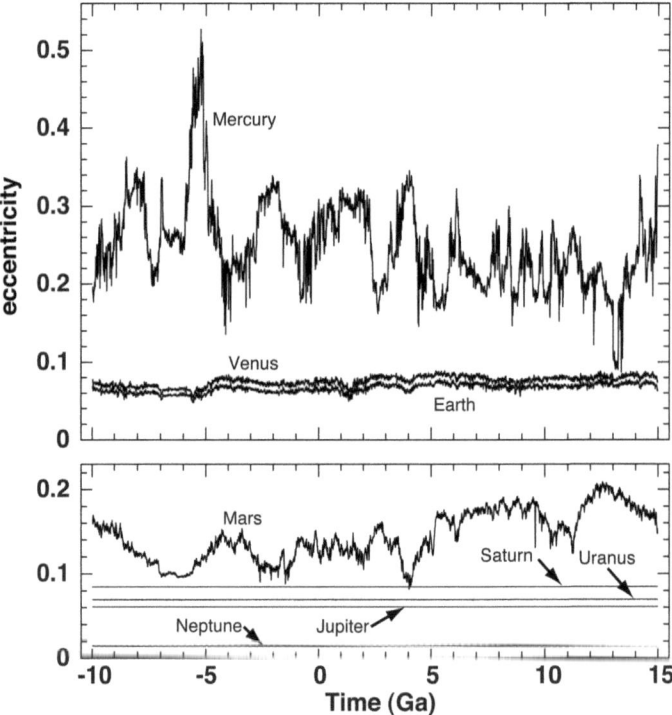

FIGURE 5. Numerical integration of the averaged equations of the Solar System from 10 to 15 Gyr. For each planet, only the maximum eccentricity reached on slices of 10 Myr is plotted. The motion of the large planets is very close to a quasiperiodic motion and the amplitude of the oscillations of their orbital elements does not vary. Instead, for all inner planets, there is a significant variation of the maximum eccentricity and inclination, which reflects the chaotic diffusion of the orbits. From [47] with permission from A & A © 1994 ESO.

The remaining variations of this maximum, which appear in Figure 5 are then the results of only the chaotic diffusion of the orbits. We see that for all the external planets (Jupiter, Saturn, Uranus, and Neptune), the maximum of the eccentricity is a horizontal line. It means that the motion of these planets is very close to quasiperiodic. However, for all the inner planets, there is a significant chaotic diffusion of the eccentricities. This diffusion is moderate for Venus and the Earth, important for Mars, whose orbit can reach eccentricities of the order of 0.2 (which does not allow collision with the Earth), and very strong for Mercury which reached an eccentricity of 0.5. This value is however not sufficient to allow for a collision with Venus, which requires an eccentricity of more than 0.7 for Mercury. But it is well understood that beyond 100 Myr, the trajectories of Figure 5 only

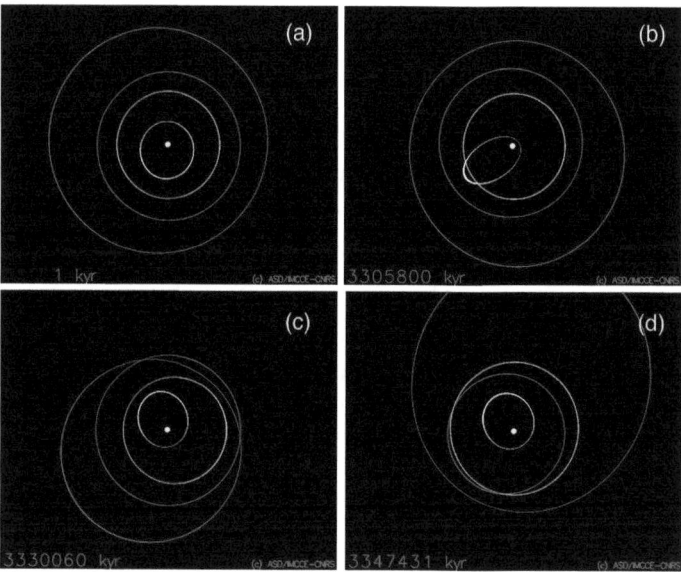

FIGURE 6. Example of long-term evolution of the orbits of the terrestrial planets: Mercury (white), Venus (green), Earth (blue), Mars (red). The time is indicated in thousands of years (kyr). (a) In the vicinity of the current state, the orbits are deformed under the influence of planetary perturbations, but without allowing close encounters or collisions. (b) In about 1% of the cases, the orbit of Mercury can deform sufficiently to allow a collision with Venus or the Sun in less than 5 Gyr. (c) For one of the trajectories, the eccentricity of Mars increases sufficiently to allow a close encounter or collision with the Earth. (d) This leads to a destabilization of the terrestrial planets which also allow collisions between the Earth and Venus (Figure adapted from the results of the numerical simulations of Laskar and Gastineau, 2009).

represent a possible trajectory of the Solar System, and a small change in initial conditions will significantly change the trajectories after 100 Myr.

To find out if collisions between Mercury and Venus are possible, it is therefore necessary to study the variations of the solutions under the influence of a small change in initial conditions. In (Laskar, 1994) I lead this study, using the secular system, showing that it was actually possible to build, section by section, an orbit of collision for Mercury and Venus. In a first step, the nominal trajectory is integrated over 500 Myr. Then, 4 additional trajectories are integrated, corresponding to small changes of 15 meters of the position reached at 500 Myr. All the trajectories are then integrated over 500 Myr and the trajectory of greater eccentricity of Mercury is retained and stopped in the neighborhood of the maximum of eccentricity. This operation is then repeated and leads to an eccentricity of Mercury of more than 0.9 in only 13 steps, in less than 3.5 Gyr, allowing thus a collisions with Venus. However, while repeating the same experiments with the trajectories of Venus, Earth and Mars, it was not possible to construct collisional solutions for these planets.

2.4. Chaotic diffusion in the Solar System

The 1994 approach however had some limitations, because the approximation obtained by the averaged equations decreases in accuracy as one approaches the collision. A study using the complete, non- averaged equations was therefore necessary to confirm these results. Despite the considerable increase in the power of computers since 1994, no complete study of this problem was conducted before 2009. Actually, because of the chaotic nature of the solutions, the only possible approach is a statistical study of a large number of solutions, with very similar initial conditions. This shows the difficulty of the problem. Indeed, before 2009, no direct integration of a single trajectory of the Solar System had yet been published using a realistic model, including the effect of the Moon and general relativity. To approach this problem, I have firstly carried out such a statistical study, using the averaged equations (whose numerical integration is about 1000 times faster than for the full equations), for 1000 different solutions that were integrated over 5 Gyr. This study (Laskar, 2008) showed that the probability to reach very high eccentricities for Mercury ($>$ 0.6) is on the order of 1%. In this same study, I could also show that over periods of time longer than 500 Myr, the distributions of the eccentricities and inclinations of the inner planets (Mercury, Venus, Earth, and Mars) followed Rice's probability densities, and behave like random walks with a very simple empirical distribution law. These results differed significantly from the results published in 2002 by Ito and Tanikawa, who had integrated 5 orbits on 5 Gyr for a purely (Non-relativistic) Newtonian model. I therefore also wanted to test the same statistics for a non-relativistic system, thinking that this system would be more stable, such as Ito's and Tanikawa's one (2002) who found for Mercury a maximum eccentricity of only 0.35. Much to my surprise, the result, on 1000 numerical solutions of the secular system with a pure Newtonian model on 5 Gyr revealed the opposite, and this system appeared far more unstable, with more than half of trajectories raising the eccentricity of Mercury up to 0.9.

To confirm these results, I have then proceeded to a direct integration, using a symplectic integrator (Laskar et al., 2004), of a pure Newtonian planetary model, for 10 trajectories with close initial conditions. The result was consistent with the results of the secular system since 4 trajectories out of 10 led to eccentricity values for Mercury larger than 0.9 (Laskar, 2008). This large excursion of the eccentricity of Mercury is explained by the presence of a resonance between the perihelion of Mercury and Jupiter, which is made easier in the absence of general relativity (GR). It is known that GR increases the precession speed of the perihelion of mercury by $0.43''/yr$. This moves it from $5.15''/yr$ to $5.58''/yr$, and thus send it further from the value of the perihelion speed of Jupiter ($4.25''/yr"$). Independently, Batygin and Laughlin (2008) published similar results shortly after. The

American team, which resumes the calculation of (Laskar, 1994) on a system of non-relativistic equations, also demonstrated the possibility of collisions between Mercury and Venus. These results were still incomplete. Indeed, as the relativistic system is much more stable than the non-relativistic system, it is much more difficult to exhibit an orbit of collision between Mercury and Venus in the realistic (relativistic) system than in the non-relativistic system taken into consideration in these two previous studies. The real challenge was therefore in the estimation of the probability of collision of Mercury and Venus for a realistic, relativistic, model. It is precisely this program that I had in mind since the writing of my 2008 paper, that allowed me to estimate the probability of success of finding a collisional orbit for a realistic model of the Solar System.

2.5. The search for Mercury-Venus collisional orbits

With M. Gastineau, we then began a massive computation of orbital solutions for the Solar System motion, under various aspects, the ambition being to confirm and extend the results obtained 15 years before with the averaged equations. For this we used a non-averaged model consistent with the short-term highly accurate INPOP planetary ephemeris that we had developed in the past years (Fienga et al., 2008). Through the previous studies of the secular system, I had estimated to 3 million hours the computing time being necessary for such a study, but at the time, no national computing center was allowing even a 10 times smaller allocation of computing time. We then used all the means which we could have access to: local cluster of workstations, computing center of Paris Observatory, and a parallel machine that had just been installed at IPGP, Paris. To search for additional CPU time, we also undertook the development of the first full-scale application in astronomy on the EGEE grid with 500 cores (Vuerli et al., 2009). These different runs, associated with multiple difficulties due to the variety of machines and operating systems, nevertheless have allowed us to recover more than 2 million hours of CPU, but at the same time, a better estimate of the necessary computing time had increased the required time to more than 5 million hours. Quite fortunately, the availability of computing resources has changed in France in 2008, with the installation of the JADE supercomputer at CINES, near Montpellier, with more than 12000 cores[9]. As we could benefit of the experimental period on this machine, we started the computations as soon as the machine was switch on, in early August 2008, using 2501 cores, with one trajectory being computed on each core. We could then finalize our computations in about 6 months.

2.6. Possibility of collisions between Mercury, Mars, Venus and the Earth

With the JADE machine, we were able to simulate 2501 different solutions of the movement of the planets of the whole Solar System on 5 billion years, corresponding to the life expectancy of the system, before the Sun becomes a red giant. The 2501 computed solutions are all compatible with our current knowledge of the Solar System. They should thus be considered as equiprobable outcomes of the future

---

[9] When this machine was installed, it was ranked 14th worldwide among supercomputer centers.

of the Solar System. In most of the solutions, the trajectories continue to evolve as in the current few millions of years: the planetary orbits are deformed and precess under the influence of the mutual perturbations of the planets but without the possibility of collisions or ejections of planets outside the Solar System. Nevertheless, in 1% of the cases, the eccentricity of Mercury increases considerably. In many cases, this deformation of the orbit of Mercury then leads to a collision with Venus, or with the Sun in less than 5 Ga, while the orbit of the Earth remained little affected. However, for one of these orbits, the increase in the eccentricity of Mercury is followed by an increase in the eccentricity of Mars, and a complete internal destabilization of the inner Solar System (Mercury, Venus, Earth, Mars) in about 3.4 Gyr. Out of 201 additional cases studied in the vicinity of this destabilization at about 3.4 Gyr, 5 ended by an ejection of Mars out of the Solar System. Others lead to collisions between the planets, or between a planet and the Sun in less than 100 million years. One case resulted in a collision between Mercury and Earth, 29 cases in a collision between Mars and the Earth and 18 in a collision between Venus and the Earth (Laskar and Gastineau, 2009). Beyond this spectacular aspect, these results validate the methods of semi-analytical averaging developed for more than 20 years and which had allowed, 15 years ago, to show the possibility of collision between Mercury and Venus (Laskar, 1994).

These results also answer to the question raised more than 300 years ago by Newton, by showing that collisions among planets or ejections are actually possible within the life expectancy of the Sun, that is, in less than 5 Gyr. The main surprise that comes from the numerical simulations of the recent years is that the probability for this catastrophic events to occur is relatively high, of the order of 1%, and thus not just a mathematical curiosity with extremely low probability values. At the same time, 99% of the trajectories will behave in a similar way as in the recent past millions of years, which is coherent with our common understanding that the Solar System has not much evolved in the past 4 Gyr. What is more surprising is that if we consider a pure Newtonian world, the probability of collisions within 5 Gyr grows to 60%, which can thus be considered as an additional indirect confirmation of general relativity.

Acknowledgment

This work has benefited from assistance from the Scientific Council of the Observatory of Paris, PNP-CNRS, and the computing center of GENCI-CINES.

References

[1] Applegate, J.H., Douglas, M.R., Gursel, Y., Sussman, G.J. and Wisdom, J.: The Solar System for 200 million years. Astron. J. **92**, 176–194 (1986).

[2] Arnold, V.: Proof of Kolmogorov's theorem on the preservation of quasi-periodic motions under small perturbations of the Hamiltonian. Rus. Math. Surv. **18** N6, 9–36 (1963a).

[3] Arnold, V.I.: Small denominators and problems of stability of motion in classical celestial mechanics. Russian Math. Surveys **18**, 6, 85–193 (1963b).

[4] Batygin, K., Laughlin, G.: On the Dynamical Stability of the Solar System. ApJ **683**, 1207–1216 (2008).

[5] Brechenmacher, F.: L'identité algébrique d'une pratique portée par la discussion sur l'équation à l'aide de laquelle on détermine les inégalités séculaires des planètes (1766-1874) Sciences et Techniques en Perspective, IIe série, fasc. 1, 5–85 (2007).

[6] Bretagnon, P.: Termes à longue périodes dans le système solaire. Astron. Astrophys. **30**, 341–362 (1974).

[7] Brumberg, V.A., Chapront, J.: Construction of a general planetary theory of the first order. Cel. Mech.8, 335–355 (1973).

[8] Carpino, M., Milani, A. and Nobili, A.M.: Long-term numerical integrations and synthetic theories for the motion of the outer planets. Astron. Astrophys. **181**, 182–194 (1987).

[9] Chierchia, L., Pinzari, G.: The planetary n-body problem: symplectic foliation, reductions and invariant tori. Inventiones Mathematicae **186**, 1–77 (2011).

[10] Cohen, C.J., Hubbard, E.C., Oesterwinter, C.: Astron. Papers Am. Ephemeris **XXII**, 1 (1973).

[11] De la Lande, F.: Abrégé d'Astronomie, première édition, Paris (1774).

[12] De la Lande, F.: Abrégé d'Astronomie, seconde édition, augmentée, Paris (1795).

[13] Duriez, L.: 1979, Approche d'une théorie générale planétaire en variable elliptiques héliocentriques, *thèse*, Lille.

[14] Euler, L.: Recherches sur les irrégularités du mouvement de Jupiter et Saturne, (1752).

[15] Féjoz, J., Herman, M.: Démonstration du Théorème d'Arnold sur la stabilité du système planétaire (d'après Michael Herman). Ergodic theory and Dynamical Systems **24**, 1–62 (2004).

[16] Fienga, A., Manche, H., Laskar, J., Gastineau, M.: INPOP06. A new numerical planetary ephemeris. A&A **477**, 315–327 (2008).

[17] Giorgilli, A., Delshams, A., Fontich, E., Galgani, L., Simo, C.: Effective stability for a Hamiltonian system near an elliptic equilibrium point, with an application to the restricted three body problem. J. Diff. Equa. **77**, 167–198 (1989).

[18] Ito, T., Tanikawa, K.: Long-term integrations and stability of planetary orbits in our Solar System. MNRAS **336**, 483–500 (2002).

[19] Kinoshita, H., Nakai, H.: Motions of the perihelion of Neptune and Pluto. Cel. Mech. **34**, 203 (1984).

[20] Kolmogorov, A.N.: On the conservation of conditionally periodic motions under small perturbation of the Hamiltonian. Dokl. Akad. Nauk. SSSR **98**, 469 (1954).

[21] Lagrange, J.-L.: *Solution de différents problèmes de calcul intégral.* Miscellanea Taurinensia, t. III, 1762-1765, Oeuvres t. I, p. 471 (1766).

[22] Lagrange, J.-L.: *Sur l'altération des moyens mouvements des planètes.* Mém. Acad. Sci. Berlin, 199 (1776), Oeuvres complètes **VI** 255 Paris, Gauthier-Villars (1869).

[23] Lagrange, J.-L.: *Recherches sur les équations séculaires des mouvements des nœuds et des inclinaisons des planètes*. Mémoires de l'Académie des Sciences de Paris, année 1774, publié en (1778).

[24] Lagrange, J.-L.: *Théorie des variations séculaires des éléments des planètes*. Première partie, Nouveaux Mémoires de l'Académie des Sciences et Belles-Lettres de Berlin, Œuvres, t. V, p. 125 (1781).

[25] Lagrange, J.-L.: *Théorie des variations séculaires des éléments des planètes*. Seconde partie contenant la détermination de ces variations pour chacune des planètes principales, Nouveaux Mémoires de l'Académie des Sciences et Belles-Lettres de Berlin, Œuvres, t. V, p. 211 (1782).

[26] Lagrange, J.-L.: *Théorie des variations périodiques des mouvements des planètes*. Première partie, Nouveaux Mémoires de l'Académie des Sciences et Belles-Lettres de Berlin, Œuvres, t. V, p. 347 (1783a).

[27] Lagrange, J.-L.: *Sur les variations séculaires des mouvements moyens des planètes*. Nouveaux Mémoires de l'Académie des Sciences et Belles-Lettres de Berlin, Œuvres, t. V, p. 381 (1783b).

[28] Lagrange, J.-L.: *Théorie des variations périodiques des mouvements des planètes*. Seconde partie, Nouveaux Mémoires de l'Académie des Sciences et Belles-Lettres de Berlin, Œuvres, t. V, p. 417 (1784).

[29] Lagrange, J.-L.: *Mémoire sur la théorie des variations des éléments des planètes et en particulier des variations des grands axes de leurs orbites*. Mémoires de la première classe de l'Institut de France, Œuvres, t. VI, p. 713 (1808).

[30] Lagrange, J.-L.: *Mémoire sur la théorie générale de la variation des constantes arbitraires dans tous les problèmes de la mécanique*. Mémoires de la première classe de l'Institut de France, Œuvres, t. VI, p. 771 (1808).

[31] Lagrange, J.-L.: *Second Mémoire sur la théorie générale de la variation des constantes arbitraires dans les problèmes de mécanique*. Mémoires de la première classe de l'Institut de France, Œuvres, t. VI, p. 809 (1809).

[32] Laplace, P.S.: Mémoire sur les solutions particulières des équations différentielles et sur les inégalités séculaires des planètes. Oeuvres complètes **9** 325 (1772), Paris, Gauthier-Villars (1895)

[33] Laplace, P.S.: Mémoire sur les solutions particulières des équations différentielles et sur les inégalités séculaires des planètes. Mémoires de l'Académie des Sciences de Paris, année 1772, Œuvres, t. VIII, p. 325 (1775).

[34] Laplace, P.S.: *Sur le principe de la Gravitation Universelle, et sur les inégalités séculaires des planètes qui en dépendent*. Mémoires de l'Académie des Sciences de Paris, Savants étrangers, année 1773, t. VII, Œuvres, t. VIII, p. 201 (1776a).

[35] Laplace, P.S.: *Mémoire sur l'Inclinaison moyenne des orbites des comètes, sur la figure de la Terre et sur leur fonctions*. Mémoires de l'Académie des Sciences de Paris, Savants étrangers, année 1773, t. VII, 1776, Œuvres, t. VIII, p. 279 (1776b).

[36] Laplace, P.S.: *Mémoire sur les inégalités séculaires des planètes et des satellites*. Mém. Acad. royale des Sciences de Paris, Oeuvres complètes **XI** 49 (1784), Paris, Gauthier-Villars (1895).

[37] Laplace, P.S.: *Théorie de Jupiter et Saturne*. Mémoires de l'Académie Royale des Sciences de Paris, année 1785, 1788, Œuvres, t. XI, p. 95 (1785).

[38] Laskar, J.: *Théorie Générale Planétaire. Eléments orbitaux des planètes sur 1 million d'années.* Thèse, Observatoire de Paris (1984), http://tel.archives-ouvertes.fr/tel-00702723

[39] Laskar, J.: Accurate methods in general planetary theory. Astron. Astrophys. **144**, 133–146 (1985).

[40] Laskar, J.: Secular terms of classical planetary theories using the results of general theory. Astron. Astrophys. **157**, 59–70 (1986).

[41] Laskar, J.: A numerical experiment on the chaotic behaviour of the Solar System. Nature **338**, 237–238 (1989).

[42] Laskar, J.: The chaotic motion of the Solar System. A numerical estimate of the size of the chaotic zones. Icarus **88**, 266–291 (1990).

[43] Laskar, J.: *La stabilité du Système Solaire.* In Chaos et Déterminisme, A. Dahan *et al.*, eds., Seuil, Paris (1992).

[44] Laskar, J.: *Lagrange et la stabilité du Système Solaire.* In Sfogliando La Mécanique Analitique, Edizioni Universitarie di Lettere Economia Diritto, Milano, pp. 157–174 (2006).

[45] Laskar, J.: Chaotic diffusion in the Solar System. Icarus **196**, 1–15 (2008).

[46] Laskar, J., Quinn, T., Tremaine, S.: Confirmation of Resonant Structure in the Solar System. Icarus **95**, 148–152 (1992a).

[47] Laskar, J.: Large scale chaos in the Solar System. Astron. Astrophys. **287** L9–L12 (1994).

[48] Laskar, J., Robutel, P., Joutel, F., Gastineau, M., Correia, A.C.M., Levrard, B.: A long term numerical solution for the insolation quantities of the Earth. A&A **428**, 261–285 (2004).

[49] Laskar, J., Gastineau, M.: Existence of colisional trajectories of Mercury, Mars and Venus with the Earth. Nature **459**, 817–819 (2009).

[50] Laskar, J., Fienga, A., Gastineau, M., Manche, H.: La2010: a new orbital solution for the long-term motion of the Earth. A&A **532**, A89 (2011).

[51] Le Monnier: *Sur le Mouvement de Saturne et sur l'inégalité de ses révolutions périodiques, qui dépendent de ses diverses configurations à l'égard de Jupiter.* Première partie, Mémoires de l'Académie Royale des Sciences, Paris, 30 avril 1746a, publié en 1751.

[52] Le Monnier: *Sur le Mouvement de Saturne.* Seconde Partie, Mémoires de l'Académie Royale des Sciences, Paris, 7 mai 1746b, publié en 1751.

[53] LeVerrier U.J.J.: *Mémoire sur les variations séculaires des éléments des orbites pour les sept planètes principales, Mercure, Vénus, la Terre, Mars, Jupiter, Saturne et Uranus.* Presented at the Academy of Sciences on September 16, 1839, Additions à la Connaissance des temps pour l'an 1843 Paris, Bachelier, pp. 3–66 (1840).

[54] LeVerrier U.J.J.: *Mémoire sur les inégalités séculaires des planètes.* Presented at the Academy of Sciences on December 14, 1840, Additions à la Connaissance des temps pour l'an 1844 Paris, Bachelier, pp. 28–110 (1841).

[55] Lochak, P.: Hamiltonian perturbation theory: periodic orbits, resonances and intermittency. Nonlinearity **6**, 885–904 (1993).

[56] Morbidelli, A., Giorgilli, A.: Superexponential stability of KAM tori. J. Stat. Phys. **78**, 1607–1617 (1995).

[57] Moser, J.: On invariant curves of area-preserving mappings of an annulus. Nach. Akad. Wiss. Göttingen, Math. Phys. Kl. II **1**, 1–20 (1962).

[58] Murray, C.D.: Book review of *The stability of planetary systems*. Edited by R.L. Duncombe, R. Dvorak, and P.J. Message. Reidel, Dordrecht, 1984. 476 pp., Icarus, **73**, 191–92 (1988).

[59] Nekhoroshev, N.N.: An exponential estimates for the time of stability of nearly integrable Hamiltonian systems. Russian Math. Surveys **32**, 1–65 (1977).

[60] Newhall, X.X., Standish, E.M., Williams, J.G.: DE102: a numerically integrated ephemeris of the Moon and planets spanning forty-four centuries. Astron. Astrophys. **125**, 150–167 (1983).

[61] Newton, I.: *Opticks: Or, A Treatise of the Reflections, Refractions, Inflexions and Colours of Light*. The Fourth Edition, corrected, 4th edition, London (seconde édition "with Additions" en 1717) (1730).

[62] Niederman, L.: Stability over exponentially long times in the planetary problem. Nonlinearity **9**, 1703–1751 (1996).

[63] Nobili, A.M., Milani, A. and Carpino, M.: Fundamental frequencies and small divisors in the orbits of the outer planets. Astron. Astrophys. **210**, 313–336 (1989).

[64] Poincaré, H.: *Les Méthodes Nouvelles de la Mécanique Céleste*. Tomes I–III, Gauthier–Villars, Paris 1892–1899, reprinted by Blanchard, 1987.

[65] Poincaré, H.: *Sur la stabilité du Système Solaire*. Annuaire du Bureau des Longitudes pour l'an 1898, Paris, Gauthier–Villars, B1-B16 (1897).

[66] Poisson: *Mémoire sur les inégalités séculaires des moyens mouvments des planètes*. Lu à l'Académie le 20 juin 1808, Journal de l'Ecole Polytechnique, Cahier XV, t. VIII, 1809, pp. 1–56.

[67] Quinn, T.R., Tremaine, S., Duncan, M.: A three million year integration of the Earth's orbit. Astron. J. **101**, 2287–2305 (1991).

[68] Robutel, P.: Stability of the planetary three-body problem. II KAM theory and existence of quasiperiodic motions. Celes. Mech. **62**, 219–261 (1995).

[69] Sussman, G.J., and Wisdom, J.: Numerical evidence that the motion of Pluto is chaotic. Science **241**, 433–437 (1988).

[70] Sussman, G.J., and Wisdom, J.: Chaotic evolution of the solar system. Science **257**, 56–62 (1992).

[71] Wilson, C.: The great inequality of Jupiter and Saturn: from Kepler to Laplace. Archive for History of Exact Sciences **33**, 15–290 (1985).

Jacques Laskar
ASD, IMCCE-CNRS UMR8028
Observatoire de Paris, UPMC
77 avenue Denfert-Rochereau
F-75014 Paris, France
e-mail: `laskar@imcce.fr`

MIX
Papier aus verantwortungsvollen Quellen
Paper from responsible sources
FSC® C105338

If you have any concerns about our products,
you can contact us on
ProductSafety@springernature.com

In case Publisher is established outside the EU,
the EU authorized representative is:
**Springer Nature Customer Service Center GmbH
Europaplatz 3, 69115 Heidelberg, Germany**

Printed by Libri Plureos GmbH
in Hamburg, Germany